Undergraduate Texts in Mathematics

Undergraduate Texts in Mathematics

Apostol: Introduction to Analytic Number Theory. Second edition.
Armstrong: Groups and Symmetry.
Armstrong: Basic Topology.
Bak/Newman: Complex Analysis.
Banchoff/Wermer: Linear Algebra Through Geometry. Second edition.
Berberian: A First Course in Real Analysis.
Brémaud: An Introduction to Probabilistic Modeling.
Bressoud: Factorization and Primality Testing.
Bressoud: Second Year Calculus.
Readings in Mathematics.
Brickman: Mathematical Introduction to Linear Programming and Game Theory.
Cederberg: A Course in Modern Geometries.
Childs: A Concrete Introduction to Higher Algebra.
Chung: Elementary Probability Theory with Stochastic Processes. Third edition.
Cox/Little/O'Shea: Ideals, Varieties, and Algorithms.
Curtis: Linear Algebra: An Introductory Approach. Fourth edition.
Devlin: The Joy of Sets: Fundamentals of Contemporary Set Theory. Second edition.
Dixmier: General Topology.
Driver: Why Math?
Ebbinghaus/Flum/Thomas: Mathematical Logic.
Edgar: Measure, Topology, and Fractal Geometry.
Fischer: Intermediate Real Analysis.
Flanigan/Kazdan: Calculus Two: Linear and Nonlinear Functions. Second edition.
Fleming: Functions of Several Variables. Second edition.
Foulds: Optimization Techniques: An Introduction.
Foulds: Combinatorial Optimization for Undergraduates.
Franklin: Methods of Mathematical Economics.
Halmos: Finite-Dimensional Vector Spaces. Second edition.
Halmos: Naive Set Theory.
Hämmerlin/Hoffmann: Numerical Mathematics.
Readings in Mathematics.
Iooss/Joseph: Elementary Stability and Bifurcation Theory. Second edition.
James: Topological and Uniform Spaces.
Jänich: Topology.
Klambauer: Aspects of Calculus.
Kinsey: Topology of Surfaces.
Lang: A First Course in Calculus. Fifth edition.
Lang: Calculus of Several Variables. Third edition.
Lang: Introduction to Linear Algebra. Second edition.
Lang: Linear Algebra. Third edition.
Lang: Undergraduate Algebra. Second edition.
Lang: Undergraduate Analysis.
Lax/Burstein/Lax: Calculus with Applications and Computing. Volume 1.
LeCuyer: College Mathematics with APL.

(continued after index)

George Pedrick

A First Course in Analysis

With 62 Illustrations

Springer-Verlag
New York Berlin Heidelberg London Paris
Tokyo Hong Kong Barcelona Budapest

George Pedrick
6231 Fairlane Drive
Oakland, CA 94611
USA

Mathematics Subject Classifications (1991): 26-01

Library of Congress Cataloging-in-Publication Data
Pedrick, George.
A first course in analysis / George Pedrick.
p. cm. — (Undergraduate texts in mathematics)
Includes bibliographical references.
ISBN 0-387-94108-8
1. Mathematical analysis. I. Title. II. Series.
QA300.P38 1994
515—dc20 93-5141

Printed on acid-free paper.

Production coordinated by Brian Howe and managed by Bill Imbornoni; manufacturing supervised by Vincent Scelta.
Typeset by Asco Trade Typesetting Ltd., Hong Kong.
Printed and bound by Edwards Brothers, Inc., Ann Arbor, MI.
Printed in the United States of America.

9 8 7 6 5 4 3 2 1

ISBN 0-387-94108-8 Springer-Verlag New York Berlin Heidelberg
ISBN 3-540-94108-8 Springer-Verlag Berlin Heidelberg New York

To
Katie and Alexandra

Preface

The first course in Analysis, which follows calculus, along with other courses, such as differential equations and elementary linear algebra, in the curriculum, presents special pedagogical challenges. There is a change of stress from computational manipulation to "proof." Indeed, the course can become more a course in Logic than one in Analysis. Many students, caught short by a weak command of the means of mathematical discourse and unsure of what is expected of them, what "the game" is, suffer bouts of a kind of mental paralysis.

This text attempts to address these problems in several ways:

First, we have attempted to define "the game" as that of "inquiry," by using a form of exposition that *begins with a question* and proceeds to analyze, ultimately to answer it, bringing in definitions, arguments, conjectures, examples, etc., as they arise naturally in the course of a narrative discussion of the question. (The true, historical narrative is too convoluted to serve for first explanations, so no attempt at historical accuracy has been made; our narratives are completely contrived.)

Second, we have kept the logic informal, especially in the course of preliminary speculative discussions, where common sense and plausibility—tempered by mild skepticism—serve to energize the inquiry. The role of formal logic is meant to emerge gradually, as the means to fulfill the *felt need for certainty*, which arises in the course of such tentative, plausible arguments. Having a view of the broad outlines and recognizing the gaps in the clarity and logical support, the stage of careful formulation of definitions and theorems, perhaps after testing and revision, leads to the substance of a concise "Definition–Theorem–Proof" exposition of the result. The reader should be able to extract such an exposition from ours, primarily by deleting most of the expository discussion.

Thus, we regard the habit of clarity and conciseness as a *goal*, to be developed gradually. A student's impatience with the style of the text could be taken as a measure of progress.

Third, we aim toward increasing facility in *all* aspects of mathematical discourse, not just step-by-step formal reasoning. (One who has followed every step in a proof does not, as a result, understand it necessarily; he is only convinced of its validity.) We include things like: fancier calculation; selection and use of appropriate notation and terminology; conjecture; awareness of applicability of previously proved results; counterexample/proof; suitable generalization; critical reading and explaining; as aspects of the maturing process which the text seeks to promote (along with the accumulation of information, of course).

It makes sense to speak of the "march of the reasoning" at two levels: the overall development of the subject, and the internal reasoning from step to step, from hypothesis to conclusion, in a proof. We have embraced the first of these more explicitly than other texts in the conviction that it is necessary to the understanding of the material and the maturing of the mind. We are also convinced that Analysis is the proper subject for the siting of the material in such a broader landscape.

A further pedagogically challenging aspect of this course is the collection of suitable exercises. Routine, scales and arpeggios type, exercises for the mastery of computational techniques are rarely available or of great interest; it is, rather, the skills of mathematical discovery and discourse that need to be practiced.

Within the confines of any single topic, only a small number of results exist that are in the first line of interest and some of those are too difficult to prove, so they are left as exercises. We have tried to preserve a good number of these for the purposes of practice. Of course, that means we have omitted the proofs at many places in our exposition. That would be a fault in a careful logical exposition, which must place heavy weight on completeness. We claim it as a virtue of our text because many good exercises result and the need to contrive exercises by calling for proofs of more obscure facts is lessened.

If the reader is able to fill a good number of the gaps we leave (whether or not we ask him to in an exercise) and if he is persuaded that he could fill out most of the rest with a good effort, the loss of logical completeness is a small price to pay.

The text can be used in several ways, both by a teacher and by an individual reader. The portion entitled Background—a narrative development of the number systems beginning with counting and arriving at the rationals—is meant to be read for notations and terminology and perhaps revisited as the work progresses. (There is a danger of bogging down in this material and not getting very far with the Analysis, the subject of the book.)

The course proper can begin either with the axioms of ordered fields (p. 19) or with Part I.

The standard material for a minimal background in Analysis: completeness, compactness, continuity, etc., is treated in Part I. An exceptional feature is that different questions are pursued which lead to several formulations of the completeness property, which are then shown to be equivalent. Completeness, rather than having the appearance of an additional axiom from "somewhere," is seen to solve several interesting problems.

The inclusion of some initial sections from Part II along with most of Part I will achieve a good one-course introduction to Real Analysis, either as a terminal course or as background to more advanced Analysis (Lebesgue integration, complex variables, differential equations, manifolds, etc.).

The complete text is enough for a two-course sequence and, it is hoped, will be found to be enjoyable "further reading," whether in a formal course or not. A good treatment of the Stieltjes integral can easily be found, in, e.g., Ross [23], Widder [27], or Hille [14]. For that reason, we have left it out here, even though it would fit admirably in Part II.

Portions of the text evolved as classroom notes over several years of teaching the shorter course. In this shorter version my colleague Russell Merris taught from the notes and made helpful suggestions. Marion Quaas typed them and Ann Cambra typed the final entire text. I owe them all my sincere thanks.

Finally, I am deeply indebted to Don Albers for his help and encouragement.

Hayward, CA, 1993 GEORGE PEDRICK

Contents

Notations and Conventions

The Text

Comparative terms are understood in the wide sense; "greater," for example, is understood to include the case of equality. In cases where the equality is to be excluded, the terminology used makes it explicit, e.g., "strictly greater" ($>$ or $\gneqq$ replaces $\geq$).

Definitions, which are sometimes presented formally, as they would be in a "Definition–Theorem–Proof" exposition, are also frequently embedded in more informal exposition (*even in exercises*). In such cases, the term is printed in **boldface** in the sentence which serves to define it precisely.

Quotation marks are used frequently (as in the above "Definition–Theorem–Proof" usage) to invoke concepts that are ill-defined or tentative, in the course of an informal exposition. In particular, discussions of the progress toward a precise, formal definition of a term might use this device.

This typeface is used to identify some parts of the exposition as being strictly motivational, parts that would surely be deleted if one were to seek to extract from the text a closer approximation of a "Definition–Theorem–Proof" presentation.

The appelation **Theorem** is reserved for those results that are judged to be part of the *goals* of the theory, as contrasted to **Lemmas** (technical results—used to consolidate proofs) and **Propositions** (everything else, including less central results, perhaps just pointing up relations between terms, e.g., open set = complement of closed set).

The Exercises

Some exercises are utterly trivial. The exercise is to recognize the fact. More generally, to assess the difficulty of each exercise is the first step in addressing it. An exercise that simply states something is understood to call for a proof of that statement.

An *overall exercise* is to rewrite portions of the text, with a view to achieving more "elegance" (brevity with clarity!), or a more clearly unassailable logic in the "march of the reasoning" as in a "Definition–Theorem–Proof" presentation.

A few exercises call for the formulation of precise definitions, or for the introduction of suitable notations, as reminders of the importance of these steps in the achievement of satisfactory theories.

Notations

The notations of set theory are by now quite standard and no lengthy explanation is deemed necessary. We only mention them briefly:

$\in$: "is a member of" or "belongs to."

$\notin$: is not a member of (use of / to negate).

$\subset$ "included in" or "is a subset of" (wide sense!).

$\{x \in S: ----\}$ "the set of all x belonging to S such that $----$"

$\cup$: union.

$\cap$: intersection.

$-$: complement: $X - S = \{x \in X: x \notin S\}$.

ϕ: empty set.

disjoint: having no members in common.

$\times$: Cartesian product: $X \times Y = \{\text{ordered pairs } (x, y): x \in X \text{ and } y \in Y\}$.

$f: X \to Y$: f is a function of $x \in X$ with values $y \in Y$.

$y = f(x)$ or $x \to f(x)$: abbreviation of above.

dom f: domain of the function $f = \{x \in X: f(x) \text{ defined}\}$.

ran f: range of the function $f = \{y \in Y: y = f(x) \text{ for some } x\}$.

graph f: graph of the function f:

$$\text{graph } f = \{(x, y) \in X \times Y: y = f(x)\}.$$

$F|_S$: restriction of the function f to S:

$$F|_S = \{(x, y) \in X \times Y: x \in S \cap \text{dom } f \text{ and } y = f(x)\}.$$

$f \circ g$: composite of functions: g followed by f

$$z = [f \circ g](x) \quad \text{if} \quad z = f(g(x)).$$

References

To formulas and figures, by numbering them within a discussion, are regarded as distractions and are kept to a minimimum, by either repeating the entire formula or leading the reader to hold the proper figure or formula in mind.

To other parts of the text are by chapter and section number, if not page.

To the Bibliography are by author's name and item number, e.g., Lebesgue [20] refers to item 20 in the Bibliography, a book by H. Lebesgue.

Introduction

Ideas of infinity—the "infinitely large" and the "infinitely small" (or "infinitesimal")—are traceable in the history of human thought to the earliest times. It is both natural and useful to imagine procedures of a kind that can be repeated over and over (bisecting a line segment, for example, or extending one to twice its length) and this evokes at once the notion of repetition "infinitely often." In turn, notions of "infinite extent" and "infinitely divisible" arise.

The usefulness of such notions is tempered by the paradoxes to which their uncritical application can lead. Examples were recorded as early as the fifth century B.C. in Zeno's famous paradoxes. One of these argues that motion is impossible, via the infinite repetition of the reasoning, that one can only move from A to B by first moving from A to the half-way point, which requires moving from A to the quarter-way point, etc.; the infinite repetition postpones indefinitely the possibility of leaving A.

To cite another example, it is a compelling intuition to think of a cone as

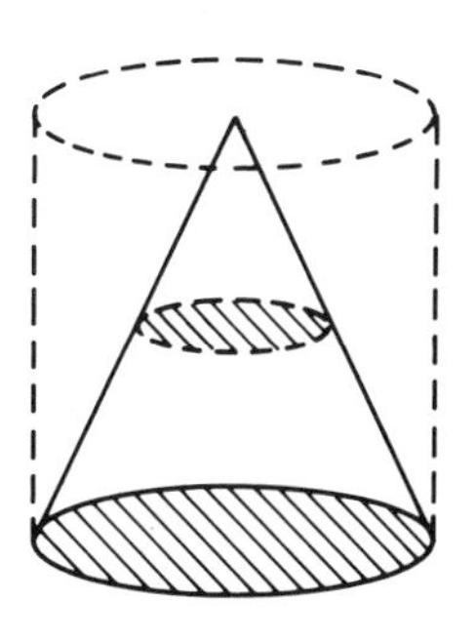

a stack of infinitely many infinitely thin disks. Ancient thinkers used this intuition to discover (i.e., guess, then prove) that the volume of the cone is one-third that of the circumscribed cylinder—a substantial achievement. But surely two "adjacent" disks have the same radius, which implies—via the same compelling intuition—that a cone cannot exist!

Surrounded by mystery and paradox, the uses of infinity expanded through the centuries, a historical process which reached its culmination in the late-seventeenth century with the development of The Calculus, and in the eighteenth century with spectacular applications of the new techniques. The lingering mysteries loomed larger as more delicate matters were investigated. It came to be recognized that the resolution of these issues depended on a better clarification of the two most fundamental concepts in mathematics: number and function.

By the end of the nineteenth century, all of the difficulties had been cleared up and it could be said that "infinity" was well understood. This set the stage for the extraordinary explosion of knowledge witnessed by the twentieth century.

It is this nineteenth-century work which this book undertakes to discuss. It was a crucial transitional body of material in history and it is a crucial transitional subject in the modern mathematics curriculum. The student will find that the focus of attention is shifted from the learning and applying of computational techniques to careful reasoning from hypothesis to conclusion, in its various forms. (It should be pointed out, however, that even the most routine computation is just a highly abbreviated presentation of an argument, or "proof," e.g., the "solution"

$$x^2 - 2x - 3 = 0,$$

$$(x + 1)(x - 3) = 0,$$

$$x + 1 = 0 \quad \text{or} \quad x - 3 = 0,$$

$$x = -1 \quad \text{or} \quad x = 3,$$

is just a sketch of an argument which proves that the only numbers x for which $x^2 - 2x - 3$ is zero are the numbers -1 and 3. The transition, therefore, is one of stress; the arguments, being less routine, demand more attention individually.)

In the old days, a function was a formula or a temperature or a velocity. The concept of function was loaded with all of these extraneous connotations. The clarification of this concept was simply the dropping of the extraneous ideas until there remained the abstract idea of a many-to-one correspondence. Modern students are taught the function concept in its full abstract form from their early years, so what was once a stumbling block is no longer.

The number concept is more complicated. Each person has an extensive familiarity with numbers and their uses, perhaps some uncertainties as well.

One of the achievements of the nineteenth century was the recognition that the proper objects of attention were not "numbers," but "number systems." From the modern perspective, we can imagine a process of evolution of "number systems," each stage being a natural response to clearly perceived defects in the previous stage. We provide a sketch of such an evolution as background for our text, at least to the point at which the perceived defects—various forms of "incompleteness" of the number systems—become central to our subject, and their correction the first major topic of that subject: Analysis.

Of course, these number systems contain no numbers that are "infinite" or "infinitesimal"; it is not in that direction that the new understanding was to be found. It is, instead, *sets* of numbers that provide the key, together with the use of *quantified* statements about the numbers in the sets. There are two kinds of quantified statements about a member x of a set of things S:

The universal quantifier—expressed as "For every x in S ..." or "For any x in S ..." or "For each x in S ..." (the ... being filled by an assertion about x).

The existential quantifier—expressed as "There exists an x in S such that ..." or "For some x in S ...".

It turns out that every useful statement about the infinite or infinitesimal has a precise formulation as a quantified statement about a number in a set that contains, respectively, arbitrarily large or arbitrarily small numbers. Much of the content of this book consists of a setting forth of the details of how this "turns out" to be true.

Finally, it can be claimed that the arguments that achieved these peaks of nineteenth-century clarification are well on their way to becoming the standard devices for new developments in mathematics and its applications ... to becoming clichés in engineering, numerical analysis, and similar areas, just as the quadratic equation or integration-by-parts once did.

Background: Number Systems

§1. Counting: The Natural Numbers

Counting is an innate mental process. Its development and use in all human cultures is evidenced by the many systems of **numerals** that there are: signs and sounds for communicating about counting. For example,

I, II, III, IV, V, . . . ;

1, 2, 3, 4, 5, . . . ;

one, two, three, . . . ;

first, second, third, . . . ;

un, deux, trois, . . . ; and

|, ||, |||, ||||, ~~||||~~, ...

§1.1. There are two aspects of counting, namely:

the ordinal: concerned with arranging things in order, or making lists; and
the cardinal: responding to the question, How many?

They are sometimes expressed by separate systems of numerals (one, two, . . . and first, second, . . .). More often, the distinction is drawn from the context and a single system of numerals is employed.

The act of counting a set of objects amounts to the establishment of a one-to-one correspondence between the objects and a set of numerals. Since each system of numerals is only one means of expression of the counting process, the *essence* of counting is to be found in that which is common to all systems of numerals. That essence consists of the making of steps from a numeral to the next or "successor" numeral, together with a starting or "first" numeral. Thus, counting is embodied abstractly in a set of symbols or objects for which a notion of **successor** is defined, which satisfies:

(a) every object has a unique successor;
(b) different objects have different successors; and
(c) there is an object that is not the successor of any object: the "first" object.

The above list of examples only suggests each of the systems of numerals. The continuation and the successor rules must be learned in each case. Thus, for example, an English-speaking person presumably knows the successor word for each numeral word in English. Similarly, one who knows the decimal system can produce, for each string of the digits $\{0, 1, 2, 3, 4, 5, 6, 7, 8, 9\}$, the string that is its successor.

EXERCISE 1. Write a description of the successor rule for the system |, ||, |||, ||||, 𝍸, ...

EXERCISE 2. Write a description of the successor rule for the decimal system.

EXERCISE 3. Observe that the decimal system is a place-value system, i.e., it assigns a certain significance to the positions of the symbols. This complicates the description of the successor rule in Exercise 2 as compared to Exercise 1. Is it worth it? (Hint: Write your age in both systems; your father's age.)

EXERCISE 4. Look up how to count in some foreign language and compare the successor rules with those of English, noting major differences, if any.

EXERCISE 5. Describe the successor rule for the system that begins

$$1, 2, 10, 11, 12, 20, 21, 22, 100, 101, \ldots,$$

that is, the three-digit, place-value system (the **ternary** system) using digits $\{0, 1, 2\}$.

EXERCISE 6. If a four-digit, place-value system employs the symbols 0, †, §, ε in that order, write the first dozen numerals in the system.

EXERCISE 7. Describe the "first" object and successor rules in the "variations" on counting that follow:

counting from 8:	$8, 9, 10, \ldots;$
counting by threes:	$3, 6, 9, \ldots;$
counting by squares:	$1, 4, 9, \ldots.$

We pursue the goal of characterizing counting abstractly, that is, by removing all irrelevant associations, such as: Roman, Arabic, French, English, ordinal, cardinal, and scratches on cell walls. We arrive at the idea of an abstract set of objects $\mathbb{N}$ in which there is an abstract successor operation * (if $n \in \mathbb{N}$ then n^* denotes the successor of n) satisfying (a), (b), and (c) above.

We require of the abstract system $\{\mathbb{N}, *\}$ that each concrete system of numerals be a realization of it. This means that if we identify the "first" elements of $\mathbb{N}$ and the concrete system with each other, then identify their successors with each other, and in general identify the successors of any pair that are already identified, there results a *complete matching* of the two sets of objects. For this to hold, the various systems must satisfy the further condition ("the matching exhausts them simultaneously").

(d) A subset of $\mathbb{N}$ is *all* of $\mathbb{N}$ if it includes the first element of $\mathbb{N}$ and includes the successor of each element it includes.

Applying this to the subsets that are covered by the identification process just described insures that both $\mathbb{N}$ and the concrete system are completely identified.

The description of counting by the conditions (a) through (d) is due to G. Peano in 1889. The abstract **Natural Number System** $\{\mathbb{N}, *\}$ is characterized by these

Peano Axioms:

(a) Every $n \in \mathbb{N}$ has a unique successor $n^* \in \mathbb{N}$.
(b) $n \neq m$ implies $n^* \neq m^*$.
(c) There is an element, denoted 1, in $\mathbb{N}$ such that $1 \neq n^*$ for all $n \in \mathbb{N}$.
(d) If $S \subset \mathbb{N}$ satisfies:

$$1 \in S \quad \text{(the \textbf{One-Case})},$$

and

$$n \in S \quad \text{implies} \quad n^* \in S \quad \text{(the \textbf{Inductive Step})},$$

then $S = \mathbb{N}$.

§1.2. The last Peano axiom is the basis for **mathematical induction**, a method both for defining concepts and for proving assertions, concerning natural numbers. Thus, the set S of all natural numbers for which a concept has been defined, or an assertion proved, is *all of* $\mathbb{N}$ provided $1 \in S$ and the **induction hypothesis**, $n \in S$, yields the inference $n^* \in S$.

As an example, we define the **addition** of a natural number n to a fixed natural number m by

$$m + 1 = m^*,$$

$$m + n^* = (m + n)^*.$$

The set S of natural numbers for which "addition to m" is defined includes 1 and the successor of each of its members, by these formulas. The induction axiom yields $S = \mathbb{N}$; addition of n to m is defined for all $n \in \mathbb{N}$. Since $m \in \mathbb{N}$ is arbitrary, it follows that $m + n$ is defined for all pairs $m, n \in \mathbb{N}$.

The definition, incidentally, provides the alternative notation $m + 1$ for m^*; the successor of m is $m + 1$.

The addition can be proved to satisfy the **associative law**

$$(m + n) + k = m + (n + k),$$

and the **commutative law**

$$m + n = n + m.$$

Briefly, "associating in pairs" and "reversing the order" do not affect the results of additions in $\{\mathbb{N}, *\}$. (In particular, indicated sums of several natural numbers, symbols like $m + n + k$, have unambiguous meaning.)

The definition having been given by induction on the second addend, it is reasonable to prove the associative law by induction on k.

Proof. 1 (One-Case).

$$(m + n) + 1 = m + (n + 1) \qquad \text{for all} \quad m, n \in \mathbb{N}.$$

This is just the second formula of the definition, expressed in terms of $+1$ rather than $*$ notation.

2 (Inductive Step). Supposing $(m + n) + k = m + (n + k)$ for all $m, n \in \mathbb{N}$, we must show that $(m + n) + (k + 1) = m + [n + (k + 1)]$ for all $m, n \in \mathbb{N}$. We have, for any $m, n \in \mathbb{N}$,

$$\begin{aligned} (m + n) + (k + 1) &= [(m + n) + k] + 1 && \text{by the one-case} \\ &= [m + (n + k)] + 1 && \text{by the induction hypothesis} \\ &= m + [(n + k) + 1] && \text{by the one-case} \\ &= m + [n + (k + 1)] && \text{by the one-case.} \end{aligned}$$

Thus, if S is the set of all $k \in \mathbb{N}$ for which

$$(m + n) + k = m + (n + k) \qquad \text{for all} \quad m, n \in \mathbb{N},$$

then $1 \in S$ and ($k \in S$ implies $k + 1 \in S$), so $S = \mathbb{N}$ and the associative law is true for all $m, n, k \in \mathbb{N}$.[1]

EXERCISE 8. Prove the one-case of the commutative law, $1 + n = n + 1$, for all $n \in \mathbb{N}$, by induction on n.

EXERCISE 9. Prove the commutative law.

Remark. The associative and commutative laws for addition are facts that we each receive with our cultural heritage; indeed we are hardly ever conscious of using them when we do, so automatic are they. Why then give proofs of the obvious? The point of the proof here is not to convince, but rather to reveal connections: that the familiar addition flows logically

[1] Summary statements like this are usually omitted from written proofs.

from the counting process. Even this fact is not surprising from the perspective of speculation about human cultural development: Addition must have emerged culturally as a systematic aid to counting; to count a bag of beans, separate it into disjoint small sets, count them and add the results. Once a person has acquired the basic addition facts in his numeration system and some facility in their use, his efficiency and accuracy at counting are clearly greatly enhanced. So addition flows culturally from counting. If it didn't flow logically as well we would regard our counting axioms as inadequate. Thus the proof provides not only clarity about connections but confirmation of our choice of axioms.

§1.3. **Multiplication** is also a form of efficient counting (the number of seats in a classroom as the product of the number of rows by the number of seats in each row). The precise definition is by induction on the second factor, with an arbitrary $m \in \mathbb{N}$ as first factor:

$$m \cdot 1 = m,$$

$$m \cdot (n + 1) = m \cdot n + m.$$

EXERCISE 10. Prove that $1 \cdot n = n$ for all $n \in \mathbb{N}$.

We say that 1 is a **neutral element under multiplication** since $1 \cdot m = m \cdot 1 = m$ for all $m \in \mathbb{N}$.

That the two operations addition and multiplication can be performed in either order with the same result is the content of the **distributive law**:

$$m \cdot (n + k) = m \cdot n + m \cdot k,$$

"add, then multiply" = "multiply, then add."

The one-case of an induction on k is just the second formula of the definition, and the rest of the proof is then easy; the very definition of multiplication contains the seed of the distributive law.

EXERCISE 11. Prove similarly that $(n + k) \cdot m = n \cdot m + k \cdot m$.

EXERCISE 12. State and prove the associative and commutative laws of multiplication.

A natural **ordering** of the elements of $\mathbb{N}$ is present: by definition, $m < n$ is taken to mean

$$m + k = n \qquad \text{for some} \quad k \in \mathbb{N}.$$

The following order properties hold:

Transitive law: $l < m$ and $m < n$ implies $l < n$; and
Trichotomy: For any $m, n \in \mathbb{N}$, precisely one of the following is true:

$$m = n \quad \text{or} \quad m < n \quad \text{or} \quad n < m.$$

EXERCISE 13. For each $n \in \mathbb{N}$ the Peano axioms imply that $\{m \in \mathbb{N}: m < n\} \cup \{n\} \cup \{m \in \mathbb{N}: n < m\}$ is a decomposition of $\mathbb{N}$ into disjoint sets. Using this, prove trichotomy.

Addition preserves order: $m < n$ implies $m + k < n + k$.
Multiplication preserves order: $m < n$ implies $m \cdot k < n \cdot k$.

EXERCISE 14. Prove these.

The symbols $>$, $\leq$, and $\geq$ are given the obvious meanings.

With the introduction of addition, multiplication, and order, the natural number system is better denoted $\{\mathbb{N}, +, \cdot, <\}$ than $\{\mathbb{N}, *\}$. In fact, for simplicity, it is usually denoted simply by $\mathbb{N}$. It is the most basic mathematical system.

Remark. It is a common practice to omit the details of inductions, giving instead a clear indication of how the definition or proof proceeds and suggesting the rest of the details by an ellipsis, ..., or an etcetera. For example, the definition of $n!$, ***n*-factorial**, is suggested by

$$n! = n \cdot (n-1) \cdot \ldots \cdot 3 \cdot 2 \cdot 1$$

or by

$$1! = 1, \qquad 2! = 2 \cdot 1, \qquad 3! = 3 \cdot 2 \cdot 1, \quad \text{etc.}$$

The precise definition is inductive:

$$1! = 1 \quad \text{and} \quad (n+1)! = (n+1) \cdot n!.$$

EXERCISE 15. Give a careful definition of **exponent notation**, m^n, to embody the idea of repeated multiplication suggested by

$$m^n = m \cdot m \cdot \ \ldots \ \cdot m \quad (n\text{-factors}).$$

EXERCISE 16. Use your definition to prove the **laws of exponents**

$$m^n \cdot m^k = m^{n+k},$$
$$(m \cdot n)^k = m^k \cdot n^k \qquad \text{for all} \quad m, n, k \in \mathbb{N},$$
$$(m^n)^k = m^{n \cdot k}.$$

§1.4. The reader will realize that the definitions and proofs we have either given or proposed as exercises, taken together, would constitute a derivation from the Peano Axioms of all of Arithmetic (of whole numbers). The completion of this program does not exhaust the uses of the induction technique. Indeed, it is essential that the technique be mastered for its use in a variety of contexts. We offer a number of examples and exercises to that end. (We make free use of knowledge about number systems, geometry, algebra, calculus, etc., in these examples.)

EXAMPLE. $|\sin nx| \le n|\sin x|$ for all x and $n \in \mathbb{N}$.

Proof. $n = 1$ gives $|\sin x|$ on both sides.

For the inductive step, we assume $|\sin kx| \le k|\sin x|$ for all x and compute:

$$\begin{aligned}|\sin(k+1)x| &= |\sin kx \cos x + \cos kx \sin x| \\ &\le |\sin kx||\cos x| + |\cos kx||\sin x| \\ &\le |\sin kx| + |\sin x| \\ &\le k|\sin x| + |\sin x| \\ &= (k+1)|\sin x|.\end{aligned}$$

Remark. The inductive step is frequently shown by a straightforward calculation, as in this example. The calculation is guided by the statement of the induction hypothesis, which must be used, and the statement of the $(k+1)$-case, which is to be proved. The use of different symbols n and k to state the proposition to be proved and the inductive step, as above, may help avoid a logical error—the confusing of the *general* statement being proved

$$|\sin nx| \le n|\sin x| \qquad \text{for all} \quad x \text{ and } n,$$

with the *fixed* case of the inductive step

$$|\sin kx| \le k|\sin x|, \qquad k \text{ fixed, all } x.$$

Once this logical point is clear, the use of the same letter, say n, in both statements causes no difficulty and is common practice.

EXAMPLE.

$$1 + 2 + 3 + \cdots + n = \frac{n(n+1)}{2} \qquad \text{for all} \quad n \in \mathbb{N}.$$

Proof. The one-case is just $1 = 1$.

For the inductive step, suppose

$$1 + 2 + 3 + \cdots + n = \frac{n(n+1)}{2} \quad (n \textit{ fixed!})$$

and compute:

$$\begin{aligned}1 + 2 + 3 + \cdots + n + (n+1) &= \frac{n(n+1)}{2} + (n+1) \\ &= \left(\frac{n}{2} + 1\right)(n+1)\end{aligned}$$

$$= \frac{(n+2)(n+1)}{2}$$

$$= \frac{(n+1)[(n+1)+1]}{2}.$$

EXERCISE 17. Prove.

$$(1+x)^n \geq 1 + nx \qquad \text{for} \quad x \geq -1 \quad \text{and} \quad n \in \mathbb{N}.$$

EXERCISE 18. Prove.

$$1^2 + 2^2 + \cdots + n^2 = \frac{n(n+1)(2n+1)}{6} \qquad \text{for} \quad n \in \mathbb{N}.$$

EXERCISE 19. Prove.

$$1 + 3 + 5 + \cdots + (2n-1) = n^2 \qquad \text{for} \quad n \in \mathbb{N}.$$

EXERCISE 20. Prove.

$$1^3 + 2^3 + \cdots + n^3 = (1 + 2 + \cdots + n)^2 \qquad \text{for} \quad n \in \mathbb{N}.$$

The one-case is so frequently obvious that one is tempted to overlook it. It is an essential part of every proof by induction, as is shown by

EXERCISE 21. The statement "$n^2 + 3n + 1$ is even" is false for every $n \in \mathbb{N}$. Show that the inductive step of its "proof" is valid.

EXERCISE 22. $2^n > n$ for $n \in \mathbb{N}$.

EXERCISE 23. $11^n - 4^n$ is divisible by 7 for $n \in \mathbb{N}$.

The one-case may be false, but an induction can begin at another integer.

EXERCISE 24. $n! > 2^n$ for $n = 4, 5, 6, \ldots$.

The inductive step in a proof is always an argument, even if the expression of that argument is a computation. In the next exercise, a geometric construction achieves the inductive step.

EXERCISE 25. A closed polygon in the plane is said to be **convex** if all of its interior angles are less than π.

(a) Show there exist n-sided convex polygons for all $n \geq 3$.
(b) Use induction to guess and prove a formula for the sum of the interior angles of an n-sided convex polygon.

Sometimes a formula too complicated to be of use is replaced by the formula for the inductive step of its proof, called a **reduction formula**.

EXERCISE 26. Prove the reduction formula

$$\int x^n e^x \, dx = x^n e^x - n \int x^{n-1} e^x \, dx, \quad n = 0, 1, 2, \ldots.$$

Contemplate the task of writing the general formula for the value of $\int x^n e^x \, dx$.

The following statement is called **complete induction**:

$S \subset \mathbb{N}$ is all of $\mathbb{N}$ provided $1 \in S$ and $\{1, 2, \ldots, k\} \subset S$ implies $k + 1 \in S$.

It is obvious that complete induction is valid in $\mathbb{N}$ since the "complete-induction hypothesis," $\{1, 2, \ldots, k\} \subset S$, includes the induction hypothesis, $k \in S$.

Suppose, on the other hand, that the complete induction statement is taken in conjunction with the first three Peano Axioms. In the resulting system, it is not hard to prove the usual induction axiom by means of a "complete induction" argument. Thus the induction and the complete induction can replace each other. Sometimes a complete induction argument is simpler or more natural.

Suppose a set $A \subset \mathbb{N}$ has no first member. Then $1 \notin A$ and it follows from $1 \notin A$ and $2 \notin A$ and ... and $k \notin A$ that $k + 1 \notin A$. Hence by complete induction A must be empty. This proves the useful

Proposition. *Every nonempty set of natural numbers has a first member.*

§1.5. The natural numbers serve as the basis for the notion of a sequence; intuitively, an unending succession of things, such as: the sequence of regular polygons △, □, ⬠, ⬡, ...; the sequence of Fibonacci numbers 1, 1, 2, 3, 5, 8, ... (after 1, 1, each number is the sum of the preceding pair); the sequence of statements 1 is odd, 2 is odd, 3 is odd, ...; and the sequence of sevens 7, 7, 7, ...

If S is any set of objects of any kind, a **sequence** of members of S is defined to be a function with domain $\mathbb{N}$ and values in S. For each $n \in \mathbb{N}$, the object that the sequence assigns to n is usually denoted with a subscript notation, e.g., x_n, rather than with the usual function notation $x(n)$. This is the ***n*th value** of the sequence. The sequence itself is indicated by writing its typical value in brackets: $\{x_n\}$. Thus, $\{1/n\}$ is a symbol for $1, \frac{1}{2}, \frac{1}{3}, \ldots$, the latter being the suggestion of the inductive definition $x_1 = 1, x_{n+1} = ((1/x_n) + 1)^{-1}$. Similarly, $\{n(n + 1)/2\}$ and "$x_1 = 1$ and $x_{n+1} = x_n + n$" both define the sequence we suggest by $1, 1 + 2, 1 + 2 + 3, \ldots$, in view of the identity $1 + 2 + \cdots + n = n(n + 1)/2$.

The inductive presentation of $\{x_n\}$ is also called recursive or iterative. The inductive step is an example of an "algorithm."

Statements about sequences can be proved directly or by induction. Thus, "the terms of $\{n(n + 1)\}$ are even" is directly evident, while "the terms of $\{5^n - 3^n\}$ are even" is most easily verified inductively.

EXERCISE 27. Give details

EXERCISE 28. Try proving by induction that $a^n - b^n$ is **divisible** by $a - b$, for $n \in \mathbb{N}$, i.e., that

$$a^n - b^n = (a - b) \cdot K_n \qquad \text{for some} \quad K_n \in \mathbb{N}.$$

Compare with the task of verifying the identity

$$a^n - b^n = (a - b)(a^{n-1} + a^{n-2} \cdot b + \cdots + a \cdot b^{n-2} + b^{n-1})$$

directly, even including the task of guessing the form of K_n as given. Isn't the latter simpler?

The natural numbers are commonly visualized as a succession of equally spaced points on a half-line extending endlessly to the right, e.g., using ordinary numerals to label the points, the picture

$$\begin{array}{ccccc} 1 & 2 & 3 & 4 & \ldots \\ \cdot & \cdot & \cdot & \cdot & \cdots\cdot \end{array}$$

The ordering is just

$$m < n \text{ means } n \text{ lies to the right of } m.$$

Addition is readily visualized as well.

A sequence in a set S, being a function, is then suggested by the arrow in

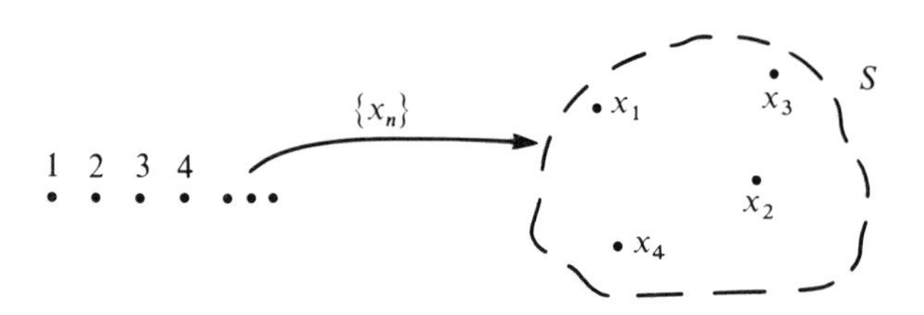

The value x_k of $\{x_n\}$ at $k \in \mathbb{N}$ is a point of S, and the **value set** of $\{x_n\}$ consists of all such points (the range of the function).

Remark. Confusion can arise from the failure to distinguish the sequence from its typical value, or from its value set. This is especially so since the same index is often used in denoting all three: $\{x_n\}$, x_n, and $\{x_n\colon n \in \mathbb{N}\}$. Indeed, the practice is not uncommon to abbreviate these simply x_n and burden the reader with drawing the distinction from the context.

The value set of $\{x_n\}$ can consist of a single point of S, in which case $\{x_n\}$ is called a **constant sequence**, or it can be another nonempty subset of S. Different sequences can have the same value set, as well as different value sets.

A sequence, as any function, can be restricted to a subset of its domain, to create a new function. Any restriction of a sequence $\{x_n\}$ to an *infinite* subset of $\mathbb{N}$ is called a **subsequence** of $\{x_n\}$. Here, infinite means not contained in $\{1, 2, \ldots, m\}$ for any $m \in \mathbb{N}$.

The best notation for a subsequence of $\{x_n\}$ arises from the fact that any infinite subset K of $\mathbb{N}$ can be put in one–one correspondence with $\mathbb{N}$ by inductively matching each $k \in \mathbb{N}$ with the first integer n_k that is as yet unmatched, starting with $n_1 =$ the first element of K. The notation is $\{x_{n_k}\}$.

EXAMPLE. (a) 1, 3, 1, 3, 1, 3, ... is a subsequence of 1, 3, 5, 1, 3, 5, 1, 3, 5, ... but not of 1, 3, 5, 7, 9, ...

(b) The odd primes 3, 5, 7, 11, ... form a subsequence of the odd integers due to the (nonobvious) existence of infinitely many primes.

EXERCISE 29. Look up a proof that there are infinitely many primes. Euclid gave one in the *Elements*.

§1.6. The practical uses of counting include the making of change, keeping of balances, recording of assets and liabilities, etc.; what might be called basic "accounting." These uses give rise to the need for zero and the negative integers, and their incorporation into a broader number system $\{\mathbb{Z}, +, \cdot, <\}$, the *integers*.

In the context of formal developments in $\mathbb{N}$, the same needs arise as soon as one poses addition problems in which one addend is unknown:

given $m, n \in \mathbb{N}$, find x such that $m + x = n$ (briefly, "solve the equation" $m + x = n$).

A solution x exists in $\mathbb{N}$ precisely when $m < n$; indeed, this is exactly the definition of $m < n$.

In case $n \leq m$, there is no solution x in $\mathbb{N}$ and the task is to provide one!, i.e., to adjoin objects to $\mathbb{N}$, forming a set $\mathbb{Z}$, and to extend the operations $+$, $\cdot$, $<$ to the new objects so that their properties are preserved. We illustrate this extension process by the

EXAMPLE. Adjoin the object 0 to $\mathbb{N}$ and define the extensions of $+$, $\cdot$ and $<$ to $\mathbb{N} \cup \{0\}$ as follows:

$$0 + n = n + 0 = n \qquad \text{for all} \quad n \in \mathbb{N} \cup \{0\},$$

(0 is said to be **neutral** under $+$),

$$0 \cdot n = n \cdot 0 = 0 \qquad \text{for all} \quad n \in \mathbb{N} \cup \{0\},$$

and

$$0 < n \qquad \text{for all} \quad n \in \mathbb{N}.$$

The resulting system $\{\mathbb{N} \cup \{0\}, +, \cdot, <\}$ preserves the properties (associative, commutative, etc.) and includes the solution 0 for all of the equations

$$n + x = n, \qquad n \in \mathbb{N} \cup \{0\}.$$

Generally, for $n < m$ there is a $k \in \mathbb{N}$ giving $n + k = m$, and we introduce a new object, denoted $-k$, with the intention of defining $m + (-k)$ to be n, so that $-k$ solves $m + x = n$. There results the set

$$\mathbb{Z} = \{-k: k \in \mathbb{N}\} \cup \{0\} \cup \mathbb{N}$$

and the task of defining $+$, $\cdot$, and $<$ throughout $\mathbb{Z}$ so as to preserve their properties.

It is, of course, possible to do this and arrive at the familiar **system of integers** $\mathbb{Z}$ in which every equation

$$m + x = n \qquad \text{for} \quad m, n \in \mathbb{Z}$$

has a unique solution. The solution is denoted $n - m$, thus introducing **subtraction** in $\mathbb{Z}$. The details are tedious and will be omitted except for a few interesting samples described in the exercises.

For any $z \in \mathbb{Z}$, the solution x of $z + x = 0$ is called the **negative**, or the **additive inverse**, of z.

EXERCISE 30. Show that the *only* meaning that can be given to $(-k)\cdot l$ for $k \in \mathbb{N}$, $l \in \mathbb{Z}$, if the distributive law is to be preserved, is the following:

$$(-k)\cdot l = \text{the additive inverse of } k \cdot l.$$

EXERCISE 31. Taking l (above) of the form $-m$, show that $(-k)\cdot(-m) = k \cdot m$.

This rule of familiar arithmetic is thus forced upon us by our desire to preserve the distributive law.

EXERCISE 32. Show that the **cancellation law for addition** holds in $\mathbb{Z}$:

$$m + k = n + k \quad \text{implies} \quad m = n.$$

EXERCISE 33. A crucial fact for solving equations in algebra is

$$m \cdot n = 0 \qquad \text{if and only if} \quad m = 0 \text{ or } n = 0.$$

Show that this fact is logically equivalent in $\mathbb{Z}$ to the **cancellation law for multiplication**

$$m \cdot k = n \cdot k \quad \text{with} \quad k \neq 0 \quad \text{implies} \quad m = n.$$

It is clear that the definition of $<$ in $\mathbb{Z}$ should be the same as in $\mathbb{N}$:

$$m < n \quad \text{means for some} \quad k \in \mathbb{N}, \quad m + k = n.$$

EXERCISE 34. Prove $m < n$ implies $-m > -n$ and that any inequality $m < n$ is "reversed" under multiplication by a negative number; $ml > nl$ if $l < 0$, for $m < n$.

Recall the natural way of visualizing the integers,

$$\begin{array}{ccccccccc} \ldots & -3 & -2 & -1 & 0 & 1 & 2 & 3 & \ldots \\ \ldots & \cdot & \cdot & \cdot & \cdot & \cdot & \cdot & \cdot & \ldots, \end{array}$$

in which "negation" appears as reflection about 0, and the result in Exercise 34 is exhibited clearly.

EXERCISE 35. Define $0! = 1$ and

$$\binom{n}{k} = \frac{n!}{k!\,(n-k)!} \qquad \text{for} \quad n, k = 0, 1, 2, \ldots.$$

These are called the **Binomial Coefficients**. Prove that

$$\binom{n}{k} + \binom{n}{k-1} = \binom{n+1}{k}.$$

We shall frequently need the **Binomial Theorem**

$$(a+b)^n = \sum_{k=0}^{n} \binom{n}{k} a^{n-k} b^k.$$

EXERCISE 36. Prove the Binomial Theorem. (Hint: Use induction on n. In the inductive step, the formula in Exercise 35 will be useful.)

EXERCISE 37. Discover and prove a formula for the number of points of intersection of n lines in the plane, assuming no three lines intersect in the same point and no two are parallel.

EXERCISE 38. Define $x_1 = 1$, $x_{n+1} = 1/(x_n + 1)$. Write a few terms, then guess and prove a connection with the Fibonacci sequence (see p. 9).

§2. Measurement: The Rational Numbers

The system of integers is adequate for counting and simple accounting. Beyond those needs and uses for numbers one encounters the more sophisticated task of *measurement*: to use numbers to express the idea that a given amount of something, such as length, weight, or temperature, contains a chosen *unit amount* a certain "number" of times (using + and − to express amounts above and below a fixed reference level, as in assets versus liabilities).

To express quantities of sand, for example, one would choose unit amounts depending on the context: single grain, pound, bucketful, truckload. In this case, the system $\mathbb{Z}$ would be adequate only if the single grain were the unit, and then the measures of common amounts would require absurdly large numbers. Measurement demands *flexibility in the choice of unit.* Of course, any other choice of unit in this example leaves most amounts of sand inexpressible by means of $\mathbb{Z}$. The need for a richer number system is clear.

§2.1. If u denotes some unit amount of something, then integer multiples of u have clear intuitive meanings: we write nu, $n \in \mathbb{Z}$. We agree to call two amounts of the thing **commensurable** if there is a unit amount u in terms of which they are both expressible, say as mu and nu, for $m, n \in \mathbb{Z}$. In terms of this concept, the desired flexibility in choice of unit can be expressed precisely: From any collection of mutually commensurable amounts, any positive amount may be chosen as the unit and then all other members of the collection are expressible in that unit.

It is useful to focus attention on a single typical quantity as we proceed with the discussion of measurement. The obvious choice, since it encourages visual interpretation, is that of lengths of line segments. We include single points as segments (length 0 for all units) and we create, out of each segment having more than one point, a pair of **directed segments**, by assigning an arrow in each of the two possible ways: $\underset{\text{left}}{\longleftarrow}$ and $\underset{\text{right}}{\longrightarrow}$. Right-directed segments will be assigned their lengths as measures and left-directed segments will be given the negatives of their lengths as measures.

All directed segments can be visualized as lying on a single line with their initial endpoints at a single reference point, or **origin**, o:

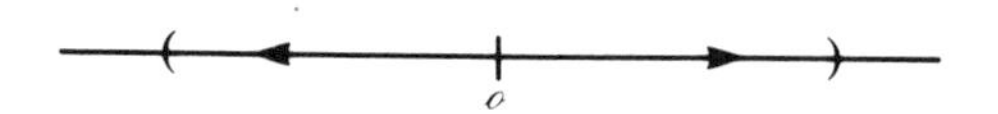

Choosing a unit segment u and marking the endpoints of the segments nu for $n \in \mathbb{Z}$ yields our visual model of $\mathbb{Z}$.

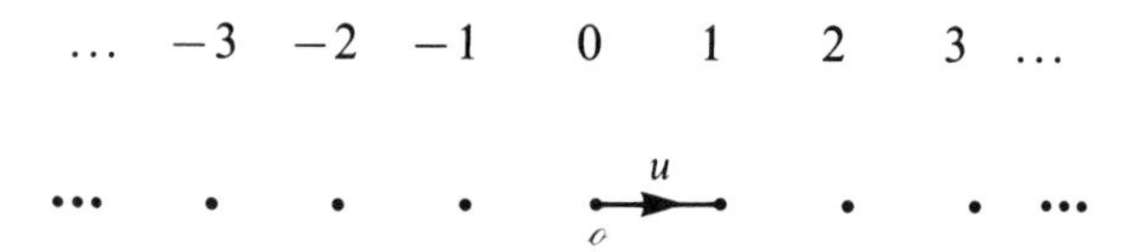

We can refer interchangeably to a point on the line and the segment terminating at that point. This leads us to the concept of mutually commensurable points on the line. If α and β denote two of these, and if u is the unit for which they have representations, respectively, $\alpha = nu$ and $\beta = mu$, for $m, n \in \mathbb{Z}$, then the possibility of expressing β in terms of α as the unit calls for a number x giving $\beta = x\alpha$, or $mu = x \cdot nu$, or

$$m = x \cdot n.$$

The criterion of flexibility for measurement is seen to translate into the requirement for a number system in which all of the equations $n \cdot x = m$ have solutions. These equations express *division* problems in $\mathbb{Z}$.

§2.2. The question of division in $\mathbb{Z}$ is an interesting one in the abstract, aside from its importance for measurement. Clearly, some division problems in $\mathbb{Z}$ have solutions in $\mathbb{Z}$ but others do not. The task is one of extending the system $\mathbb{Z}$ to a larger system $\{\mathbb{Q}, +, \cdot, <\}$ in which division is generally possible. Of course, the $+, \cdot$, and $<$ in $\mathbb{Q}$ should have all the usual properties and reduce to the corresponding things in $\mathbb{Z}$ when applied to elements of $\mathbb{Z}$.

We want division problems to have solutions and we want unique solutions. This *excludes division by zero*, i.e., it excludes the equations $nx = m$ with $n = 0$. For, if $m = 0$ any x would be a solution ($0 \cdot x = 0$ would hold for any x in the extended system), violating the uniqueness, whereas if $m \neq 0$ no solution could exist, for the same reason.

We recognize among the remaining equations

$$n \cdot x = m, \qquad n \neq 0,$$

a natural equivalence as in

$$1 \cdot x = 2, \qquad 2 \cdot x = 4, \qquad -3x = -6, \ldots$$

and

$$1 \cdot x = -4, \qquad -3x = 12, \qquad 6x = -24, \ldots$$

as well as in

$$3x = 2, \qquad 6x = 4, \qquad -15x = -10, \ldots$$

and

$$-7x = 4, \qquad 28x = -16, \qquad 14x = -8, \ldots.$$

Here, the first two lists can serve as stand-ins for the numbers 2 and -4, respectively, in $\mathbb{Z}$. Clearly, every member of $\mathbb{Z}$ has such a stand-in, while every one of our equations yields an object of the same kind—namely, the collection of all equations mutually "equivalent" to it—which can serve as the "solution" of that equation. Thus the second two lists above are, respectively, the newly created "numbers" we shall call $2/3$ and $-4/7$.

To carry out these ideas formally, we begin with the

Definition. The equations $n \cdot x = m$, $n \neq 0$, and $q \cdot x = p$, $q \neq 0$, in $\mathbb{Z}$ are **equivalent** if $m \cdot q = p \cdot n$.

Exercise 1. Verify this property for each of the above lists. Identify all equivalent equations among

$$\begin{array}{llll} 6 \cdot x = 12, & 3 \cdot x = 12, & 2 \cdot x = 3, & 4 \cdot x = 8, \\ 8 \cdot x = 12, & 7 \cdot x = 28, & -6 \cdot x = -9, & 3 \cdot x = 0. \\ -9 \cdot x = -18, & 9 \cdot x = -18, & -9 \cdot x = -18. & \end{array}$$

Notation. An equation in $\mathbb{Z}$, $n \cdot x = m$, $n \neq 0$, and the division problem it poses, can be abbreviated by the symbol m/n, called a **fraction**, in which $n \neq 0$ is understood. We call m the **numerator** and n the **denominator** of m/n. The equivalence of equations provides the meaning of **equality** for fractions:

$$\frac{m}{n} = \frac{p}{q} \quad \text{means} \quad m \cdot q = p \cdot n \quad \text{in } \mathbb{Z}.$$

By the **negative** of a fraction, $-\frac{p}{q}$, we mean either of the equal fractions $\frac{-p}{q}$ and $\frac{p}{-q}$. Fractions of the form $m/1$ are written simply m. In this way, the stand-in for an integer as described above is denoted by the same symbol as the integer itself.

This idea of representing each equivalence class, i.e., collection of mutually equal fractions, by a special one of its members can be carried over to all fractions. For it is easy to see that any fraction is equal to a unique fraction **in lowest terms**; having positive denominator and no common factors between the numerator and denominator.

Exercise 2. What rule of integer arithmetic is central to this observation?

The fractions in lowest terms can be taken to constitute the set of objects $\mathbb{Q}$ of our extended number system, which is the same, taking account of the equality of fractions, as the set of all equivalence classes of fractions. There remains the task of extending the meanings of $+$, $\cdot$, and $<$ and verifying that their properties continue to hold and that any division problem formulated in the extended system has a unique solution there. Once again, we choose to point out the details in the form of exercises, the results being completely familiar.

Definition. The **sum** of two fractions is

$$\frac{m}{n} + \frac{p}{q} = \frac{m \cdot q + n \cdot p}{n \cdot q}.$$

(Note: $n \cdot q \neq 0$.)

This is an allowable definition of sum in $\mathbb{Q}$ because if m/n and p/q are replaced by any equal fractions $\bar{m}/\bar{n}$ and $\bar{p}/\bar{q}$, the sums are equal.

Exercise 3. Prove this.

Exercise 4. Addition in $\mathbb{Q}$ is associative and commutative, 0 is neutral, and $-p/q$ is the additive inverse of p/q.

Definition. The **product** of two fractions is

$$\frac{m}{n} \cdot \frac{p}{q} = \frac{m \cdot p}{n \cdot q}.$$

Exercise 5. Prove that this is an allowable definition.

Exercise 6. Multiplication in $\mathbb{Q}$ is associative and commutative, 1 is neutral, and if $p/q \neq 0$, i.e., $p \neq 0$ as well as $q \neq 0$, then q/p is a multiplicative inverse of p/q.

Exercise 7. Prove the distributive law in $\mathbb{Q}$.

Theorem. *Any equation in $\mathbb{Q}$ of the form*

$$\frac{m}{n} \cdot x = \frac{p}{q} \qquad \textit{with} \quad m \neq 0$$

has a unique solution in $\mathbb{Q}$.

Exercise 8. Show that only $(n \cdot p)/(m \cdot q)$ can solve this equation, and that it does, i.e., prove the theorem.

Definition. The **order** between fractions is given by

$$\frac{m}{n} < \frac{p}{q} \quad \text{means} \quad mq < np \quad \text{in } \mathbb{Z},$$

where expressions are chosen in which the denominators are positive. $\Big($Note: $\dfrac{2}{-3} < \dfrac{-1}{2}$ is true, even though $4 < 3$ is false, hence we need to take care with the optional notation $\dfrac{-p}{q} = \dfrac{p}{-q}$ in this definition.$\Big)$

Exercise 9. The transitive and trichotomy laws hold in $\mathbb{Q}$.

Exercise 10. Order is preserved under addition and under multiplication by a positive fraction; for $x, y, z \in \mathbb{Q}$,

$$x < y \quad \text{implies} \quad x + z < y + z$$

and

$$x < y \quad \text{and} \quad z > 0 \quad \text{implies} \quad zx < zy.$$

The number system $\{\mathbb{Q}, +, \cdot, <\}$ achieved by this construction, abbreviated simply $\mathbb{Q}$, derives its name from the word ratio: the **Rational Number System**.

Exercise 11. Check that the definitions of $+$, $\cdot$, and $<$ reduce to those in $\mathbb{Z}$ when applied to fractions of the form $m/1$.

Exercise 12. Prove the following three useful computational facts in $\mathbb{Q}$: For $x, y, z \in \mathbb{Q}$,

1. $x^2 \geq 0$ (squares are nonnegative);
2. $x < y$ and $z < 0$ implies $zx > zy$ (multiplication by a negative number *reverses* order); and
3. $0 < x < y$ implies $0 < 1/y < 1/x$ (inversion reverses order, for positive numbers).

Remark. We comment, finally, on the question of measurement, particularly of directed segments.

Suppose α and β are any two commensurable segments, with β not a single point; that is,

$$\alpha = mu, \qquad \beta = nu, \qquad n \neq 0, \quad \text{for some unit segment } u.$$

Then using numbers from $\mathbb{Q}$ we can write u as $(1/n)\beta$ and hence $\alpha = (m/n)\beta$, i.e., α can be expressed in terms of β using $\mathbb{Q}$; $\mathbb{Q}$ is adequate under our flexibility criterion for measurement.

On the other hand, if α and β are such that one is expressed in terms of the other using $\mathbb{Q}$, say,

$$\alpha = \frac{m}{n}\beta, \qquad m \in \mathbb{Z} \quad \text{and} \quad 0 < n \in \mathbb{Z},$$

where β is not a single point, then by dividing β in n equal parts we find a unit u for which $\alpha = mu$ and $\beta = nu$. Thus, *only* commensurable segments can be expressed, one in terms of the other, using $\mathbb{Q}$.

EXERCISE 13. Recall or rediscover a ruler-and-compass construction for dividing a given segment in a given number of equal parts.

The measurement process must be flexible in another sense: it must not impose a prior limit on the amount of a thing that can be measured; there must be no positive amount of the thing, so large that it is not exceeded by some multiple of the unit amount. This **Archimedean property** must hold for a number system that is to serve the measurement function. We assert the

Theorem. *If $0 < y \in \mathbb{Q}$ then there is an $m \in \mathbb{N}$ such that $y \leq m$.*

Proof. Let the expression of y in lowest terms be m/n so that $n \geq 1$. Then $1/n \leq 1$ and the result follows upon multiplication by $m > 0$;

$$y = \frac{m}{n} \leq m.$$

Corollary. *For $0 < x \in \mathbb{Q}$ there is an $n \in \mathbb{N}$ for which $1/n < x$.*

If we paraphrase the Archimedean property as "there are arbitrarily large natural numbers," then in the same spirit we can paraphrase the corollary as, "the reciprocals of the natural numbers can be arbitrarily small."

The Archimedean property is often stated in a seemingly more general form:

Archimedean Property. *For any $x > 0$, no y exceeds all the integer multiples of x.* (The theorem states the case $x = 1$.)

EXERCISE 14. Prove that the general $x > 0$ case follows from the $x = 1$ case.

We shall see presently that there exist noncommensurable pairs of line segments, indeed, many such. This fact forces us to accept the existence of points on the line that are *not* measurable using $\mathbb{Q}$; from the point of view of the *ideal* objective, to be able to measure *every* segment, the number system $\mathbb{Q}$ is inadequate. We shall investigate the inadequacy of $\mathbb{Q}$ from several points of view in the next chapter, with an eye to the possibility of making another extension, to a number system, say $\{F, +, \cdot, <\}$, that *is* adequate for the measurement of *any* segment.

We summarize the properties of $\mathbb{Q}$ that should be preserved in such an extension F as follows:

The Axioms of Ordered Fields

Addition is associative and commutative; there is an element $0 \in F$ that is neutral under addition; each $x \in F$ has an additive inverse $-x \in F$.

Multiplication is associative and commutative; there is an element $1 \in F$ that is neutral under multiplication; each nonzero $x \in F$ has a multiplicative inverse $1/x \in F$.

Multiplication and Addition are connected by the distributive law.

Order is transitive and trichotomy holds.

Addition of $x \in F$ preserves order.

Multiplication by $x > 0$ preserves order.

Any number system satisfying these axioms is called an **ordered field**. Of course, $\mathbb{Q}$ is an ordered field.

There are examples[2] of ordered fields in which the Archimedean property fails to hold, so we must distinguish one, such as $\mathbb{Q}$, that does have the additional property as an **Archimedean ordered field**.

Any ordered field F contains 1, hence $1 + 1$, hence $1 + 1 + 1$, etc., and the negatives of these, along with 0, that is, a copy of $\mathbb{Z}$. But the existence of multiplicative inverses in F provides all of the quotients m/n, $m \in \mathbb{Z}$, $0 \neq n \in \mathbb{Z}$, as members of F; F contains a copy of $\mathbb{Q}$. In this sense, $\mathbb{Q}$ is the smallest

[2] See Gelbaum and Olmsted [8].

ordered field. We shall seek our extension of $\mathbb{Q}$ among the collection of all Archimedean ordered fields. We exhibit, first, an ordered field other than $\mathbb{Q}$.

EXAMPLE. Consider the set of all ordered pairs of rationals, (r, s) with $r, s \in \mathbb{Q}$ and $(r, s) = (r', s')$ to mean $r = r'$ and $s = s'$ in $\mathbb{Q}$. Define

$$(r, s) + (r', s') = (r + r', s + s'),$$

$$(r, s) \cdot (r', s') = (rr' + 2ss', rs' + r's),$$

$$(r, s) < (r', s') \qquad \text{if} \quad r < r' \text{ and } s \le s' \text{ or } r \le r' \text{ and } s < s'.$$

This is an Archimedean ordered field. The subset consisting of all pairs $(r, 0)$ is a copy of $\mathbb{Q}$. Since $(0, 1) \cdot (0, 1) = (2, 0)$, the element $(0, 1)$ can be said to be a square root of 2. We shall see that $\mathbb{Q}$ itself contains no such element; this is an Archimedean ordered field different from $\mathbb{Q}$; a *proper* extension of $\mathbb{Q}$.

Since $(r, s) = (r, 0) + (0, s) = (r, 0) + (s, 0)(0, 1)$, the notation $r + s\sqrt{2}$ suggests itself as an alternative to the pair notation for elements of this field. The operations defined above take more familiar forms in this notation.

EXERCISE 15. Verify the multiplication axioms.

EXERCISE 16. Verify the distributive law.

EXERCISE 17. Verify the Archimedean property.

EXERCISE 18. Replace the 2 in the definition of multiplication by an integer ≥ 3. Is the resulting system an Archimedean ordered field? Is it a proper extension of $\mathbb{Q}$?

There is a practical way of viewing measurement, in which each measurement is seen as the result of an action, using an instrument, that leads to the reporting of a number ("with this micrometer, this length is measured to be x inches"). Any such procedure has a limited *precision* (± 0.005 inches for a micrometer) and it is enough to be able to report a number within the range determined by that precision. A number system F admits the reporting of a rational number for all such acts of measurement provided $\mathbb{Q}$ is dense in F in the sense of the

Definition. $\mathbb{Q}$ is **dense** in F if, for any $x < y$ in F there is a member of Q between x and y.

EXERCISE 19. Show that if $\mathbb{Q}$ is dense in F then there are "infinitely many" members of $\mathbb{Q}$ between any given two members in F in the sense that no natural number can count all of them.

Theorem. *$\mathbb{Q}$ is dense in any Archimedean ordered field F.*

The idea of the proof is to lay off positive integer multiples of a certain $1/m$, insuring that one of them falls between x and y, i.e., that a step of size $1/m$ cannot jump from a point $< x$ to a point $> y$.

Proof. If x and y have opposite signs, then 0 is between them. We pursue the case $0 \le x < y$. The remaining case follows from this one by changing signs.

By the corollary to the Archimedean property of $\mathbb{Q}$, there is an $m \in \mathbb{N}$ such that

$$\frac{1}{m} < \frac{y-x}{2} \quad \text{(half the distance from } x \text{ to } y\text{)}.$$

Consider the positive integer multiples of $1/m$. That there is a multiple k/m larger than x is trivial if $x = 0$, and is a consequence of the Archimedean property if $x > 0$. Thus the set of all $k \in \mathbb{N}$ for which $k/m > x$, as a nonempty set of natural numbers, has a first member n. We have, since $(n-1)/m \le x$,

$$\frac{n}{m} \le x + \frac{1}{m} < x + \frac{y-x}{2} = \frac{x+y}{2} < \frac{y+y}{2} = y.$$

Hence

$$x < \frac{n}{m} < y.$$

If we represent an Archimedean ordered field geometrically, using a line (that is, an origin and unit point are chosen on the line—a "coordinate system"), the numbers of F are represented by the points they measure. The Archimedean property conforms with our intuition of the "infinite extent" of the line. The density of $\mathbb{Q}$ in F certainly shows that an abundance of the points on the line are measured by rational numbers. Whether or not the density conforms to intuition is impossible to say since the intuitions of individuals as to the nature of a line probably vary on this point—of it being "infinitely divisible." The notion of "line" is not a part of the formal mathematics being discussed here; it is only a visual aid, albeit an extremely useful one. It is perhaps at this point that the formal mathematics begins to inform the intuition.

Although it is essential to use intuition as a guide to ideas and arguments, it is equally essential to finally accept only those conclusions that are provable in the formal context, in our case, derivable from the axioms of Archimedean ordered fields.

§2.3. The computational rules we have cited for rational numbers are all either axioms or formally derivable from the axioms of ordered fields. Hence they hold in any ordered field F, even if it is larger than $\mathbb{Q}$. We now collect some additional familiar notations and rules that hold in ordered fields, mostly in the form of exercises.

EXERCISE 20. Every *linear equation* $ax + b = 0$, $a \neq 0$ (x actually is present) has a unique solution.

EXERCISE 21. Define x^n for $x \in F$, $n \in \mathbb{N}$, and prove the laws of exponents.

EXERCISE 22. The aim is to extend the x^n notation to include exponents from $\mathbb{Z}$, in such a way that the laws of exponents continue to hold. For $0 \neq x \in F$, show that the only possible meanings of x^0 and x^{-n}, $n \in \mathbb{N}$, are $x^0 = 1$ and $x^{-n} = 1/x^n$. These definitions are then accepted, 0^0 and 0^{-n} are left undefined, and the laws of exponents do hold.

EXERCISE 23. The aim is to further extend exponent notation, for $0 \neq x \in F$, to include x^y for any $y \in \mathbb{Q}$. Show that the laws of exponents will continue to hold *only* if $x^{p/q}$ denotes an element of F whose qth power is x^p.

It is not clear whether F contains points of the required kind, for various $x \neq 0$, p and $q \neq 0$ in $\mathbb{Z}$. If it does, we represent them also through the use of radical notations:

for q odd, $x^{1/q} = \sqrt[q]{x}$ denotes the unique number whose qth power is x; and

for q even, if $y^q = x$, then also $(-y)^q = x$ and the notations $x^{1/q}$ and $\sqrt[q]{x}$ refer to the *positive* number whose qth power is x.

We shall have frequent occasion to refer to **intervals** in F; the subsets defined using the order in F and denoted as follows: for $a \leq b$ in F,

$(a, b) = \{x \in F: a < x < b\}$ **open** interval,

$[a, b) = \{x \in F: a \leq x < b\}$ left-closed, right-open interval,

$(a, b] = \{x \in F: a < x \leq b\}$ left-open, right-closed interval,

$[a, b] = \{x \in F: a \leq x \leq b\}$ **closed** interval.

The closed interval in case $a = b$ reduces to a single point; the others are empty in this case.

The **length** of any of these is $b - a$.

A similar notation is used for "half-lines" and the entire "line" F, using the symbol ∞:

$$(-\infty, a) = \{x \in F: x < a\},$$

$$[a, \infty) = \{x \in F: a \leq x\},$$

and

$$(-\infty, \infty) = F.$$

We define the **absolute value** of $x \in F$ by

$$x = \begin{cases} x & \text{if } x \geq 0, \\ -x & \text{if } x < 0. \end{cases}$$

EXERCISE 24. $|xy| = |x||y|$.

Intuitively, $|x|$ is just the distance of x from the origin. More generally, the distance between any two points, a, $b \in F$ is $|b - a|$. This intuition helps us "see" facts and find formal proofs for them.

Clearly, for each $x \in F$

$$-|x| \leq x \leq |x|,$$

and one of the equality signs holds.

EXERCISE 25. $|x| \leq y$ if and only if $-y \leq x \leq y$.

By adding the inequalities

$$-|x| \leq x \leq |x|$$

and

$$-|y| \leq y \leq |y|$$

we get

$$-(|x| + |y|) \leq x + y \leq |x| + |y|$$

and, by Exercise 25, we obtain

The **triangle inequality**: $|x + y| \leq |x| + |y|$.

EXERCISE 26. $||x| - |y|| \leq |x - y|$. (Hint: $x = (x - y) + y$.)

EXERCISE 27. The interval $[-a, a]$, $a > 0$, centered at 0, is the same as the set $\{x \in F: |x| \leq a\}$. What is the interval $(-a, a)$?

EXERCISE 28. From the interval $|x| \leq 1$ remove the open middle third $|x| < \frac{1}{3}$. From each remaining piece remove the open middle third. Repeat ad infinitum. Contemplate the set of points that remains.

We shall have frequent occasion to discuss subsets of F from the point of view of whether or not they have "arbitrarily large" members. The careful definition and terminology follows.

Definition. A set $S \subset F$ is **bounded above** if there is a $b \in F$ for which

$$x \in S \quad \text{implies} \quad x \leq b.$$

Any such b is called an **upper bound** of S. Clearly, any number larger than an upper bound of S is also an upper bound of S.

The Archimedean property of F is just the statement that $\mathbb{N}$ is not bounded above in F, or, in its other form,

$$\text{for } x > 0, \{nx: n \in \mathbb{N}\} \text{ is not bounded above.}$$

The negative integers in an Archimedean ordered field provide an example of a set that is not "bounded below."

EXERCISE 29. Define S **is bounded below** and l **is a lower bound of** S, for $S \subset F$.

A set is **bounded** if it is both bounded above and bounded below.

EXERCISE 30. S is bounded if and only if there is an $M \in F$ such that

$$x \in S \quad \text{implies} \quad |x| \leq M, \qquad \text{i.e.,} \quad S \subset [-M, M].$$

EXERCISE 31. Show that $\{2^n : n \in \mathbb{N}\}$ is not bounded above in $\mathbb{Q}$.

EXERCISE 32. For what values of x is $\{x^n : n \in \mathbb{N}\}$ bounded above, below, in an Archimedean ordered field?

EXERCISE 33. Show that, in $\mathbb{Q}$, the set

$$\left\{1 + \frac{1}{2} + \frac{1}{3} + \cdots + \frac{1}{n}; n \in \mathbb{N}\right\}$$

is not bounded above. Hint: $1 + \frac{1}{2} > 1$, $1 + \frac{1}{2} + (\frac{1}{3} + \frac{1}{4}) > 2$, $1 + \frac{1}{2} + (\frac{1}{3} + \frac{1}{4}) + (\frac{1}{5} + \frac{1}{6} + \frac{1}{7} + \frac{1}{8}) > 2\frac{1}{2}, \ldots$.

EXERCISE 34. Let

$$S_n = 1 + \frac{1}{2} + \frac{1}{2^2} + \frac{1}{2^3} + \cdots + \frac{1}{2^n}, \qquad n = 1, 2, 3, \ldots.$$

Is the set $\{S_n\}$ bounded above? Hint: Compute $S_n - \frac{1}{2}S_n$ and solve for S_n. Then use the resulting expression for S_n to see the answer.

EXERCISE 35. Let

$$T_n = 1 + \frac{1}{2!} + \frac{1}{3!} + \cdots + \frac{1}{n!}.$$

Compare T_n with S_n in Exercise 34 and decide whether $\{T_n\}$ is bounded above.

EXERCISE 36. Give an inductive definition of the set of numbers suggested by

$$\sqrt{2}, \qquad \sqrt{2 + \sqrt{2}}, \qquad \sqrt{2 + \sqrt{2 + \sqrt{2}}}, \ldots$$

(assuming an Archimedean ordered field that contains such numbers) and determine whether it is bounded above.

We shall have use for the process of "division with remainder" in an Archimedean ordered field (as in: 4 goes into 14 three times with a remainder of 2), called the **division algorithm**. For a **dividend** $n \in F$ and a **divisor** $0 < d \in F$, this asserts the existence in F of a **quotient** $q \in \mathbb{Z}$, and a **remainder** r which satisfies $0 \leq r < d$, giving

$$n = qd + r.$$

By the Archimedean property, $\{kd: k \in \mathbb{Z}\}$ is not bounded below. In particular, n is not a lower bound for this set; $kd < n$ for some integers k. Choose for q the largest such integer, so that $qd < n$ and $(q+1)d \geq n$. Then the remainder $r = n - qd$ falls in the required range, $0 \leq r < d$. If n and d are integers then so is r.

EXERCISE 37. Prove there is a largest integer q as asserted, by showing in the contrary case ($kd < n$ implies $(k+1)d < n$) that the Archimedean law is violated.

§3. Decimal Representation. Irrationals

The decimal system of representation of numbers employs as **digits** the symbols in the set $\{0, 1, 2, 3, 4, 5, 6, 7, 8, 9\}$. We take the word *number* to mean an element of a fixed Archimedean ordered field F, and adhere to the common practice of using the words number and point interchangeably, as we examine the decimal representation process.

§3.1. Given a number x, there is a largest integer that is $\leq x$, which we denote $[x]$.

EXERCISE 1. Prove this assertion.

We shall call $[x]$ the **integer part** of x and $x - [x]$ the **fractional part** of x. (Fractional in the sense $0 \leq x - [x] < 1$.)

For $x \geq 0$, each part will be assigned a decimal representation, i.e., a list of digits:

$$[x] = b_k b_{k-1} \ldots b_1 b_0, \text{ each, } b_i \text{ a digit, and } b_k \neq 0,$$

$$x - [x] = a_1 a_2 a_3 \ldots, \text{ each } a_j \text{ a digit,}$$

and then $b_k b_{k-1} \ldots b_1 b_0 . a_1 a_2 a_3 \ldots$ will be the decimal representing x.

For $x < 0$, the decimal for $|x|$ is preceded by a minus sign to get the decimal for x.

The process begins with the following observation: The set $\{10^n: n \in \mathbb{N}\}$ is not bounded above. Hence, given $x \geq 0$, it contains a first member 10^k that exceeds $[x]$.

EXERCISE 2. Prove these assertions.

Divide $[0, 10^k)$ into ten equal subintervals

$$[0, 10^{k-1}), [10^{k-1}, 2 \cdot 10^{k-1}), \ldots, [9 \cdot 10^{k-1}, 10^k)$$

and associate them respectively with the digits 0, 1, 2, ..., 9 (the multiplier of 10^{k-1} in the expression for the left endpoint). Choose as b_k the digit of the interval that contains $[x]$;

$$[x] \in [b_k 10^{k-1}, (b_k + 1)10^{k-1}).$$

Next, divide this interval in ten equal parts and choose b_{k-1} as the digit associated with the one containing $[x]$. Continuing, we arrive at intervals of length $1 = 10^0$ and a choice of b_0.

If all digits so chosen are 0, then $[x] = 0$ and the expansion is taken to be 0, simply. With this exception, we drop any initial digits 0 in the list before writing the decimal

$$[x] = b_k b_{k-1} \dots b_1 b_0 . , \qquad b_k \neq 0.$$

Note that the use of half-open intervals insures at each stage that exactly one subinterval contains $[x]$, hence that each b_i is well determined.

For the fractional part $x - [x]$, we choose the digits $a_1, a_2, a_3, \dots$ by induction as we now indicate.

Divide the interval $[0, 1)$, which contains $x - [x]$, into ten equal parts

$$[0, \tfrac{1}{10}), [\tfrac{1}{10}, \tfrac{2}{10}), \dots, [\tfrac{9}{10}, 1)$$

and take as a_1 the digit (numerator of left endpoint) corresponding to the subinterval that contains $x - [x]$.

Next, divide $[a_1/10, (a_1 + 1)/10)$ into ten equal parts

$$\left[\frac{a_1}{10}, \frac{a_1}{10} + \frac{1}{10^2}\right), \left[\frac{a_1}{10} + \frac{1}{10^2}, \frac{a_1}{10} + \frac{2}{10^2}\right), \dots, \left[\frac{a_1}{10} + \frac{9}{10^2}, \frac{a_1 + 1}{10}\right)$$

and choose as a_2 the digit (multiplier of $1/10^2$ for left endpoint) of the one that contains $x - [x]$.

Continue, to get $a_1 a_2 a_3 \dots$ and hence the **decimal of** $\boldsymbol{x}$:

$$x = b_k b_{k-1} \dots b_1 b_0 . a_1 a_2 a_3 \dots .$$

If x happens to be an endpoint of one of the subintervals, the digits for x are all 0 from some stage on, and conversely. In this case, x is just the finite sum

$$x = b_k 10^k + b_{k-1} 10^{k-1} + \cdots + b_1 10 + b_0 + \frac{a_1}{10} + \frac{a_2}{10^2} + \cdots + \frac{a_l}{10^l}$$

and the decimal can be viewed as an abbreviation for this sum. If x is not an endpoint, the decimal has infinitely many nonzero digits and a sum interpretation of the decimal for x requires a dicusssion of infinite series.

Proposition. *Distinct numbers have distinct decimal representations.*

Proof. Suppose $x < y$. Since $\{10^k\}$ is unbounded, by the Archimedean property of F, there is a k for which $0 < 1/(y - x) < 10^k$, i.e.,

$$\frac{1}{10^k} < y - x.$$

Clearly, the choices of a_k for x and y differ.

Every number has a decimal uniquely assigned to it. It is not clear whether every decimal stands for a number, or, stated more precisely, whether F can be taken large enough to insure *there is* a number for every decimal. We consider this later.

The endpoints of the intervals of subdivision are assigned decimals that we might describe as **terminating**, i.e., they conclude with an infinite *repetition of the digit* 0. For example

$$\frac{1}{4} = \frac{25}{100} = 0.25000\ldots.$$

We could just as well have settled on left-open, right-closed intervals for the subdivisions. The only differences in the decimal representations would be at the endpoints, where a decimal of the form xxx999... would then result, such as

$$\tfrac{1}{4} = 0.24999\ldots.$$

Either method sets up a one–one correspondence between the numbers in F and all decimals of a certain kind.

A further alternative is to use closed subintervals and accept an ambiguity in the decimal representations of some numbers; each endpoint would belong to two subintervals at some stage, and the two choices $a_k000\ldots$ and $(a_k - 1)999\ldots$ would result. This is a tolerable degree of ambiguity and it is commonly accepted, in the form of statements like

$$0.25000\ldots = 0.24999\ldots.$$

The rationals are present in any Archimedean ordered field, so it is natural to ask if there is a way to recognize whether a given decimal represents a rational. We could write down the decimals for a large number of rationals and search for common features, formulate conjectures and then try to prove them, refining the conjecture until the right one—the provable one—is found.

EXERCISE 3. Try some of this before reading what follows.

Consider the decimal $0.735421421421\ldots$ that has the 421 group of digits repeating: we can write it $0.735\overline{421}$. If x is the number it represents then

$1000x = 735.\overline{421}$ and

$$1000x - 735 = 0.\overline{421}.$$

Hence

$$1000[1000x - 735] = 421 + 0.\overline{421}$$

and, equating two versions of $0.\overline{421}$,

$$1000[1000x - 735] - 421 = 1000x - 735,$$

$$999[1000x - 735] = 421,$$

$$999000x = 421 + 999 \cdot 735,$$

which shows that x is rational.

EXERCISE 4. Define "recurring decimal" and show a notation for an arbitrary one. Show that every recurring decimal represents a rational.

We show next the converse: the decimal for any rational is recurring. It is enough to show this for a rational p/q in $(0, 1)$.

Divide $10p$ by q using the division algorithm:

$$10p = a_1 q + r_1 \qquad \text{where} \quad r_1 \in \{0, 1, 2, \ldots, q - 1\} \quad (\text{since } 0 \leq r_1 < q).$$

The quotient a_1 and remainder r_1 have interpretations, which are revealed by dividing through by $10q$:

$$\frac{p}{q} = \frac{a_1}{10} + \frac{r_1}{10q} \quad \text{with} \quad 0 \leq \frac{r_1}{10q} < \frac{1}{10}.$$

Clearly, a_1 is the first digit in the expansion of p/q, in view of the inequality.

Now divide $10r_1$ by q

$$10r_1 = a_2 q + r_2, \qquad \text{where} \quad r_2 \in \{0, 1, 2, \ldots, q - 1\} \quad (\text{since } 0 \leq r_2 < q).$$

Again, the interpretations of a_2 and r_2 are found by division, this time by $10^2 q$:

$$\frac{r_1}{10q} = \frac{a_2}{10^2} + \frac{r_2}{10^2 q} \qquad \text{with} \quad 0 \leq \frac{r_2}{10^2 q} < \frac{1}{10^2}.$$

For we have then

$$\frac{p}{q} = \frac{a_1}{10} + \frac{a_2}{10^2} + \frac{r_2}{10^2 q} \qquad \text{with} \quad 0 \leq \frac{r_2}{10^2 q} < \frac{1}{10^2}$$

in which it is clear that a_2 is the second digit in the expansion of p/q. Continuing this process, we obtain the sequence of equations

$$\begin{aligned} 10p &= a_1 q + r_1, \\ 10r_1 &= a_2 q + r_2, \\ 10r_2 &= a_3 q + r_3, \\ &\vdots \end{aligned}$$

in which the quotients form the decimal for p/q

$$\frac{p}{q} = 0.a_1 a_2 a_3 \ldots$$

and the remainders all belong to $\{0, 1, 2, \ldots, q-1\}$.

It is clear that there must be a recurrence of the same value for the remainder before the "q plus second" equation is reached, which means some group of the equations in the sequence recurs, hence that a group of the digits recurs in the expansion of p/q.

EXERCISE 5. Make up a nonrecurring decimal.

The entire discussion of decimal representations of numbers in an Archimedean ordered field can be repeated with the following modifications. Instead of dividing intervals in ten equal parts and assigning digits from the set $\{0, 1, \ldots, 9\}$, make the divisions into B equal parts ($B \geq 2$) and use the digits $0, 1, \ldots, B-1$. The result is the **base *B* expansion** of numbers. None of our major observations is changed. Every number has a base B representation and distinct numbers have distinct representation. Half-open intervals taken in the two ways lead, for the endpoints of the intervals of subdivision, either to a terminating expansion (left-closed) or one that repeats the last digit ($B-1$) (right-closed). Closed intervals give ambiguous expansions at those points. Rationals are characterized by having recurring expansions.

EXERCISE 6. Consider the ternary (base 3, digits 0, 1, 2) expansions of numbers in $[0, 1]$. Suppose those numbers of $[0, 1]$ for which both the expansions (terminating and nonterminating) have first digit 1 are removed from $[0, 1]$. What set remains? (Note $\frac{2}{3} = 0.200\ldots$ remains, as does $\frac{1}{3} = 0.0222\ldots$.) Remove, in addition, those numbers whose second digit is 1 in both versions. What set remains? Continue in this way ad infinitum. Describe the set that remains. It is called the **Cantor Set**.

We may regard a base-B expansion

$$\pm b_k \ldots b_2 b_1 b_0 . a_1 a_2 a_3 \ldots,$$

where each $a_i, b_j \in \{0, 1, \ldots, B-1\}$, as a series of *instructions* for finding a number. Supposing the ambiguous closed interval method is understood, these successive digits designate a succession of closed intervals in which the number is to be sought. The question whether F contains a number for the expansion to represent is seen to be that of whether the intersection of a certain sequence of closed intervals is empty.

Some definitions help to formulate this idea.

Definition. A sequence, $\{x_n\}$, of numbers in an Archimedean ordered field F

converges to 0, written $x_n \to 0$, if for every $\varepsilon > 0$ in F there is an index N_ε such that

$$|x_m| < \varepsilon \qquad \text{whenever} \quad m > N_\varepsilon.$$

EXAMPLES.

$$\frac{1}{n} \to 0, \qquad \frac{1}{B^n} \to 0 \qquad \text{if} \quad B > 1.$$

Remark. The Archimedean property insures that examples exist.

EXERCISE 7. Prove that if $\{y_n\}$ is a strictly increasing sequence that is not bounded above, then $1/y_n \to 0$.

Definition. A **nest** of closed intervals is a sequence $\{I_n\}$ of closed intervals satisfying

$$I_1 \supset I_2 \supset I_3 \ldots \quad \text{and} \quad \text{length } I_n \to 0.$$

Each base B expansion yields a nest of closed intervals if $B > 1$, as noted above.

EXERCISE 8. For a nest of closed intervals $\{I_n\}$, the intersection $\bigcap_{n=1}^{\infty} I_n$ cannot contain two distinct points. Hint: If $x < y$ are any two distinct points, consider $\varepsilon = (y - x)/2$ in the definition of length $I_n \to 0$.

Thus for any nest of closed intervals in the Archimedean ordered field F the intersection $\bigcap_{n=1}^{\infty} I_n$ is either empty or a single point. The property of F that insures the existence of a number in F for every base B expansion to represent is just the

Nested Intervals Property: For every nest of closed intervals in F the intersection $\bigcap_{n=1}^{\infty} I_n$ is nonempty.

The Archimedean ordered fields that have this property are the ones in which the use of decimal, or other base B, representations of numbers is completely satisfactory.

EXERCISE 9. The use of closed intervals is essential, in order that the Nested Intervals Property say what we want it to say; taking $I_n = (0, 1/10^n]$ gives $\bigcap_{n=1}^{\infty} I_n$ empty, even though the number it leads to, namely 0, is present in F. Give an example of a "nest" of left-open, right-closed intervals whose intersection is the number 0.

§3.2. The side and diagonal of any square are noncommensurable segments. For, suppose some square had side su and diagonal du, for some unit segment u, and $s, d \in \mathbb{N}$. The Pythagorean Theorem gives $d^2 = s^2 + s^2$, or $d^2 = 2s^2$. Thus d/s solves $x^2 = 2$ in $\mathbb{Q}$. But we have the

Proposition. $x^2 = 2$ *has no rational solution.*

Proof. We show that $(m/n)^2 = 2$ forces both m and n to be even, which would prevent the fraction m/n from having a lowest-terms form, which is impossible.

Now, $(m/n)^2 = 2$ means $m^2 = 2n^2$, so m^2 is even. But m^2 can only be even if m is even, so $m = 2l$, for some $l \in \mathbb{N}$. Then $4l^2 = 2n^2$, or $n^2 = 2l^2$ is even and so n is even.

EXERCISE 10. Show that m^2 even implies m is even.

Definition. If F is any ordered field, the members of $F - \mathbb{Q}$ are called **irrational** elements of F.

The set $I = F - \mathbb{Q}$ of all irrational elements of F may be empty, but if F contains a solution of $x^2 = 2$ it must be in I; $\sqrt{2}$ *is irrational.* (Here, as always, $\sqrt{2}$ means the positive solution. The negative solution is $-\sqrt{2}$.)

The sum of two irrationals can be rational ($\sqrt{2} + (-\sqrt{2}) = 0$) but the sum of a rational and an irrational must be irrational.

EXERCISE 11. Prove this.

EXERCISE 12. Discuss products.

Definition. A subset $D \subset F$ is **dense** if there is a member of D between any two members of F (hence infinitely many members of D between them).

Now suppose F is Archimedean, so that $\mathbb{Q}$ is known to be dense. If there is an irrational $\alpha \in F$ then the set of irrationals

$$\{\alpha + r : r \in \mathbb{Q}\}$$

is seen to be dense, because if $x < y$ in F then $x - \alpha < y - \alpha$ and so $x - \alpha < r < y - \alpha$ for some $r \in \mathbb{Q}$, i.e., $x < \alpha + r < y$. Thus, *the set of irrationals is either empty or dense* in an Archimedean ordered field F.

This reveals the nature of any proper extension of $\mathbb{Q}$ to an Archimedean ordered field F: there must be infinitely many numbers inserted between each pair of rationals to achieve such an extension.

We have seen that $\mathbb{Q}$ is adequate for measurement from a practical point of view (see p. 20). From an ideal point of view, however, we must consider the existence of noncommensurable segments as constituting a defect of $\mathbb{Q}$ for purposes of measurement. Stated otherwise, the "line" as represented by all points with rational "coordinates" has "gaps"; no matter what the unit segment is taken to be, the diagonal of the corresponding square does not end at a point of the rational "line."

From the point of view of decimal (and other base) expansions, the existence of nonrecurring expansions can also be considered a defect of $\mathbb{Q}$.

Finally, the lack of a solution to $x^2 = 2$ in $\mathbb{Q}$ is a defect of $\mathbb{Q}$ from the algebraic point of view.

A more general way to generate irrationals is seen when we replace $x^2 = 2$ by any polynomial equation formulated in $\mathbb{Q}$ and consider its solutions.

If $n \in \mathbb{N}$, a **polynomial equation of degree n in $\mathbb{Q}$** is any equation

$$a_n x^n + a_{n-1} x^{n-1} + \cdots + a_1 x + a_0 = 0$$

with

$$a_n \neq 0 \quad \text{and} \quad a_i \in \mathbb{Q}, \qquad i = 0, 1, \ldots, n.$$

(If the equation holds, we can multiply both sides by all of the denominators of the fractions a_i, $i = 0, 1, \ldots, n$, and the resulting equation with integer coefficients holds, and conversely, i.e., we can assume $a_i \in \mathbb{Z}$, $i = 0, 1, \ldots, n$.)

Definition. An **algebraic number** in an ordered field F is any $x \in F$ that satisfies a polynomial equation in F with integer coefficients.

Algebraic numbers can be irrational; $x^2 - 2 = 0$, but all rationals are algebraic; $qx - p = 0$. We show that a rational number bears a special relation to every polynomial equation it solves (integer coefficients).

Proposition. *If p/q in lowest terms solves $a_n x^n + a_{n-1} x^{n-1} + \cdots + a_1 x + a_0 = 0$, $a_i \in \mathbb{Z}$, $i = 0, 1, \ldots, n$ and $a_n \neq 0$, then*

$$p \text{ divides } a_0; \qquad a_0 = p \cdot \text{an integer}$$

and

$$q \text{ divides } a_n; \qquad a_n = q \cdot \text{an integer}.$$

Proof. The hypothesis is

$$a_n \left(\frac{p}{q}\right)^n + a_{n-1} \left(\frac{p}{q}\right)^{n-1} + \cdots + a_1 \frac{p}{q} + a_0 = 0.$$

Multiplying by q^n,

$$a_n p^n + a_{n-1} p^{n-1} q + \cdots + a_1 p q^{n-1} + a_0 q^n = 0$$

or

$$\begin{aligned} a_n p^n &= q(-a_{n-1} p^{n-1} - \cdots - a_1 p q^{n-2} - a_0 q^{n-1}) \\ &= q \cdot \quad \text{an integer}. \end{aligned}$$

Thus q divides $a_n p^n$. But q does not divide p^n, so it must divide a_n. For the other conclusion, solve for $a_0 q^n$ and argue similarly.

This allows us to check algebraic numbers for irrationality by showing they are not among the possible rational solutions of some equation they satisfy. Thus, the only possible rational solutions of $x^2 - 2 = 0$ being ± 2, there are none; $\sqrt{2}$ is irrational.

EXAMPLE. $\sqrt{3 + \sqrt{2}}$ is irrational because, calling it x,

$$x^2 = 3 + \sqrt{2},$$

$$(x^2 - 3)^2 = 2,$$

$$x^4 - 6x^2 + 7 = 0,$$

and only ± 1, ± 7 are candidates for rational solutions.

EXERCISE 13. Give the details of the rest of the proof of the proposition.

EXERCISE 14. The proof uses the fact about $\mathbb{Z}$, that a divisor of ab that is not a divisor of a must be a divisor of b. A proof can be devised using the division algorithm and occurs in books on Modern Algebra and/or Number Theory. Find a proof.

EXERCISE 15. Make up some numbers involving two radicals and see if you can decide whether they are irrational.

EXERCISE 16. Make up an irrational number that involves three radicals (reduced, i.e., $\sqrt{\sqrt{\sqrt{2}}}$ is one radical). Prove it is irrational.

Part I

Analysis

The three broad categories of Mathematics are Algebra, Geometry, and Analysis. There is no sharp definition of these categories or clear distinction between them, but, generally speaking, it is the idea of approximation which characterizes Analysis.

Many of the central concepts of Analysis are technical in nature; some seem artificial, but each appeared originally in the course of a search for answers to fundamental questions. We shall endeavor to present them in such contexts.

CHAPTER 1

Approximation: The Real Numbers

The fixing of a reference point, or origin, and a unit point on a line constitutes the establishment of a *coordinate system* on that line. For each choice of a coordinate system on a line there is determined the set of rational points; those measurable using $\mathbb{Q}$, and the set of irrational points such as $\sqrt{2}$. We have mentioned the idea of using the fact that $\mathbb{Q}$ is dense, to "approximate" irrationals by rationals, a notion that we must make more precise.

If one is working in a specific practical context (e.g., building a house), that context includes the means to arrive at a judgment of what constitutes "good enough" approximation, or "small enough" error ε. Factors affecting this judgment include: the uses to which numbers will be put; the precision of any instruments being used, and the units employed.

EXAMPLES. 1000 is a *small* number of grains of sand on a beach, but a *large* number in a serving of spinach. Context!

0.01 is a *small* error in a paycheck expressed in dollars, but a *large* error in a bank transfer expressed in millions of dollars. Units!

It is easy to imagine circumstances in which 1.4 is an adequate $\sqrt{2}$ and 22/7 is a satisfactory π ... the errors would be "small enough."

Our point is that the meaning of "small enough" error depends critically on context, including choice of unit.

This observation is to be taken together with the observation that

For mathematics to be applicable it is essential that it be free of special context, i.e., that it be abstract.

In the absence of a context, the only possible meaning of "small enough error" is that *arbitrarily small* error be achievable: error less than ε for any $\varepsilon > 0$.

(Note: For this to say what it intends, the error bound ε should be taken from an Archimedean ordered field, where there *are* arbitrarily small positive numbers: $\mathbb{Q}$, for example.)

These comments lead us to a precise, context free, meaning for: *a number x is approximated by numbers in a set A*. Namely, for any $\varepsilon > 0$ there is an $a_\varepsilon \in A$ such that $|x - a_\varepsilon| < \varepsilon$. Of course the approximating number a_ε might change as the error bound ε changes: the notation a_ε stresses this. The statement expresses the possibility of achieving "good enough" accuracy regardless of the context.

One could say that the subject matter of "Analysis" is the study of approximation in this sense. The first step in such a study would be the extension of $\mathbb{Q}$ to an Archimedean ordered field that includes all numbers that *can be* approximated by sets of rationals. Using such "numbers" one could analyze *any* particular practical context, confident that whatever amounts of various quantities are encountered, they either are rational or are approximable by rationals.

The discussion of how to extend $\mathbb{Q}$ in this way must begin with a description *in terms of* $\mathbb{Q}$ of those numbers that can be approximated by sets of rationals. We begin by so describing $\sqrt{2}$, to point the way to more general cases.

§1. Least Upper Bound

If an Archimedean ordered field F contains $\sqrt{2}$, we can describe $\sqrt{2}$ naturally in terms of rationals by use of the set

$$A = \{0 < r \in \mathbb{Q}: r^2 < 2\}.$$

Surely, $\sqrt{2}$ can be approximated by this set.

We shall examine the relationship of $\sqrt{2}$ to A, to express it in a form that applies generally.

First, $\sqrt{2}$ is an upper bound of A since $s > \sqrt{2}$ implies $s \notin A$.

EXERCISE 1. Prove this by multiplying $s > \sqrt{2}$ by appropriate positive quantities in F.

The error in approximating $\sqrt{2}$ by $r \in A$ is $\sqrt{2} - r$. To say that this error is less than $\varepsilon > 0$ is to say

$$r > \sqrt{2} - \varepsilon \qquad \text{with} \quad r \in A,$$

that is, $\sqrt{2} - \varepsilon$ is *not* an upper bound of A. The statement that $\sqrt{2}$ can be approximated by A is the same as this statement:

$\sqrt{2}$ is an upper bound of A and no smaller number $\sqrt{2} - \varepsilon$ is an upper bound of A.

Since rational values of ε suffice to describe arbitrarily small numbers, this gives a characterization of $\sqrt{2}$ in terms of $\mathbb{Q}$. The point is this: If we wished to adjoin $\sqrt{2}$ to $\mathbb{Q}$ we could describe the required number within $\mathbb{Q}$, as an upper bound of A that is its least upper bound, i.e., no smaller number is an upper bound of A.

The idea generalizes to the important

Definition. The number l is the **least upper bound** of a set A if l is an upper bound of A and no smaller number is an upper bound of A; $l > a$ for all $a \in A$ and $\varepsilon > 0$ implies $l - \varepsilon < a_\varepsilon$ for some $a_\varepsilon \in A$.

It is obvious that a set can have a least upper bound only if it is bounded above.

EXERCISE 2. Prove that l is the least upper bound of A if and only if l is an upper bound of A and $l \leq u$ whenever u is an upper bound of A. (In other words, l is an upper bound of A and a lower bound of the set of all upper bounds of A.)

EXERCISE 3. Define "g is the **greatest lower bound** of a set A" and imitate Exercise 2.

The terms **supremum** and **infimum** are commonly used instead of least upper, respectively, greatest lower bound. These terms give rise to the standard notations

$$\sup A \quad \text{and} \quad \inf A.$$

Clearly, $\inf A \leq \sup A$ for any bounded nonempty set A.

EXERCISE 4. If $A \neq \phi$ is bounded above then

$$\sup A = \inf\{u\colon u \geq a \text{ for all } a \in A\}.$$

EXERCISE 5. Exhibit sets in $\mathbb{Q}$ that have:

(a) both an inf and a sup in $\mathbb{Q}$;
(b)–(c) one but not the other; and
(d) neither.

Exhibit bounded sets of each kind.

EXERCISE 6. For a set A having a supremum, $\sup A$ may or may not belong to A. Give examples in $\mathbb{Q}$.

If $\sup A$ exists and belongs to A, it is the largest member of A and is also denoted $\max A$, the **maximum** of A. Similarly for $\min A$, the **minimum** of A, when it exists.

EXERCISE 7. For each of the following subsets of $\mathbb{Q}$ identify (including "doesn't exist") the inf, sup, min, and max:

(a) $\mathbb{N}$;
(b) the value set of $\{(-1)^n\}$;
(c) the value set of $\{\tan n\}$;
(d) the value set of $\{n^{(-1)^n}\}$; and
(e) $\{x \in \mathbb{Q}: x^3 < 8\}$.

We complete the discussion of the $\sqrt{2}$ example as follows.

Let $x = \sup\{0 < r \in \mathbb{Q}: r^2 < 2\}$ exist in an extension F of $\mathbb{Q}$. We show that $x^2 = 2$ by showing that $x^2 < 2$ and $x^2 > 2$ cannot hold.

First, $x^2 < 2$ cannot hold, for if it did we could increase x slightly, to $x + h$, $h > 0$, while retaining a square less than 2, which is impossible since then some $r \in \mathbb{Q}$ satisfies $x < r < x + h$ and $r^2 < (x + h)^2 < 2$; r would be a member of the set, exceeding the sup of the set.

We can in fact figure out a specific h that does this, i.e., for which $(x + h)^2 < 2$ and $h > 0$. If $h < 2$,

$$(x + h)^2 = x^2 + 2xh + h^2 = x^2 + h(2x + h) < x^2 + h(2x + 2).$$

The last expression is equal to 2 provided

$$h = \frac{2 - x^2}{(2x + 2)},$$

and this h works, i.e., $h > 0$ since $x^2 < 2$ and $h < 2$, obviously.

Second, $x^2 > 2$ cannot hold, for if it did we could lower x slightly, to $x - h$, $h > 0$, retaining a square greater than 2, which is impossible since this $x - h$ would be an upper bound of the set, smaller than the least upper bound x.

We conclude $x^2 = 2$ because of trichotomy.

EXERCISE 8. Supply the details of the second part of the argument, including a specific h.

§2. Completeness. Nested Intervals

The example $\sqrt{2}$ and the general ideas to which it leads suggest how to adjoin numbers to $\mathbb{Q}$ in order to achieve an extension F that includes all "numbers" that can be approximated by rationals: namely, provide a least upper bound for every set that should have one.

The sets that should have least upper bounds are the sets that are bounded above, except for the empty set.[1]

[1] Any number exceeds all members of the empty set since there are none. Thus, all numbers are upper bounds ... no least upper should exist.

Definition. An ordered field F is **complete** if it includes a least upper bound for every nonempty subset that is bounded above.

Any complete ordered field is then an extension of $\mathbb{Q}$ of the desired kind.

Proposition. *A complete ordered field is Archimedean.*

EXERCISE 1. Proof? Hint: Suppose $\mathbb{N}$ has an upper bound b. Consider $b - 1$.

EXERCISE 2. Show that F is complete if and only if every nonempty set that is bounded below has a greatest lower bound.

Recall (see p. 30) that for an extension of $\mathbb{Q}$ to be satisfactory as regards decimal expansions (and expansions in other bases) it should have the nested intervals property. We have the

Nested Intervals Theorem. *A complete ordered field has the nested intervals property.*

Proof. Let $I_n = [a_n, b_n]$, $n = 1, 2, 3, \ldots$, be any nest of closed intervals. Then

$$a_1 \leq a_2 \leq a_3 \leq \cdots \leq b_3 \leq b_2 \leq b_1.$$

We must show $\bigcap_{n=1}^{\infty} I_n \neq \phi$.

The set $A = \{a_1, a_2, \ldots\}$, of all left endpoints is nonempty and bounded above; indeed, each right endpoint is an upper bound. Thus sup A exists in F by the hypothesis. But clearly,

$$a_n \leq \sup A \leq b_n \qquad \text{for all} \quad n,$$

i.e.,

$$\sup A \in \bigcap_{n=1}^{\infty} I_n.$$

On the other hand, we also have the

Theorem. *An Archimedean ordered field with the nested intervals property is necessarily complete.*

Proof. Consider any nonempty set A that is bounded above. If A has a maximum, it is sup A. Assuming A has no maximum, we find sup A as the common point of a nest of intervals as follows. Let a_1 and b_1 be any nonupper[2] bound, respectively, upper bound of A. Such points exist by the assumptions on A.

Either the midpoint $(a_1 + b_1)/2$ is an upper bound of A or it isn't. If it is, put $a_2 = a_1$ and $b_2 = (a_1 + b_1)/2$. If it isn't, put $a_2 = (a_1 + b_1)/2$ and $b_2 = b_1$.

[2] We choose *nonupper bound* to express "a number that is not an upper bound" over other awkward expressions.

Repeat this definition (induction) to form successive midpoints, each of which is adjoined to the "a sequence" or the "b sequence," yielding

$$a_1 \le a_2 \le a_3 \le \cdots \le b_3 \le b_2 \le b_1$$

with these properties:

(i) each a_n is not an upper bound of A;
(ii) each b_n is an upper bound of A; and
(iii) $b_n - a_n \to 0$. (Since $b_1 - a_1$ is divided by 2 at each stage, i.e., $b_n - a_n = (b_1 - a_1)/2^{n-1} \to 0$.)

By hypothesis, there is an $x \in \bigcap_{n=1}^{\infty} [a_n, b_n]$. We show $x = \sup A$ in two steps:

1. $x \ge a$ for all $a \in A$. Otherwise, i.e., $x < a$ for some $a \in A$, choosing the positive number $\varepsilon = a - x$ in the definition of $b_n - a_n \to 0$ yields an index N such that

$$b_N - a_N < a - x,$$

or

$$b_N < a + (a_N - x).$$

But $a_N - x < 0$, so $b_N < a$, a contradiction since b_N is an upper bound of A.

2. $x \le b$ if b is an upper bound of A. Suppose, to the contrary, that $x - b > 0$ for some upper bound b of A. Choose N so that

$$b_N - a_N < x - b,$$

or

$$a_N - b_N > b - x.$$

But $x \le b_N$ or $-x \ge -b_N$. Hence

$$a_N - b_N > b - b_N,$$

or

$$a_N > b.$$

This is absurd since a_N is a nonupper bound, while b is an upper bound.

Thus the requirements for satisfactory extensions of $\mathbb{Q}$ from two distinct points of view coincide. We consider another, slightly different, point of view next—approximation by sequences.

§3. Bounded Monotonic Sequences

The task of approximating a number is perhaps most naturally carried out recursively. That is, a sequence of numbers $x_1, x_2, x_3, \ldots, x_n, \ldots$ is generated in a way that produces better and better approximations as n increases, so that any prescribed degree of approximation $\varepsilon > 0$ is achieved, once n is large enough. For example, considering $\sqrt{2}$ again, suppose $x_1 > 0$ is any guess for $\sqrt{2}$. Then x_1 and $2/x_1$ have product 2 and the closer to equal these factors are, the closer to $\sqrt{2}$ each is. The idea of averaging the two factors suggests itself at once, as a means of improving the estimate, to

$$x_2 = \frac{1}{2}\left(x_1 + \frac{2}{x_1}\right).$$

EXERCISE 1. Show that the new pair of factors is indeed closer to equal than the old pair, i.e.,

$$\left|x_2 - \frac{2}{x_2}\right| \leq \left|x_1 - \frac{2}{x_1}\right|.$$

By induction,

$$\begin{aligned} x_1 &> 0, \\ x_{n+1} &= \frac{1}{2}\left(x_n + \frac{2}{x_n}\right), \end{aligned}$$

we define a sequence $\{x_n\}$ which we expect to yield arbitrarily good approximations to $\sqrt{2}$.

To test this expectation we need to define it precisely (in an Archimedean ordered field F).

Definition. A number x is the **limit** of a sequence $\{x_n\}$ if, for every $\varepsilon > 0$, there is an index N such that $n \geq N$ gives $|x_n - x| < \varepsilon$.

Thus, the arbitrary degree of approximation $\varepsilon > 0$ is achieved by all terms of the sequence from some index N on.

EXERCISE 2. At most one number x can be the limit of a given sequence $\{x_n\}$ (*limits are unique*). Hint: Suppose x and y are both limits of $\{x_n\}$. Show that $|x - y| < \varepsilon$ for any $\varepsilon > 0$ (hence that $x = y$) by using the estimate

$$|x - y| = |x - x_n + x_n - y| \leq |x - x_n| + |y - x_n|.$$

This justifies the usage "*the* limit" in the definition, as well as the introduction of notations:

$$x = \lim x_n, \quad \text{or} \quad \lim_{n\to\infty} x_n = x$$

and

$$x_n \to x, \quad \text{or} \quad x_n \to x \quad \text{as} \quad n \to \infty.$$

The definition includes that of $x_n \to 0$ as we stated it on p. 29.

If a limit exists for $\{x_n\}$, we say $\{x_n\}$ **converges** or is **convergent**. Otherwise, we say $\{x_n\}$ **diverges**.

Remark. The definition of $x_n \to x$ says that x can be approximated by A, where A is the value set of the sequence $\{x_n\}$, but it says more. It restricts how the *function* $n \to x_n$, of $\mathbb{N}$ into A, is involved in making this approximation hold.

Definition. A sequence $\{x_n\}$ is **bounded above** if its value set $\{x_n: n \in \mathbb{N}\}$ is bounded above. Similarly for $\{x_n\}$ **bounded below** and **bounded**.

Proposition. *Every convergent sequence is bounded.*

EXERCISE 3. Proof? Hint: Fix $\varepsilon > 0$. All but a finite number of the values x_n lie in $(x - \varepsilon, x + \varepsilon)$.

Definition. A sequence $\{x_n\}$ is **increasing** if $x_n \leq x_{n+1}$ for all $n \in \mathbb{N}$ and **strictly increasing** if $x_n < x_{n+1}$ for all $n \in \mathbb{N}$. Similarly, $\{x_n\}$ is **decreasing (strictly decreasing)** if $x_n \geq x_{n+1}$ ($x_n > x_{n+1}$) for all $n \in \mathbb{N}$.

All four kinds of special sequences are called **monotonic**.

The intuition about numbers that we get from thinking of the line suggests very strongly that an increasing sequence (x_n moves to the right or stays fixed as n increases) which is bounded above (x_n stays to the left of some b) must accumulate at some $x \leq b$; i.e., $x_n \to x$ must hold for some $x \leq b$.

$x_1 \quad x_2 \quad x_3 \quad \to \quad x$

b

Similarly for decreasing and bounded below.

We return to the example

$$x_1 > 0 \quad \text{and} \quad x_{n+1} = \frac{1}{2}\left(x_n + \frac{2}{x_n}\right) = \frac{x_n^2 + 2}{2x_n}.$$

EXERCISE 4. Show that $a^2 + b^2 \geq 2ab$, for any $a, b \in F$.

EXERCISE 5. Show that $2 \leq x_{n+1}^2$ for $n \geq 1$. Hint: Use Exercise 4 to calculate. Avoid writing $\sqrt{2}$.

EXERCISE 6. Show $\{x_n\}$ for $n \geq 2$ is decreasing. Use Exercise 5.

The intuition above suggests that $x_n \to x$ for some x with $2 \leq x^2$, which we expect is the number $\sqrt{2}$, if F contains such a number. Further support for this expectation results if we suppose x_n converges to *some* $l \in F$ and that it is permissible to let $n \to \infty$ in the equation

$$x_{n+1} = \frac{1}{2}\left(x_n + \frac{2}{x_n}\right).$$

For then l must satisfy

$$l = \frac{1}{2}\left(l + \frac{2}{l}\right),$$

or

$$l^2 = 2.$$

These considerations suggest that a way of expressing the statement "$\sqrt{2} \in F$" is to say that $\{x_n\}$ has a limit in F, for the above bounded, monotonic sequence. More generally, we might (boldly!) conjecture that F is complete if it contains limits for *all* of its bounded, monotonic sequences.

Theorem. *An Archimedean ordered field is complete if and only if each of its bounded, monotonic sequences converges.*

Proof. Consider a complete F and a sequence $\{x_n\}$ in F that is increasing and bounded (same as bounded above since x_1 is a lower bound). By completeness $x = \sup x_n$ exists. We show $x_n \to x$.

For any $\varepsilon > 0$, $x - \varepsilon$ is not an upper bound of $\{x_n\}$, hence

$$x_N > x - \varepsilon \qquad \text{for some } N.$$

If $n \geq N$, then $x_n \geq x_N > x - \varepsilon$, so $|x_n - x| < \varepsilon$. The case of a decreasing sequence is similar.

Conversely, suppose every bounded, monotonic sequence in F converges. Consider a nonempty set A that is bounded above. We need only pursue the case in which A has no maximum. In that case, a nonupper bound of A exists.

Define a_1 to be any nonupper bound of A and b_1 to be any upper bound of A. The midpoint $(a_1 + b_1)/2$ is either an upper bound of A or it isn't. If it is, call it b_2 and let $a_2 = a_1$. If not, call it a_2 and put $b_2 = b_1$. Repeat the process using the midpoint $(a_2 + b_2)/2$ to define a_3 and b_3, and so on by induction.

There results $\{a_n\}$, nonupper bounds, and $\{b_n\}$, upper bounds, for which

$$a_1 \leq a_2 \leq a_3 \leq \cdots \leq b_3 \leq b_2 \leq b_1$$

and $b_n - a_n \to 0$.

By hypothesis $\{a_n\}$ and $\{b_n\}$ have limits, clearly equal, say

$$x = \lim a_n = \lim b_n.$$

We show $x = \sup A$ in the usual two steps.

1. x is an upper bound of A.

Suppose not; $x < y$ for some $y \in A$. Take $y - x$ for ε in the definition of $b_n \to x$ to get

$$b_N - x < y - x \qquad \text{for some } N.$$

This is impossible since $y \le b_N$.

2. $x \le b$, for any upper bound b of A.

If, for some upper bound b, we had $x > b$ then $x - a_N < x - b$ would hold for some N, i.e., $a_N > b$, which is impossible.

EXERCISE 7. For any $c > 0$ exhibit a sequence that converges to $\sqrt{c}$ in a complete ordered field. Give proofs.

EXERCISE 8. Assume that the sequence

$$1,\quad \sqrt{2},\quad \sqrt{2\sqrt{2}},\quad \sqrt{2\sqrt{2\sqrt{2}}},\quad \ldots$$

has a limit. What must the limit be?

EXERCISE 9. Prove that the sequence in Exercise 8 has a limit, in any complete ordered field containing the sequence.

EXERCISE 10. Which of the following sequences has a limit in a complete ordered field? Give proofs.

(a) $$x_n = 1 + \frac{1}{2} + \frac{1}{3} + \cdots + \frac{1}{n};$$

(b) $$x_n = 1 + \frac{1}{2!} + \frac{1}{3!} + \cdots + \frac{1}{n!}; \quad \text{and}$$

(c) $$x_n = 1 + r + r^2 + \cdots + r^n \qquad \text{for} \quad 0 < r < 1.$$

EXERCISE 11. Consider an arc of a circle $\widehat{PQ}$. For any finite set of points in order on $\widehat{PQ}$, $\{P_1, P_2, \ldots, P_n\}$, add the lengths of the segments joining these points:

$$l\{P_1, P_2, \ldots, P_n\} = \sum_{i=1}^{n-1} \text{length } \overline{P_iP_{i+1}}.$$

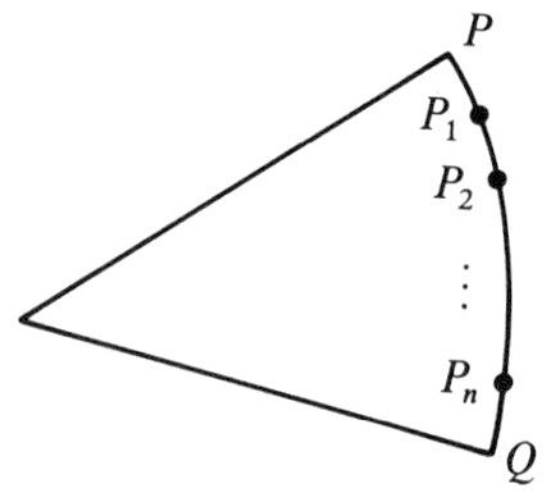

Show that $\sup[l\{P_1, P_2, \ldots, P_n\}: P_1, P_2, \ldots, P_n \in \widehat{PQ}]$ exists. It is called the **length** of the arc $\widehat{PQ}$.

§4. Cauchy Sequences

Intuition supports the result that every bounded, monotone sequence should converge, in a complete ordered field. The converse, that completeness follows from the existence of limits for just these very special sequences, is surprising. One might expect to have to describe quite generally the sequences $\{x_n\}$ that *ought to* have limits, and then require that all such sequences *do* have limits, in order to infer the completeness.

This approach to the question was taken by A. L. Cauchy in 1821. His preliminary thinking might have developed along the following lines.

A first attempt at a meaning for "$\{x_n\}$ ought to converge," a notion that must be expressed only in terms of the values x_n themselves, is

$$x_{n+1} \text{ and } x_n \text{ differ by amounts that} \to 0 \text{ as } n \to \infty.$$

Every sequence that does converge has this property.

EXERCISE 1. Proof?

However, unbounded sequences ought not to converge, and some of them have this property (for example, $\{1 + 1/2 + \cdots + 1/n\}$). A stricter requirement is needed.

One soon arrives at a stricter looking condition: For each $k \in \mathbb{N}$, $|x_{n+k} - x_n| \to 0$ as $n \to \infty$.

The same example, however, is still present.

$$|x_{n+k} - x_n| = \left(1 + \frac{1}{2} + \cdots + \frac{1}{n} + \frac{1}{n+1} + \cdots + \frac{1}{n+k}\right)$$
$$- \left(1 + \frac{1}{2} + \cdots + \frac{1}{n}\right)$$
$$= \frac{1}{n+1} + \cdots + \frac{1}{n+k} \to 0 \quad \text{as} \quad n \to \infty, \quad \text{for each} \quad k \in \mathbb{N}.$$

Examining this more closely, we notice that for a given $\varepsilon > 0$, the index N beyond which $|x_{n+k} - x_n| < \varepsilon$ holds, is *dependent on k*.

A stronger statement would be:

For $\varepsilon > 0$ there is an N, *independent of k*, such that for every $k \in \mathbb{N}$

$$n \geq N \quad \text{gives} \quad |x_{n+k} - x_n| < \varepsilon.$$

We say that

$$|x_{n+k} - x_n| \to 0 \qquad \text{as} \quad n \to \infty,$$

uniformly in $k \in \mathbb{N}$, when this holds.

EXERCISE 2. This does not hold for the above example, $\{1 + \frac{1}{2} + \cdots + 1/n\}$.

Cauchy gave a slightly different formulation.

Definition. $\{x_n\}$ is a **Cauchy sequence** if, for any $\varepsilon > 0$, there is an N such that

$$m, n \geq N \quad \text{gives} \quad |x_m - x_n| < \varepsilon.$$

EXERCISE 3. Prove that $\{x_n\}$ is a Cauchy sequence if and only if

$$|x_{n+k} - x_n| \to 0 \qquad \text{as} \quad n \to \infty,$$

uniformly in $k \in \mathbb{N}$.

Proposition. *Every Cauchy sequence is bounded.*

EXERCISE 4. Proof? Hint: Fix $\varepsilon > 0$ (say $\varepsilon = 1$) and choose N as in the Cauchy condition. Some interval I contains $x_1, x_2, \ldots, x_N$, a finite set. Now enlarge I to include all of $\{x_n\}$.

This shows that no unbounded sequence "ought to converge" in the Cauchy sense. Furthermore, we have the

Proposition. *Every convergent sequence is a Cauchy sequence.*

EXERCISE 5. Proof?

We shall prove the

Cauchy Completeness Theorem. *An Archimedean ordered field is complete if and only if every Cauchy sequence in it converges.*

The proof makes use of a lemma which is interesting in itself.

Lemma. *Every sequence has a monotonic subsequence.*

Proof. We need only consider sequences that have no increasing subsequence and show that they each necessarily have a decreasing subsequence.

Suppose $\{x_n\}$ has no increasing subsequence. We argue that there is a largest value among $\{x_n: n \in \mathbb{N}\}$, i.e., max x_n exists.

If not, x_1 is not the largest, so there is a first index n_1 giving $x_1 < x_{n_1}$. But x_{n_1} is not the largest, so there is a first index n_2 giving $x_{n_1} < x_{n_2}$, and $n_2 > n_1$. Continuing, we find an increasing subsequence, contrary to assumption.

Choose as x_{m_1} the first occurrence of the largest value (the obvious place to start constructing a decreasing subsequence).

Consider $\{x_n: n > m_1\}$, which has no increasing subsequence, and take as x_{m_2} the first occurrence of its largest value. Clearly, $x_{m_1} \geq x_{m_2}$.

Continuing by induction in this way, we arrive at a decreasing subsequence of $\{x_n\}$.

EXERCISE 6. Give the alternative argument in which you consider the sequences that have no decreasing subsequence.

Proposition. *If a sequence converges then every one of its subsequences converges to the same limit.*

EXERCISE 7. Proof?

Proposition. *If a Cauchy sequence has a convergent subsequence then the sequence itself converges to the same limit.*

Proof. Let $\{x_n\}$ be a Cauchy sequence and suppose $x_{n_k} \to x$ as $k \to \infty$. Fix $\varepsilon > 0$. To show $|x_n - x| < \varepsilon$ for n sufficiently large, we estimate $|x_n - x|$, using the triangle inequality and introducing the differences $|x_n - x_{n_k}|$ and $|x_{n_k} - x|$ which we can make small for appropriate n and k by our hypotheses:

$$|x_n - x| \le |x_n - x_{n_k}| + |x_{n_k} - x|.$$

Now, we can choose N so that

$$|x_n - x_m| < \frac{\varepsilon}{2} \quad \text{when} \quad n, m \ge N.$$

Applying this with n_k as m will give an estimate for the first term on the right above, provided $n, n_k \ge N$. Now we choose K so that

$$k \ge K \quad \text{gives} \quad n_k \ge N \quad \text{and} \quad |x_{n_k} - x| < \frac{\varepsilon}{2},$$

thus insuring the requirement $n_k \ge N$ and estimating the second term.

Combining, we see that the desired conclusion,

$$n \ge K \quad \text{gives} \quad |x_n - x| < \varepsilon,$$

follows via the triangle inequality estimate using any $k \ge K$.

(Notice that K can only be chosen *after* N has been chosen, because of the requirement that $k \ge K$ gives $n_k \ge N$. This is a subtle point in the argument which is easily missed.)

Proof of Cauchy Completeness Theorem. Suppose $\{x_n\}$ is a Cauchy sequence in a complete ordered field. Since $\{x_n\}$ is bounded, it contains a bounded, monotonic subsequence because of the lemma. The completeness provides a limit for the subsequence, which is then a limit for the Cauchy sequence. Thus every Cauchy sequence in a complete ordered field converges.

Conversely, if every Cauchy sequence in an Archimedean ordered field converges and if $\{I_n\}$ is a nest of closed intervals, $I_n = [a_n, b_n]$, then, since $b_n - a_n \to 0$, it is immediate that both $\{a_n\}$ and $\{b_n\}$ are Cauchy sequences. The set $\bigcap_{n=1}^{\infty} I_n$ is seen to consist of the common limit of these sequences, which exists by the hypothesis.

EXERCISE 8. Give another proof of the converse by considering A nonempty and bounded above and showing a Cauchy sequence whose limit is sup A.

EXERCISE 9. Every bounded, monotonic sequence is a Cauchy sequence. Infer from the theory, but also prove directly from the definitions.

EXERCISE 10. Show that $\{x_n\}$ is a Cauchy sequence if $|x_{k+1} - x_k| < 1/2^k$ for each k.

EXERCISE 11. Prove that $x = \sup A$ if and only if x is an upper bound of A with the property, some sequence $\{x_n\}$ contained in A converges to x.

EXERCISE 12. Prove the *Bolzano–Weierstrass Theorem*: Every bounded sequence has a convergent subsequence, in a complete ordered field.

§5. The Real Number System

The idea of completeness in its various forms:

Nested Intervals Property;
Bounded, monotone sequences converge;
Least upper bounds exist; and
Cauchy sequences have limits;

has been seen to provide the needed property of an Archimedean ordered field, in order that there be a satisfactory outcome of certain natural lines of investigation:

Existence of numbers for all decimals, etc.
Existence of all numbers approximable by sets of rationals.
Existence of numbers to solve certain equations.
Existence of limits for sequences.

The obvious issue that remains for us to address, is that of the *existence* of complete ordered fields. The equally obvious approach to this issue is the construction of an extension of $\mathbb{Q}$ with an eye to the completeness property. Our several lines of investigation above suggest several ways to create new "numbers" for the construction of an extension F of $\mathbb{Q}$:

1. Take the collection of all decimals as the objects of F. Identify the set of recurring decimals with $\mathbb{Q}$. Extend the meanings of $+$, $\cdot$, and $<$ to all of F so that the ordered field axioms hold. Prove the resulting F is complete.

One would expect the full elaboration of such a program to be tedious. The main difficulty is in formulating the definitions of $+$ and $\cdot$ for decimals.

2. The approximation approach via least upper bounds suggests the creation of an object to serve as sup A for every $\phi \neq A \subset \mathbb{Q}$ that is bounded above. Every such A determines a decomposition of $\mathbb{Q}$ into two nonempty sets

$$L = \text{the set of nonupper bounds of } A \text{ in } \mathbb{Q},$$

$$U = \text{the set of upper bounds of } A \text{ in } \mathbb{Q},$$

with $\mathbb{Q} = L \cup U$ and $L \cap U = \phi$.

Such a decomposition of $\mathbb{Q}$ into a pair of disjoint sets (L, U) is called a *cut* in $\mathbb{Q}$. The elements of $\mathbb{Q}$ are identified with cuts by

$$r \leftrightarrow (\{s \in \mathbb{Q}: s < r\}, \{s \in \mathbb{Q}: s \geq r\}).$$

Under the definition of order between cuts given by

$$(L, U) < (L', U') \quad \text{means} \quad L \subset L'$$

it is fairly clear that the set of all cuts will include least upper bounds for all of its nonempty sets that are bounded above.

There remains, once more, the task of defining $+$ and $\cdot$, this time for cuts, in agreement with their meanings in $\mathbb{Q}$, and to verify that this makes the set of all cuts a complete ordered field.

This approach was carried out by R. Dedekind in 1872. Details can be found in Landau [19] and Rudin [24].

3. From the point of view of Cauchy sequences, one begins by observing that two Cauchy sequences of rationals, $\{x_n\}$ and $\{y_n\}$ should have the same limit precisely when $x_n - y_n \to 0$. Lumping together all Cauchy sequences of rationals that are mutually equivalent in this sense, $x_n - y_n \to 0$, we create an object, the class of equivalence, to serve as the limit of all of those Cauchy sequences. Thus, the objects making up F from this approach are the classes of equivalence of Cauchy sequences of rationals.

The rational r yields the constant Cauchy sequence $r, r, r, \ldots$ and its class of equivalence, hence an element of F. This embeds $\mathbb{Q}$ in F and leaves the task of extending the meanings of $+, \cdot$, and $<$ from $\mathbb{Q}$ to F to obtain a complete ordered field.

The details of this approach can be found in Goffman [11].

Can we reconcile these various approaches to the "completion" of $\mathbb{Q}$? Or must we each commit ourselves to a preference between decimals, cuts, classes of equivalence of Cauchy sequences and whatever new approach may present itself? The answer to these questions is that any two complete ordered fields are **isomorphic**; there is a one–one correspondence $\varphi: F \leftrightarrow F^*$ of any complete ordered field, F, onto any other, F^*, such that the structures are preserved by φ:

$$\varphi(x + y) = \varphi(x) + \varphi(y),$$

$$\varphi(x \cdot y) = \varphi(x) \cdot \varphi(y),$$

$$x < y \quad \text{if and only if} \quad \varphi(x) < \varphi(y).$$

Such a φ is called an **isomorphism** of the complete ordered fields. Its existence means that in the abstract sense all complete ordered fields are the same—any differences between two concrete realizations of the complete ordered field axioms lie only in the interpretations of their objects, e.g., decimals versus cuts in $\mathbb{Q}$; they are the same number system.

In this sense, i.e., up to isomorphism, there is exactly one complete ordered field, which is called $\mathbb{R}$, the **Real Number System**.

The construction of a $\varphi: F \to F^*$ begins with the neutral elements, i.e., using the obvious notations, the assignments

$$\varphi(0) \text{ is defined to be } 0^*,$$

$$\varphi(1) \text{ is defined to be } 1^*,$$

and

$$\varphi(n + 1) = \varphi(n) + 1^*,$$

match up $\mathbb{Z}$ and $\mathbb{Z}^*$ in the required way. In particular $\varphi(m \cdot n) = \varphi(m) \cdot \varphi(n)$, so that the division problems are matched, allowing φ to be extended to an isomorphism of $\mathbb{Q}$ and $\mathbb{Q}^*$. Finally, $\{x_n\}$ is a Cauchy sequence in $\mathbb{Q}$ if and only if $\{\varphi(x_n)\}$ is a Cauchy sequence in $\mathbb{Q}^*$, and the completeness allows the extension of φ to all of F, carrying it onto all of F^*, by matching the limits of corresponding Cauchy sequences.

The many details of the development of the real number system that we have omitted as tedious are perhaps only tedious to read; constructing some of these details for himself can be a fascinating task for the interested student. For those who are left with a feeling of being cut off from the logical basis of the real number system by these omissions, there is a perfectly acceptable logical alternative. Instead of starting with the Peano Axioms, one starts with the axioms of a complete ordered field, i.e., it is *assumed* that a complete ordered field exists.

Finally, it should be observed that our intuitive geometric model, the line, which we would now consider as having "points" at every real number, could itself be constructed from appropriate axioms of one-dimensional geometry—after which the resulting "**Real line**" could be proved to provide just another realization of the complete ordered field axioms.

The real number system is also often referred to as **the continuum**.

§6. Countability

Further insight into $\mathbb{R}$ comes from a consideration of the plentitude of various kinds of real numbers.

Two finite sets have the same number of elements precisely when there is a one–one correspondence between them. The latter idea generalizes:

Definition. Sets A and B have the **same cardinality** if there exists a one–one function with domain A and range B.

EXAMPLES. 1. $\mathbb{N}$ and the set of all even numbers. The matching spelled out by

$$\begin{array}{ccccc} 1, & 2, & 3, & \ldots, & n, \ldots \\ \updownarrow & \updownarrow & \updownarrow & & \updownarrow \\ 2, & 4, & 6, & \ldots, & 2n, \ldots \end{array}$$

can be abbreviated by simply writing the even numbers as a sequence: 2, 4, 6, ...

2. Two nontrivial closed intervals A and B. Placing them parallel in the plane, we find a one–one correspondence as in the figure.

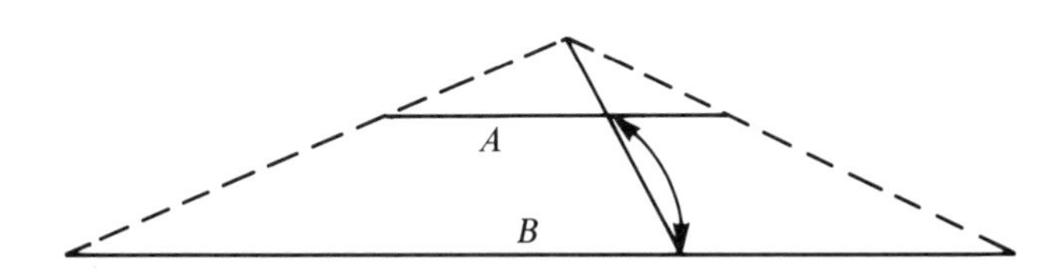

3. $(0, 1]$ and $[0, \infty)$ using

$$A = [0, \infty): x \to \frac{1}{1 + x^2} \in B = (0, 1].$$

EXERCISE 1. Any two nontrivial intervals, including $(-\infty, \infty) = \mathbb{R}$, half-lines (a, ∞), $(-\infty, a)$, $[a, \infty)$, and $(-\infty, a]$ and finite open, closed or semiclosed have the same cardinality. Draw pictures as in 2.

Definition. A **countable set** is one that has the same cardinal number as a nonempty subset of $\mathbb{N}$. A set is **finite** if it has the same cardinal number as $\{1, 2, \ldots, n\}$ for some $n \in \mathbb{N}$. A set that is countable but not finite is **countably infinite**. **Uncountable** means not countable.

EXAMPLE. $\mathbb{Z}$ is countably infinite. The list $0, 1, -1, 2, -2, 3, -3, \ldots$ of the elements of $\mathbb{Z}$ as a sequence exhibits a one–one correspondence with $\mathbb{N}$.

In general, a countable set is one that can be listed as a sequence, possibly finite.

Proposition. *If A_n is a countable set for each $n \in \mathbb{N}$, then $\bigcup_{n=1}^{\infty} A_n$ is countable.*

(We say, a countable union of countable sets is countable.)

Proof. Let $A_n = a_{n1}, a_{n2}, a_{n3}, \ldots$ be any listing of A_n as a sequence. The union can then be listed as an array

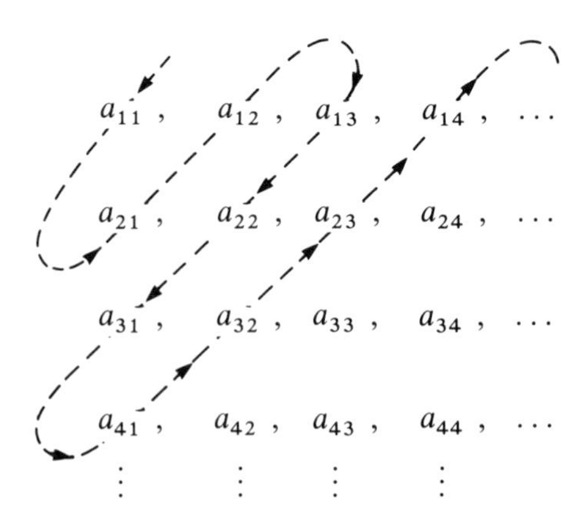

By following the indicated path and omitting any object already encountered, we can list the union as a sequence.

This is called Cantor's first diagonal procedure.

Theorem. $\mathbb{Q}$ *is countable.*

EXERCISE 2. Proof? Hint: Let A_n be the set of all fractions in lowest terms with denominator n.

EXERCISE 3. The set of all algebraic numbers is countable. Hint: The equation $a_n x^n + \cdots + a_1 x + a_0 = 0$, with $n \geq 1$, all $a_i \in \mathbb{Z}$ and $a_n \neq 0$, has only finitely many real solutions, so it is enough to show only countably many such equations exist. For each $k \in \mathbb{N}$ let A_k be the set of all equations for which

$$n + |a_n| + |a_{n-1}| + \cdots + |a_1| + |a_0| = k.$$

Theorem. $\mathbb{R}$ *is uncountable.*

We give two proofs.

***Proof* 1.** $\mathbb{R}$ is uncountable if $[0, 1)$ is. The decimal expansion process sets up a one–one correspondence between $[0, 1)$ and terminating decimals. We show that no sequence of terminating decimals

$$\begin{array}{ll} 0.\ a_{11}a_{12}a_{13}\ldots & \\ 0.\ a_{21}a_{22}a_{23}\ldots, & a_{ij}\in\{0,1,2,\ldots,9\} \\ 0.\ a_{31}a_{32}a_{33}\ldots & \\ \vdots\quad\vdots\quad\vdots & \end{array}$$

can contain all of them.

Define b_i for each $i \in \mathbb{N}$ to be any digit other than 9 and a_{ii}. Then

$$0.\ b_1 b_2 b_3 \ldots$$

is a terminating decimal (no digit 9, hence cannot have the form xxx9999...) and differs in the ith place from the ith entry on the list.

This argument is called Cantor's second diagonal method.

The Nested Intervals Theorem can be used to show that no sequence of reals $x_1, x_2, x_3, \ldots$ can contain all reals, as follows.

***Proof* 2.** Let I_1 be any closed interval of length 1 that does not include x_1. One of the closed thirds of I_1 must exclude x_2. (Notice that closed halves might not do, since x_2 could be the midpoint of I_1.) Choose a closed third of I_1 that excludes x_2 and call it I_2. Continue by induction to define $I_1 \supset I_2 \supset I_3 \supset \cdots$ such that each is closed, I_n excludes x_n for all n, and length $I_n = 1/3^{n-1} \to 0$.

By the Nested Intervals Theorem, for some $x \in \mathbb{R}$, $x \in \bigcap_{n=1}^{\infty} I_n$. Clearly, $x \neq x_n$ for all $n \in \mathbb{N}$. Thus the sequence does not include all of $\mathbb{R}$.

Corollary. *The set of irrational real numbers is uncountable.*

For, if it were countable then its union with $\mathbb{Q}$ would be countable.

EXERCISE 4. The set of all transcendental numbers, i.e., real numbers that are not algebraic, is uncountable.

EXERCISE 5. The base 2, or binary, expansion with digits 0 and 2 and terminating symbols (no xxx222...) sets up a one–one correspondence between [0, 1) and the set of all sequences of the digits 0, 2. Prove from this that the Cantor Set is uncountable (see p. 29, Exercise 6).

Appendix. The Fundamental Theorem of Algebra. Complex Numbers

We refer by the term algebra only to the question of solving polynomial equations. The lack of a solution in $\mathbb{Q}$ of $x^2 - 2 = 0$ was one of our motivations for the extension of $\mathbb{Q}$ to $\mathbb{R}$. However, there remain simple equations having no solutions, even after this extension; since $x^2 \geq 0$ in any ordered field, the equations $x^2 + c = 0$, for $c > 0$, can have no solutions in ordered fields. Typical of these is the equation $x^2 + 1 = 0$.

It is remarkable that by adjoining to $\mathbb{R}$ an object i created to solve this equaton:

$$i^2 = -1, \qquad i \text{ is the "square root" of } -1,$$

and then extending $\{\mathbb{R}, +, \cdot\}$ in a very natural way, one arrives at a number system $\{\mathbb{C}, +, \cdot\}$ in which one can prove the

Fundamental Theorem of Algebra. *Every equation*

$$a_n z^n + a_{n-1} z^{n-1} + \cdots + a_1 z + a_0 = 0, \qquad n \geq 1, \quad a_n \neq 0,$$

formulated with coefficients from $\mathbb{C}$ *has a solution* z *in* $\mathbb{C}$.

Of course, the order relation is not retained in the extension to $\mathbb{C}$, just $+$ and $\cdot$ with their usual properties. The system $\mathbb{C}$ that results, the complex number system, is modeled geometrically by the plane. It is of great importance throughout mathematics and its applications.

Since the purpose of this book is primarily to pursue the ramifications of $\mathbb{R}$, especially those flowing from the completeness, we shall confine our discussion of $\mathbb{C}$ to a sketch of its development, some of the basics of complex computation and a proof (eventually) of the Fundamental Theorem. This brings to a satisfactory conclusion the series of extensions of number systems to which we have been led.

§1. A **complex number** is just an ordered pair of real numbers (a, b) and the **equality** of two complex numbers means they have the same first entry and the same second entry:

$$(a, b) = (c, d) \quad \text{means} \quad a = c \quad \text{and} \quad b = d \quad \text{in } \mathbb{R}.$$

To **add** complex numbers, add their first and second entries:

$$(a, b) + (c, d) \quad \text{means} \quad (a + c, b + d).$$

Obviously, addition is associative and commutative.

We pause to exploit the geometric interpretation. Consider a plane in which a rectangular coordinate system has been chosen.

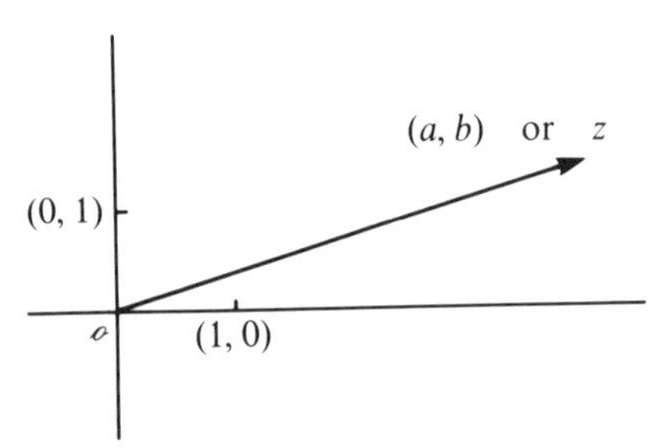

The pair (a, b) is identified with the point z whose coordinates are a and b, respectively, or with the vector (arrow) from the origin o terminating at that point. o is $(0, 0)$ and is also simply 0, the complex zero.

The addition is immediately seen to be the addition of vectors by the *parallelogram law*: if $z = (a, b)$ and $z' = (c, d)$, then $z + z' = (a + c, b + d)$ is the diagonal through o of the parallelogram with sides z and z'.

Clearly, $o = 0 = (0, 0)$ is neutral under this addition.

The vector $-z = (-a, -b)$ is seen to be an additive inverse of $z = (a, b)$.

We can write $(a, b) = (a, 0) + (0, b)$ and identify the pair $(a, 0)$ with the real number a. The horizontal axis is called the **real axis** in the plane and a is called the **real part** of $z = (a, b)$, also written $a = \text{Re } z$.

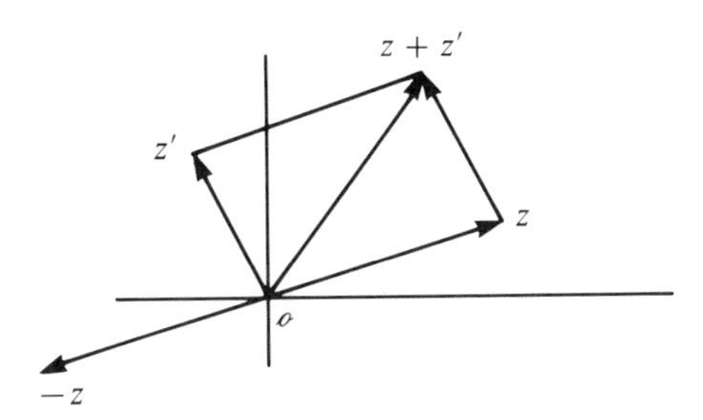

The pair $(0, b)$ is abbreviated bi, so that $z = (a, b) = a + bi$. Thus, the pair notation can be abandoned in favor of the equivalent "binomial" notation $a + bi$, in which, for now, the i is simply a tag to designate which real number is the second one in the ordered pair.

The vertical axis consists of the numbers bi, in this notation, and is called the **imaginary axis**. The **imaginary part** of the complex number $z = a + bi$ is the (real!) number b, we write $\text{Im } z = b$. So $z = \text{Re } z + i \text{ Im } z$.

To define **multiplication** we put $i^2 = -1$ and otherwise treat $(a + bi)(c + di)$ like a product of binomials:

$$\begin{aligned}(a + bi)(c + di) &= ac + adi + bci + bdi^2 \\ &= (ac - bd) + (ad + bc)i.\end{aligned}$$

This multiplication is obviously commutative, and the fact that 1 is neutral is also obvious. The associative law of multiplication and the distributive law connecting addition and multiplication are straightforward calculations.

EXERCISE 1. Give the details.

Of the standard properties, there only remains that of the existence and uniqueness of multiplicative inverses of nonzero numbers. Suppose $a + bi \neq$

0, i.e., either $a \neq 0$ or $b \neq 0$. An inspired guess for the inverse of $a + bi$ is found by once more imitating binomial calculations and recalling the trick of "rationalizing the denominator," i.e.,

$$\frac{1}{a+bi} = \frac{1}{a+bi} \cdot \frac{a-bi}{a-bi} = \frac{a-bi}{a^2+b^2} = \frac{a}{a^2+b^2} + \frac{-b}{a^2+b^2} i.$$

EXERCISE 2. Prove that if a or b is $\neq 0$ then

$$(a+bi)\left(\frac{a}{a^2+b^2} + \frac{-b}{a^2+b^2} i\right) = 1.$$

Letting $(\mathbb{C}, +, \cdot)$, or simply $\mathbb{C}$, denote the number system so constructed, the **Complex Number System**, we have a system in which the four arithmetic operations can be done and the usual rules of arithmetic hold. The actual calculations are done by treating complex numbers as binomials and replacing i^2 by -1 whenever it occurs, including "rationalizing the denominator" to do division.

The symbol bi as a "tag" now has the meaning b times i.

Multiplication of a complex number z by a real number r is easily visualized geometrically:

If $r > 0$, stretch the vector z in the ratio $r:1$. If $r < 0$, stretch z by $|r|$ and reverse its direction.

The length of the vector z is called the **absolute value** or **modulus** of the complex number z,[3]

$$|z| = \sqrt{a^2 + b^2} \qquad \text{if} \quad z = a + bi.$$

The **triangle inequality** holds

$$|z + w| \leq |z| + |w|.$$

The geometric interpretation explains the name: The length of one side of a triangle (the diagonal $z + w$ of the parallelogram) cannot exceed the sum of the lengths of the other two sides. (Sketch!)

The proof we gave in case the numbers are real made strong use of the order relation in $\mathbb{R}$; a new proof is needed. First, some useful notations and minor facts.

The **conjugate** $\bar{z}$ of a complex number z is defined by

$$\bar{z} = a - bi \qquad \text{if} \quad z = a + bi.$$

Geometrically, reflect z in the real axis to get its conjugate. Of course, $|\bar{z}| = |z|$.

[3] The existence of $\sqrt{c}$ for $c \geq 0$ was shown in Exercise 7, p. 46.

We have

$$|z|^2 = z \cdot \bar{z}$$

EXERCISE 3. Prove this.

We have

$$2 \operatorname{Re} z = z + \bar{z} \quad \text{and} \quad 2i \operatorname{Im} z = z - \bar{z},$$

$$\bar{\bar{z}} = z, \qquad \overline{(z + w)} = \bar{z} + \bar{w} \quad \text{and} \quad \overline{z \cdot w} = \bar{z} \cdot \bar{w},$$

$$\operatorname{Re} z \leq |z| \quad \text{and} \quad \operatorname{Im} z \leq |z|$$

EXERCISE 4. Proofs?

The proof of the triangle inequality is now just a computation. Since both sides are ≥ 0, it is enough to show that $|z + w|^2 \leq (|z| + |w|)^2$, which is simpler to write:

$$\begin{aligned}
|z + w|^2 &= (z + w)(\bar{z} + \bar{w}) = |z|^2 + z \cdot \bar{w} + \bar{z} \cdot w + |w|^2 \\
&= |z|^2 + 2 \operatorname{Re} z\bar{w} + |w|^2 \\
&\leq |z|^2 + 2|z| \cdot |w| + |w|^2 = (|z| + |w|)^2.
\end{aligned}$$

Circles and circular disks in the plane are easily described analytically. If the complex number z_0 is the center and r is the radius, then $|z - z_0| = r$ describes the points z on the circle and $|z - z_0| \leq r$ describes the disk. The circle $|z| = 1$ is called the **unit circle**.

If z is on the unit circle, the counterclockwise arc from 1 to z has a length[4] θ satisfying $0 \leq \theta < 2\pi$. The number θ is called the argument of z. Generally, if $z \neq 0$ then $z/|z|$ is on the unit circle and its argument is called the **argument of z**. The number 0 has no argument.

The coordinates of $z/|z|$ are determined by $\theta = \arg z$ and are called $\cos \theta$ and $\sin \theta$,

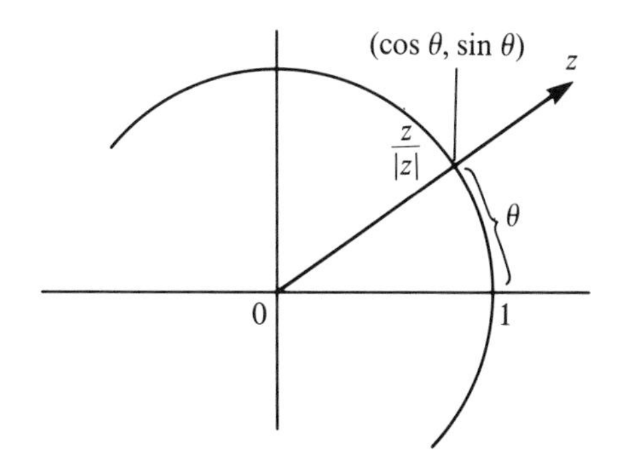

[4] See p. 46, Exercise 11.

$$\frac{z}{|z|} = (\cos\theta, \sin\theta) \quad \text{or} \quad z = |z|(\cos\theta + i\sin\theta).$$

The geometric interpretation of multiplication is seen by using this "polar" form of the complex numbers: If

$$z = |z|(\cos\theta + i\sin\theta) \quad \text{and} \quad w = |w|(\cos\varphi + i\sin\varphi)$$

then

$$\begin{aligned} z\cdot w &= |z||w|\{(\cos\theta\cos\varphi - \sin\theta\sin\varphi) + i(\sin\theta\cos\varphi + \cos\theta\sin\varphi)\} \\ &= |z||w|\{\cos(\theta+\varphi) + i\sin(\theta+\varphi)\}, \end{aligned}$$

where we have used the addition formulas for sine and cosine from Trigonometry.

The sum $\theta + \varphi$ may be outside the range $[0, 2\pi)$, so it may not be $\arg zw$, i.e., $\arg zw = \theta + \varphi - 2\pi$ might be the case. The best way to express this result, and others, is by the use of congruence modulo 2π instead of equality: **α is congruent to β modulo 2π**, written $\alpha \equiv \beta \pmod{2\pi}$, means $\alpha - \beta$ is an integer multiple of 2π. Thus our result is

$$\arg z\cdot w \equiv (\arg z + \arg w) \pmod{2\pi}.$$

EXERCISE 5. Sketch the complex numbers $-\sqrt{3} - i$ and $-\sqrt{3}/2 - \frac{3}{2}i$. Find their product.

EXERCISE 6. Sketch all the powers $[(1+i)/\sqrt{2}]^n$ for $n = 0, 1, 2, 3, \ldots$.

EXERCISE 7. Describe geometrically the effect of "multiplication by i" on the points of the plane.

The numbers z^n, $n = 1, 2, 3, \ldots$, are easily located if $z \neq 0$ is given: on the circles of center 0 and radii $|z|^n$ simply mark the points whose angles with the positive real axis are, respectively, $\arg z$, $2\arg z$, $3\arg z$, ... By reversing the process, we see how to find nth roots of complex numbers, $n = 1, 2, 3, \ldots$. We shall see later that any positive number has a positive nth root, due to the completeness of $\mathbb{R}$. Thus, if $a \neq 0$, the nth roots of a, or the solutions of $z^n = a$, lie on the circle $|z| = |a|^{1/n}$ at the points whose arguments θ satisfy $n\theta \equiv \arg a \pmod{2\pi}$. This means

$$n\theta = \arg a + 2k\pi, \qquad k \in \mathbb{Z},$$

but among these values of θ,

$$\theta = \frac{\arg a + 2k\pi}{n}, \qquad k \in \mathbb{Z},$$

the ones occurring for $k = 0, 1, \ldots, n-1$ are not congruent modulo 2π, and all others are congruent to one of these. We can say that the equation $z^n = a$

for $a \neq 0$ has exactly n distinct complex solutions, given by

$$|a|^{1/n}\left\{\cos\frac{\arg a + 2k\pi}{n} + i\sin\frac{\arg a + 2k\pi}{n}\right\}, \qquad i = 0, 1, \ldots, n-1.$$

This result is known as *De Moivre's Theorem.*

EXERCISE 8. Sketch the three cube roots of 8.

EXERCISE 9. Sketch the four fourth roots of $16i$.

EXERCISE 10. Sketch the two square roots of -1.

§2. De Moivre's Theorem provides a complete analysis of the solutions of the particular polynomial equations $z^n - a = 0$, $n = 1, 2, 3, \ldots$. The subtle part, that $\sqrt[n]{c}$ exists for $c > 0$, assumed above, is examined in Chapter 3.

Turning now to the general polynomial equation $P(z) = 0$, where $n \geq 1$ and $a_n \neq 0$ in

$$P(z) = a_n z^n + a_{n-1}z^{n-1} + \cdots + a_1 z + a_0,$$

we can write out $P(a)$ and subtract term by term to get

$$P(z) - P(a) = a_n(z^n - a^n) + a_{n-1}(z^{n-1} - a^{n-1}) + \cdots + a_1(z - a).$$

Now, direct computation verifies the identities

$$z^k - a^k = (z - a)(z^{k-1} + az^{k-2} + a^2z^{k-3} + \cdots + a^{k-2}z + a^{k-1})$$

for $k = 1, 2, \ldots$. Hence

$$\begin{aligned} P(z) - P(a) &= (z - a)\{a_n(z^{n-1} + az^{n-2} + \cdots + a^{n-2}z + a^{n-1}) \\ &\qquad + a_{n-1}(z^{n-2} + az^{n-3} + \cdots + a^{n-2}) + \cdots + a_1\} \\ &= (z - a)Q(z), \end{aligned}$$

where $Q(z)$ is the indicated polynomial of degree $n - 1$. This is the division algorithm for polynomials in the special case of the divisor of first degree, $z - a$;

$$P(z) = (z - a)Q(z) + P(a), \qquad \text{degree } Q = \text{degree } P - 1.$$

The fact that the remainder in this division is just the value of P at a is called the *Remainder Theorem*, while the *Factor Theorem* is the obvious observation that "$P(a) = 0$" and "$(z - a)$ is a factor of $P(z)$" are equivalent statements. Such numbers a are called **roots** of the polynomial P.

EXERCISE 11. If $P(z) = (z - a)^m Q(z)$ for some polynomial Q having no factor $(z - a)$, we say that a is a root of P of **multiplicity m**. Assuming (the Fundamental Theorem) that every nontrivial polynomial *has* a root, show that the number of roots, counting each as many times as its multiplicity, is exactly its degree.

Repeated application of the division algorithm gives, for any $a \in \mathbb{C}$,

$$\begin{aligned}P(z) &= (z-a)Q_1(z) + P(a)\\ &= (z-a)[(z-a)Q_2(z) + Q_1(a)] + P(a)\\ &= (z-a)^2Q_2(z) + (z-a)Q_1(a) + P(a)\\ &= (z-a)^2[(z-a)Q_3(z) + Q_2(a)] + (z-a)Q_1(a) + P(a)\\ &= (z-a)^3Q_3(z) + (z-a)^2Q_2(a) + (z-a)Q_1(a) + P(a),\end{aligned}$$

etc., giving finally, constants $A_0, A_1, \ldots, A_n$ for which

$$(*) \qquad P(z) = A_0 + A_1(z-a) + A_2(z-a)^2 + \cdots + A_n(z-a)^n.$$

Our proof of the Fundamental Theorem of Algebra consists in showing two things:

1. $|P(z)|$ assumes its minimum value; and
2. $|P(z)|$ cannot have a minimum value other than zero.

Taken together, these insure the existence of an $a \in \mathbb{C}$ for which $|P(a)| = 0$, that is, $P(a) = 0$; a is a root of P or a solution of $P(z) = 0$.

In Chapter 2 we shall examine the general question of the existence of maxima and minima of real-valued quantities. Our proof of the first statement will appear there (p. 76) as an application of the general considerations.

We conclude this chapter with a proof of the second statement, formulated as the

Lemma. *If* $|P(a)| > 0$ *there is a* z *such that* $|P(z)| < |P(a)|$.

Proof. In the expression $(*)$ for P at the point a in question, we know $A_0 = P(a) \neq 0$ and $A_n \neq 0$ because of our hypotheses ($|P(a)| > 0$ and degree $P = n$, respectively). There is, therefore, a first nonzero coefficient A_k among $A_1, A_2, \ldots, A_n$, so that $(*)$ becomes, more explicitly,

$$P(z) = P(a) + A_k(z-a)^k + \cdots + A_n(z-a)^n, \qquad A_k \neq 0.$$

The object is to show that z can be chosen to give a vector $P(z)$ shorter than the vector $P(a)$. The approach is to use the term $A_k(z-a)^k$ to directly subtract length from $P(a)$, while preventing the remaining terms from cancelling this effect, by choosing $|z-a|$ small enough (using the fact that the remaining terms involve higher powers of $(z-a)$, hence they become relatively negligible for $|z-a|$ small).

To see that this approach succeeds, we begin by rewriting $(*)$ even more explicitly

$$\begin{aligned}(**) \qquad P(z) &= P(a) + A_k(z-a)^k\\ &\quad + A_k(z-a)^k\left\{\frac{A_{k+1}}{A_k}(z-a) + \cdots + \frac{A_n}{A_k}(z-a)^{n-k}\right\}\\ &= P(a) + A_k(z-a)^k + A_k(z-a)^kQ(z)(z-a),\end{aligned}$$

where

$$Q(z) = \frac{A_{k+1}}{A_k} + \frac{A_{k+2}}{A_k}(z - a) + \cdots + \frac{A_n}{A_k}(z - a)^{n-k-1}.$$

Now, z is chosen so as to give $A_k(z - a)^k$ a direction opposite to that of $P(a)$, i.e.,

$$\arg A_k(z - a)^k - \arg P(a) = \pi,$$

$$\arg A_k + k \arg(z - a) - \arg P(a) \equiv \pi \pmod{2\pi},$$

$$\arg(z - a) \equiv \frac{\arg P(a) - \arg A_k + \pi}{k} \pmod{2\pi}.$$

For any such z, $A_k(z - a)^k = -cP(a)$ for some $c > 0$, and (**) becomes

$$P(z) = (1 - c)P(a) - cP(a)Q(z)(z - a),$$

so that

$$|P(z)| \leq |1 - c||P(a)| + c|P(a)||Q(z)||z - a|.$$

Now we restrict $z - a$ in size. First, since $c = -A_k(z - a)^k/P(a)$ and $k \geq 1$, we can insure that $c < 1$, hence that $|1 - c| = 1 - c$ in the last formula:

$$|P(z)| \leqq (1 - c)|P(a)| + c|P(a)||Q(z)||z - a|.$$

We notice that if $|Q(z)||z - a| < 1$ here, then

$$|P(z)| < (1 - c)|P(a)| + c|P(a)| = |P(a)|,$$

the desired conclusion. But if $|z - a| < 1$ then

$$\begin{aligned} |Q(z)| &= \left| \frac{A_{k+1}}{A_k} + \frac{A_{k+2}}{A_k}(z - a) + \cdots + \frac{A_n}{A_k}(z - a)^{n-k-1} \right| \\ &\leq \left|\frac{A_{k+1}}{A_k}\right| + \left|\frac{A_{k+2}}{A_k}\right| |z - a| + \cdots + \left|\frac{A_n}{A_k}\right| |z - a|^{n-k-1} \\ &< \left|\frac{A_{k+1}}{A_k}\right| + \left|\frac{A_{k+2}}{A_k}\right| + \cdots + \left|\frac{A_n}{A_k}\right| \end{aligned}$$

and we need only choose $|z - a|$ less than the reciprocal of this last number to obtain $|Q(z)||z - a| < 1$, as required.

Thus any z giving $z - a$ the required direction and a modulus meeting the three restrictions noted, gives the result.

CHAPTER 2

The Extreme-Value Problem

We present an informal discussion, to illustrate how certain concepts might naturally arise in the pursuit of an answer to a fundamental question: the question of the existence of minimum and maximum values of real-valued functions.

This is indeed a fundamental question. We have seen in Chapter 1 how a proof of the Fundamental Theorem of Algebra turns on the existence of a minimum value of a quantity, $|P(z)|$. This is not an isolated example; many important questions in Mathematics have formulations that ask for the existence of an extreme value (i.e., a minimum or a maximum value) of a real-valued function. More importantly perhaps, extreme values play central roles in the sciences: a ball thrown in the air reaches a maximum height; an economic strategy is designed to yield maximum profit; a chemical reaction follows a course that minimizes the energy consumed. Examples abound. In seeking mathematical models of physical (economic, chemical, biological, etc.) phenomena, one is frequently guided to a suitable model simply by excluding those in which certain functions fail to achieve the extreme values they must have.

All of this points to the need for a usable general theorem that gives conditions under which a real-valued function assumes extreme values: "conditions on f" implies f assumes extreme values. The task is to find such "conditions on f."

§1. Continuity, Compactness, and the Extreme-Value Theorem

We begin with the terminology. If f is a real-valued function, its domain dom f can be a set of any kind, but we shall consider the case dom $f \subset \mathbb{R}$; a function of a real variable.

(We choose to stress the fact that functions with different domains are different functions, e.g.,

$$f(x) = 4x^3 - 3x^2 + 7 \qquad \text{for} \quad -3 \leq x < 19$$

and

$$g(x) = 4x^3 - 3x^2 + 7 \qquad \text{for} \quad 0 < x \leq 12$$

are different functions. This means that we are temporarily abandoning the commonly used convention, whereby the domain of a function is not specified, but is understood to be the largest set for which the given formula has meaning, as in

$$f(x) = \sqrt{x} \qquad (\text{dom } f = \text{all } x \geq 0 \text{ implied})$$

or

$$f(x) = \frac{1}{x-3} \qquad (\text{dom } f = \text{all } x \neq 3 \text{ implied})$$

or

$$4x^3 - 3x^2 + 7 \quad (\text{all } x \in \mathbb{R} \text{ implied}).)$$

The **range**, or value set, of f

$$\text{ran } f = \{f(x) \colon x \in \text{dom } f\}$$

is a subset of $\mathbb{R}$ in the case we are considering. We say that **the function f is bounded above, bounded below** or **bounded** in case ran f is a set with the corresponding property. **Upper (lower) bound of** f means upper (lower) bound of ran f.

When they exist, the infimum, supremum, minimum, and maximum of the set ran f are called the **infimum, supremum, minimum**, and **maximum of the function**, denoted, respectively, inf f, sup f, min f, and max f.

If the function f is bounded above, so that sup f exists, the question of whether max f exists is just the question whether sup $f \in$ ran f, by the definition of the maximum of a set of reals. Thus the maximum value of f, if it exists, is a real number that:

(a) occurs as a value of f, $f(x_0)$, at some point (perhaps several) $x_0 \in$ dom f; and
(b) satisfies $f(x) \leq f(x_0)$ for all $x \in$ dom f.

Similarly for min f.

Exercise 1. Write out the details in the minimum case.

EXERCISE 2. Suppose g is a **restriction** of f: $\operatorname{dom} g \subset \operatorname{dom} f$ and $g(x) = f(x)$ for $x \in \operatorname{dom} g$. What can be said about the existence of, and relations between, $\min f$, $\max f$, $\min g$, and $\max g$?

EXERCISE 3. Give examples, via formulas and also via sketches of graphs, of functions f for which $\max f$ does, does not, exist.

Observe that f has a minimum value if and only if $-f$ has a maximum value, so that it suffices to concentrate on maximum values as the discussion proceeds.

Our procedure in seeking a theorem is a common one: we consider examples of functions for which the conclusion is false (no maximum exists) and write conditions that exclude those examples, to arrive at a conjecture

"conditions" implies maximum exists.

Then we either prove the conjecture or, failing to do so, discover additional examples (counterexamples to the conjecture) along with further conditions, to provide a new conjecture.

In our case, the obvious functions having no maximum value are those that are not bounded above, giving the conjecture

f bounded above implies $\max f$ exists.

Counterexamples are ready to hand, such as

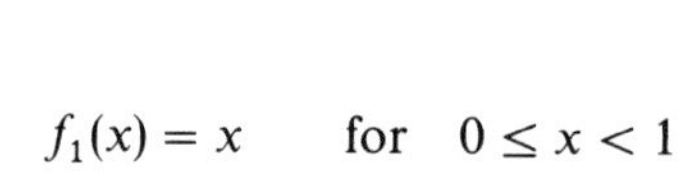

$$f_1(x) = x \qquad \text{for} \quad 0 \le x < 1$$

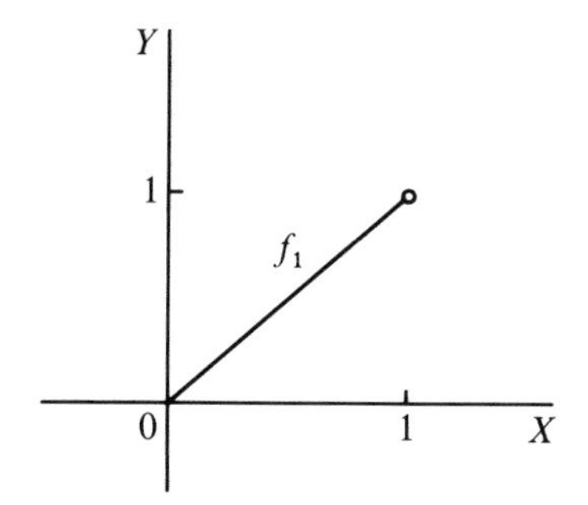

$$f_2(x) = \begin{cases} x, & 0 \le x < 1, \\ \frac{1}{2}, & x = 1. \end{cases}$$

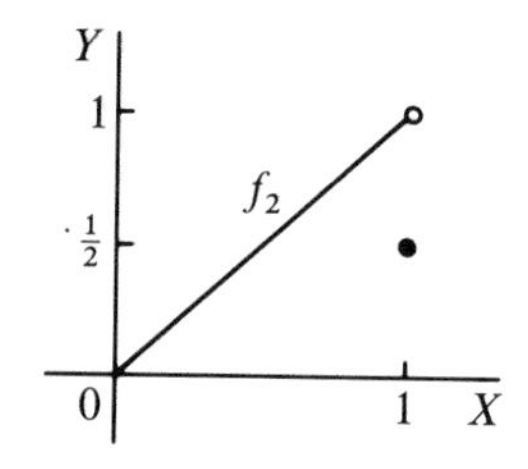

Here, the value $\sup f$, which exists by the hypothesis, is not a value of f, either because the point where it "should occur" is missing from $\operatorname{dom} f$, as with f_1, or the function has the "wrong" value there, as with f_2.

The f_1 example, or more generally, any example f of the kind illustrated,

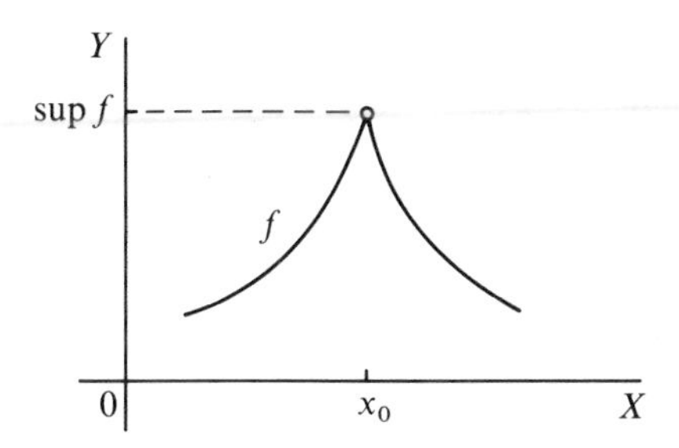

must be excluded by means of a condition that forces dom f to include certain points x_0. To attempt a description of such points, notice that, since sup f is a limit of numbers in ran f, there is a sequence $\{x_n\} \subset$ dom f giving that sequence of values $\{f(x_n)\}$, $f(x_n) \to \sup f$. There is no reason to suppose $\{x_n\}$ converges, but x_0 in our example would be the limit of a subsequence, and the examples of this kind would be excluded by the condition on dom f, that it include the limit of any convergent sequence in it (the subsequence of $\{x_n\}$ in the example).

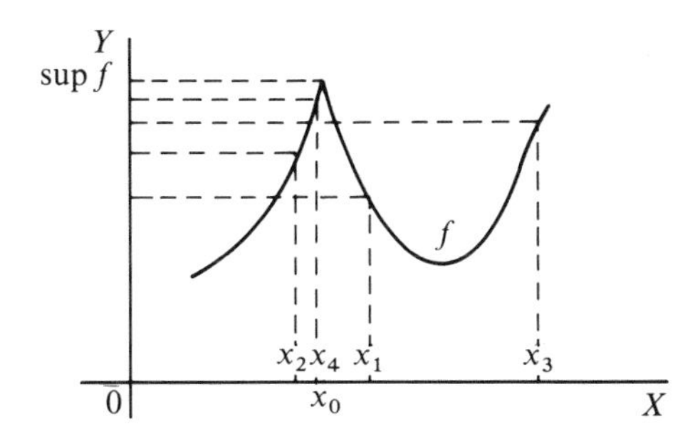

Definition. A set A of reals is **closed** if whenever $\{x_n\}$ is a convergent sequence of members of A, its limit is also a member of A.

Notice that the set $\mathbb{R}$ is closed, the empty set ϕ is closed because of logical convention (there being no convergent sequences in ϕ, the limits of all of them are in ϕ) and a closed interval $[a, b]$ is a closed set because any $x_0 \notin [a, b]$ is not the limit of a sequence in $[a, b]$.

EXERCISE 4. Prove this last assertion (take $x_0 < a$, to be specific) by applying the definition of $x_n \to x_0$, taking $(a - x_0)/2 > 0$ as ε.

EXERCISE 5. A finite union of closed intervals is a closed set. Proof? Generalization?

EXERCISE 6. An open or semiopen interval is not a closed set. Neither is $(0, \infty)$. Show that a union of countably many closed sets need not be closed. Are $\mathbb{N}$ and $\mathbb{Z}$ closed?

The condition that dom f be closed will eliminate many functions that have no maximum value. (To be sure, it will also exclude some functions that do have a maximum value, but it is not our intention to avoid that.)

EXERCISE 7. Give an example of a function whose domain is not closed, which has a maximum value.

The examples like f_2 above (more generally, as in the figure) for which dom f is closed but, at points where the value should be sup f, it is too small, must be excluded next. The points x_0 we are considering are limits of sequences $\{u_k\}$ in dom f for which $\{f(u_k)\}$ converges to sup f. A simple and natural condition that excludes such phenomena is

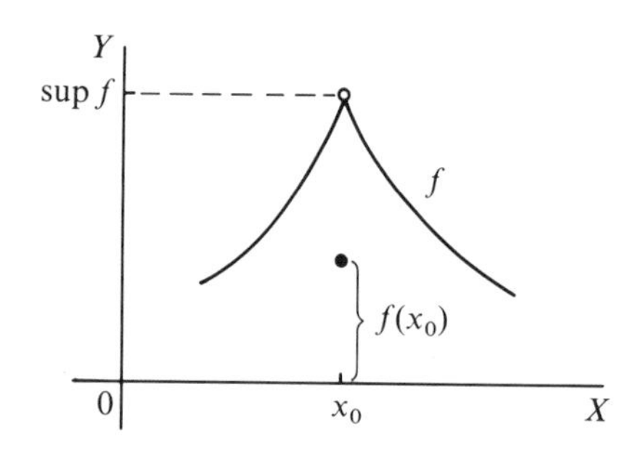

$$x_n \to x_0 \text{ in dom } f \quad \text{implies} \quad f(x_n) \to f(x_0),$$

for it says in the case of $\{u_k\}$ that

$$f(x_0) = \lim f(u_k) = \sup f.$$

We make the

Definition. A function $f: \mathbb{R} \to \mathbb{R}$ is **continuous at** $x_0 \in \text{dom } f$ if

$$\{x_n\} \subset \text{dom } f \quad \text{and} \quad x_n \to x_0 \quad \text{implies} \quad f(x_n) \to f(x_0).$$

A function is **continuous** if it is continuous at every point of its domain.

Remark. Continuity can be symbolically abbreviated

$$f(\lim x_n) = \lim f(x_n),$$

i.e., as the interchangeability of the application of f with the limit operation. This suggests how continuity might be verified: that constant functions and the identity function, $f(x) = x$ for $x \in \mathbb{R}$, are continuous is obvious, so any function built up from these by operations that are interchangeable with the limit operation will be continuous. This idea is developed later in this chapter; a theorem that assumes continuity will be seen to be a usable theorem; the hypothesis is readily verified.

We can now formulate a new conjecture: If f is bounded above and continuous and dom f is closed then f has a maximum value.

The framework for a possible proof is contained in the reasoning that led to the conjecture:

1. Since f is bounded above, sup f exists and there are points $x_1, x_2, x_3, \ldots$ in dom f for which $f(x_n) \to \sup f$.
2. *Supposing* $\{x_n\}$ has a subsequence $\{u_k\}$ that converges in dom f to a point x_0, then $x_0 \in \operatorname{dom} f$, since dom f is closed. That is, $f(x_0)$ is defined.
3. Since f is continuous, it is continuous at x_0. Hence, using the fact that $u_k \to x_0$, we have

   $$f(x_0) = \lim f(u_k) = \lim f(x_n) = \sup f,$$

   the desired conclusion; max f exists (it is $f(x_0)$).

The supposition in 2 is the only gap in the argument; we obtain a theorem if we make the ad hoc hypothesis that dom f is compact, as in the

Definition. A set A of reals is **compact** if every sequence in A has a convergent subsequence whose limit belongs to A.

EXERCISE 8. Observe that a compact set is closed, in view of "whose limit belongs to A." State and prove a theorem on the existence of maximum values.

The theorem having the hypothesis that dom f be compact is unsatisfactory, since it is not clear how to check whether a set is compact, so we return to our last conjecture and counterexamples to it.

To get a sequence with no convergent subsequence, we soon think of one that "goes to infinity in fixed steps," yielding the counterexample shown. This suggests the new hypothesis: that dom f be bounded.

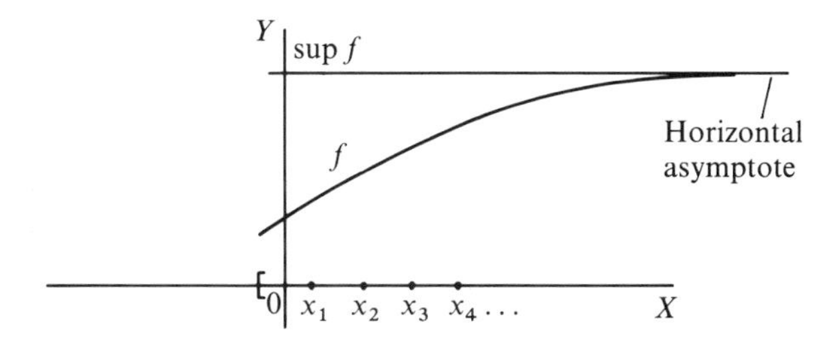

EXERCISE 9. Show that a compact set is necessarily bounded.

Our new conjecture is

If f is bounded above and continuous and if dom f is closed and bounded then f has a maximum.

The proof sketched above applies if we show the

Proposition. *A closed, bounded set is compact.*

Proof. Let $\{x_n\}$ be any sequence in the closed, bounded set A. By the Bolzano–Weierstrass Theorem, $\{x_n\}$ has a convergent subsequence (or, directly, a monotone subsequence of $\{x_n\}$, being bounded, converges). The limit of the subsequence belongs to A since A is closed.

Thus, the ad hoc property, compactness, has the more intuitive form closed and bounded.

EXERCISE 10. Which of the following sets are compact?

(a) $\mathbb{N}$.
(b) $[0, \infty)$.
(c) $[3, 5]$.
(d) A finite union of finite closed intervals.
(e) A union of closed intervals.
(f) $\bigcap_{n=1}^{\infty} (-1/n, 1 + 1/n)$.

EXERCISE 11. Suppose dom f is compact and f is not bounded above. There are points $x_1, x_2, x_3, \ldots$ in dom f such that $f(x_n) > n$ for each $n = 1, 2, 3, \ldots$. Show that f must be discontinuous at some point $x_0 \in \text{dom } f$.

In view of this exercise, the boundedness of f can be shifted from the hypothesis to the conclusion of our conjecture, in view of the presence of the properties dom f compact and f continuous.

We summarize our discussion of the extreme value problem with the statement and proof of one of the most important theorems of Analysis.

The Extreme-Value Theorem. *If f is continuous and* dom f *is compact then f is bounded and assumes its maximum and minimum values.*

Proof. We prove that f is bounded above and assumes its maximum.

Suppose f is not bounded above and choose $\{x_n\} \subset \text{dom } f$ so that $f(x_n) > n$ for each $n = 1, 2, 3, \ldots$. Choose a convergent subsequence $\{x_{n_k}\}$ and let $x_0 = \lim x_{n_k}$. Since dom f is closed, $x_0 \in \text{dom } f$. By the continuity at x_0, $f(x_0) = \lim f(x_{n_k})$. This is impossible since for k large enough n_k exceeds any prescribed value, say $f(x_0) + 1$, hence for such a k

$$f(x_{n_k}) > n_k > f(x_0) + 1.$$

Thus, sup f exists. Choose $\{x_n\} \subset \text{dom } f$ so that $f(x_n) \to \sup f$. Selecting a convergent subsequence $\{x_{n_k}\}$and denoting its limit by $x_M \in \text{dom } f$, the continuity at x_M gives

$$f(x_M) = \lim f(x_{n_k}) = \lim f(x_n) = \sup f;$$

$f(x_M)$ is the required maximum value of f.

EXERCISE 12. Supply the details of the proof that f is bounded below and assumes its minimum value.

EXERCISE 13. Prove the minimum case by showing that f continuous implies $-f$ continuous.

EXERCISE 14. Characterize those sets of reals having the property that *every* sequence in the set converges. In view of this, the simplest *nontrivial* condition on a set that insures an abundance of convergent sequences in the set is the compactness condition.

EXERCISE 15. Write a complete exposition in the Definition–Theorem–Proof style, of the extreme value theorem, in three to five pages.

It should be remarked that the notion of "continuity" arose early in various applied contexts with compelling, if vague, intuitive content:

A "continuous" motion as one that does not change abruptly, or jump.
A "continuous" curve as one that can be drawn without lifting the pencil.
A "smooth" curve (no corners) as one whose tangent turns "continuously."
A "real" physical quantity, such as the temperature of an object, varies "continuously."

It was only in the mid-nineteenth century that our precise definition of continuity (and of function) was set down, to clarify such ideas. It was found to only partially support such intuitions, indeed to admit some possibilities quite at odds with intuition, such as continuous (without quotes) curves having no tangents, and continuous curves that fill an area. A milder example is the following function, which is continuous at every irrational and discontinuous at every rational:

$$f(x) = \begin{cases} 1/q & \text{if } x = p/q \text{ in lowest terms,} \\ 0 & \text{if } x \text{ is irrational.} \end{cases}$$

This violated the intuition that points of discontinuity could be at worst "few and far between."

EXERCISE 16. Prove the above assertions about $f(x)$. Sketch a dozen or so points of the graph of $f(x)$, for $0 < x < 1$ and rational.

It is evident, however, that only continuous functions need be considered, in order to model "continuous" phenomena; the concept of continuity has a central applied relevance.

§2. Continuity of Rational Functions. Limits of Sequences

The task of navigating from one place to another has created demands for mathematical tools from ancient times to the present. It gave rise to the computation of certain ratios which we now call the sines of angles. The evolution of the uses of such computations led ultimately to our present trigonometric functions and the discipline, Trigonometry, that elaborates their properties and uses. The continuing interplay between the developing theory and emerging applied problems led to the employment of the trigonometric functions to describe (or "model") things like circular motion and alternating currents, then more complex periodic phenomena such as the motions of the planets, the vibrations of violin strings, radio waves, and light.

One feature of the hand-in-hand evolution of Mathematics and its applications which this example illustrates is a growing *need for new functions* to describe various phenomena, indeed, for sweeping general methods to define new functions—in anticipation of future needs. It is possible from the modern perspective, to describe such sweeping general methods systematically, and doing so provides us with a theme for the introduction of some major ideas of Analysis. In particular, we can trace the presence of the central property, continuity, to see when the extreme value result, and any other in which continuity is a central hypothesis, applies.

The first sweeping general method is the application of the four arithmetical operations, to build the function. It is the subject of this section.

In this section, the numbers we admit will be any complex numbers. By admitting complex numbers, we are able to complete the proof of the Fundamental Theorem of Algebra at the end.

We begin by choosing a "variable" number $z \in \mathbb{C}$ and some (complex) constants, and doing arithmetic to arrive at a value $f(z)$. If only additions and multiplications are used, the result can be assembled in the form

$$a_n z^n + a_{n-1} z^{n-1} + \cdots + a_1 z + a_0, \qquad n \in \mathbb{N},$$

for appropriate constants $a_0, a_1, a_2, \ldots, a_n \in \mathbb{C}$. The resulting function P is, of course, the general polynomial function on $\mathbb{C}$. If some division steps are included, it is easy to show that the final result can also be obtained using only one division—at the last step, i.e., the result can be assembled in the form of a **rational function**

$$R(z) = \frac{P(z)}{Q(z)} = \frac{a_n z^n + a_{n-1} z^{n-1} + \cdots + a_1 z + a_0}{b_m z^m + b_{n-1} z^{m-1} + \cdots + b_1 z + b_0},$$

the quotient of two polynomials.

EXERCISE 1. Prove the assertion in case only two division steps are involved.

EXERCISE 2. Complete the proof by an induction on the number of division steps.

Of course, the domain of $R(z)$ consists of all $z \in \mathbb{C}$ for which $Q(z) \neq 0$.

The formal definitions of limit and continuity in the complex case are identical, with z replacing x, to those in the real case:

Definition. A number z is the **limit** of a sequence $\{z_n\}$ if, for any $\varepsilon > 0$ there is an index N such that $n \geq N$ gives $|z_n - z| < \varepsilon$.

Definition. A function $f: \mathbb{C} \to \mathbb{C}$ is **continuous at a point** $z_0 \in \operatorname{dom} f$ if

$$\{z_n\} \subset \operatorname{dom} f \quad \text{and} \quad z_n \to z_0 \quad \text{implies} \quad f(z_n) \to f(z_0).$$

A function is **continuous** if it is continuous at each point of its domain.

EXERCISE 3. Let x_n and y_n be the real and imaginary parts of z_n for each $n = 0, 1, 2, 3, \ldots$;

$$z_n = x_n + iy_n.$$

Show that $z_n \to z_0$ if and only if $x_n \to x_0$ and $y_n \to y_0$. (Hint: $|z_n - z_0| \leq |x_n - x_0| + |y_n - y_0|$ and $|x_n - x_0| \leq |z_n - z_0|$.)

A set A of complex numbers is **bounded** if it is contained in some disk centered at 0, i.e., if there is an $R > 0$ such that

$$z \in A \quad \text{implies} \quad |z| < R.$$

A function is called **bounded** if its set of values is bounded. In particular, a **bounded sequence** is a sequence $\{z_n\}$ such that for some $R > 0$,

$$|z_n| < R \qquad \text{for} \quad n = 1, 2, 3, \ldots.$$

EXERCISE 4. A convergent sequence is bounded.

The theorem we need, to infer that every rational function is continuous, is one that establishes the interchangeability of each arithmetic operation with the limit operation.

Theorem. *Suppose* $z_n \to z_0$ *and* $w_n \to w_0$. *Then the following limits exist and have the indicated values.*

1. $z_n \pm w_n \to z_0 \pm w_0$.
2. $z_n \cdot w_n \to z_0 \cdot w_0$.
3. $z_n/w_n \to z_0/w_0$, *provided* $w_0 \neq 0$.

Remark. We haven't specified $w_n \neq 0$ in the quotient statement. This is because it can be inferred, at least for all large enough indices, from $w_0 \neq 0$, and so the quotient sequence

$$\frac{z_n}{w_n}$$

is defined from some index on.

EXERCISE 5. Prove that $w_n \to w_0 \neq 0$ implies there is an n_0 such that $w_n \neq 0$ for $n \geq n_0$. (Hint: Use $\varepsilon = |w_0|/2$ in the definition of $w_n \to w_0$.)

***Proof of* 1.** Consider the sum case. Fix $\varepsilon > 0$. Since

$$\begin{aligned} |(z_n + w_n) - (z_0 + w_0)| &= |(z_n - z_0) + (w_n - w_0)| \\ &\leq |z_n - z_0| + |w_n - w_0|, \end{aligned}$$

we need only make each term on the right less than $\varepsilon/2$ by choosing n large enough. By hypothesis, there exists N_1 such that

$$n \geq N_1 \quad \text{gives} \quad |z_n - z_0| < \varepsilon/2$$

and N_2 such that

$$n \geq N_2 \quad \text{gives} \quad |w_n - w_0| < \varepsilon/2.$$

Hence, if $n \geq \max\{N_1, N_2\}$ we have

$$|(z_n + w_n) - (z_0 + w_0)| < \varepsilon,$$

proving the addition result. The subtraction proof is similar, or follows by putting $z_n - w_n = z_n + (-w_n)$ and using the obvious fact that $-w_n \to -w_0$.

***Proof of* 2.** The idea is the same: to estimate $|z_n \cdot w_n - z_0 \cdot w_0|$ from above in terms of $|z_n - z_0|$ and $|w_n - w_0|$. The trick of adding and subtracting $z_0 \cdot w_n$ gives an estimate:

$$\begin{aligned} |z_n \cdot w_n - z_0 \cdot w_0| &= |z_n \cdot w_n - z_0 \cdot w_n + z_0 \cdot w_n - z_0 \cdot w_0| \\ &= |w_n(z_n - z_0) + z_0(w_n - w_0)| \\ &\leq |w_n||z_n - z_0| + |z_0||w_n - w_0|. \end{aligned}$$

Since a restriction on n that makes $|z_n - z_0|$ small can be frustrated by $|w_n|$ becoming large, we make one more step, to achieve a satisfactory estimate. The sequence $\{w_n\}$ being convergent, it is bounded. Choose $R > 0$ so that

$|w_n| < R$ for all n. Thus

$$|z_n \cdot w_n - z_0 \cdot w_0| \leq R|z_n - z_0| + |z_0||w_n - w_0|, \qquad \text{all } n.$$

For any $\varepsilon > 0$, there exist, by hypothesis, N_1 and N_2 such that

$$n \geq N_1 \quad \text{implies} \quad |z_n - z_0| < \frac{\varepsilon}{2R}$$

and

$$n \geq N_2 \quad \text{implies} \quad |w_n - w_0| < \frac{\varepsilon}{2|z_0|}, \qquad \text{if} \quad z_0 \neq 0.$$

If $z_0 = 0$, the term $|z_0||w_n - w_0|$ is zero. So in any case

$$n \geq \max(N_1, N_2) \quad \text{gives} \quad |z_n \cdot w_n - z_0 \cdot w_0| < \varepsilon,$$

completing the proof.

(Note that we could take $N_2 \geq N_1$ so that N_2 *is* the $\max\{N_1, N_2\}$.)

***Proof of* 3.** We estimate with a similar trick:

$$\begin{aligned}
\left|\frac{z_n}{w_n} - \frac{z_0}{w_0}\right| &= \left|\frac{z_n w_0 - z_0 w_n}{w_n \cdot w_0}\right| \\
&= \frac{1}{|w_n w_0|}|z_n w_0 - z_0 w_0 + z_0 w_0 - z_0 w_n| \\
&= \frac{1}{|w_n w_0|}|w_0(z_n - z_0) + z_0(w_0 - w_n)| \\
&\leq \frac{1}{|w_n|}|z_n - z_0| + \frac{1}{|w_n|} \cdot \left|\frac{z_0}{w_0}\right| |w_n - w_0|,
\end{aligned}$$

whenever n is in a range, say $n \geq N_1$, in which $w_n \neq 0$.

The possibly frustrating factor, $1/|w_n|$, can be replaced by a constant $K > 0$ (the proof is left as an exercise) to give

$$\left|\frac{z_n}{w_n} - \frac{z_0}{w_0}\right| \leq K|z_n - z_0| + \frac{K|z_0|}{|w_0|}|w_n - w_0|.$$

Now, for any $\varepsilon > 0$, choose $N_2 \geq N_1$ to give

$$n \geq N_2 \quad \text{implies} \quad |z_n - z_0| < \frac{\varepsilon}{2K}$$

and then choose $N_3 \geq N_2$ so that

$$n \geq N_3 \quad \text{implies} \quad |w_n - w_0| < \frac{|w_0|\varepsilon}{2K|z_0|}, \qquad \text{if} \quad z_0 \neq 0,$$

and $N_3 = N_2$ if $z_0 = 0$. Then

$$n \geq N_3 \quad \text{implies} \quad \left| \frac{z_n}{w_n} - \frac{z_0}{w_0} \right| < \varepsilon,$$

as was to be shown.

EXERCISE 6. If $w_n \neq 0$ for all n and $w_n \to w_0 \neq 0$, then $\{1/w_n\}$ is a bounded sequence.

Remark. The product result in the case where $\{w_n\}$ is the constant sequence $a, a, a, \ldots$ says that

$$\lim az_n = a \lim z_n;$$

the limit operation is interchangeable with that of multiplication by a constant.

The sum, difference, product, and quotient of two functions is defined point-by-point on the common part of their domains in terms of the functions being combined. Thus the continuity of the combined function, assuming that of the functions being combined, is an immediate corollary of this theorem.

Corollary. *If f and g are continuous (at x_0) then $f \pm g$, fg, and f/g are continuous (at x_0, unless $g(x_0) = 0$ in the quotient).*

The use of induction extends this sum, difference and product result to any number of functions.

Since constant functions and the identity function are obviously continuous, we have the desired

Theorem. Every rational function is continuous.

EXERCISE 7. If $f(z)$ is continuous then $|f(z)|$ is continuous. Hint: The triangle inequality gives

$$||f(z)| - |f(z_0)|| \leq |f(z) - f(z_0)|.$$

Appendix: Completion of the Proof of the Fundamental Theorem of Algebra

The continuity of $|P(z)|$ for any polynomial P is now established. By the Extreme Value Theorem, $|P(z)|$ restricted to any compact subset K of $\mathbb{C}$ assumes its minimum value; $\min_{z \in K} |P(z)|$ exists.

To complete the proof of the Fundamental Theorem of Algebra (Chapter 1,

Appendix) we must show that for any polynomial of degree at least 1,

$$P(z) = a_n z^n + a_{n-1} z^{n-1} + \cdots + a_1 z + a_0, \qquad n \geq 1, \quad a_n \neq 0,$$

the minimum over *all* of $\mathbb{C}$, $\min_{z \in \mathbb{C}} |P(z)|$, exists. Since $\mathbb{C}$ is not compact, some additional argument is needed.

The polynomial assumes the value a_0 at $z = 0$, hence if a compact set K contains 0,

$$\min_{z \in K} |P(z)| \leq |a_0|.$$

If K can be chosen so that

$$|P(z)| \geq |a_0| \qquad \text{whenever} \quad z \notin K,$$

then, clearly, $\min_{z \in K} |P(z)|$ is the required $\min_{z \in \mathbb{C}} |P(z)|$. A set K in the form of a closed disk centered at 0 suggests itself, because $|P(z)|$ should behave outside a large disk, i.e., for $|z|$ large, like its dominant term $|a_n z^n|$ (the lower powers are "comparatively negligible" when $|z|$ is large) and this term is easily made to exceed $|a_0|$ for $|z|$ large.

Exercise 1. A closed disk in $\mathbb{C}$ is a compact set.

These thoughts would be made precise by means of an inequality of the form

$$|P(z)| \geq \text{constant}\, |z^n|, \qquad \text{for } |z| \text{ large},$$

so we estimate $|P(z)|$ from below:

$$|P(z)| = |a_n|\, |z^n + b_{n-1} z^{n-1} + \cdots + b_1 z + b_0|,$$

where

$$b_i = \frac{a_i}{a_n}, \qquad i = 0, 1, \ldots, n-1,$$

$$|P(z)| \geq |a_n| (|z|^n - |b_{n-1} z^{n-1} + \cdots + b_1 z + b_0|)$$

$$\geq |a_n| \left(|z|^n - \sum_{i=0}^{n-1} |b_i|\, |z|^i \right),$$

the last two steps both making use of the triangle inequality. Replacing each $|b_i|$ by the largest among them, $\beta = \max_{0 \leq i \leq n-1} |b_i|$, continues the string of inequalities and simplifies the expression:

$$|P(z)| \geq |a_n| \left(|z|^n - \beta \sum_{i=0}^{n-1} |z|^i \right) = |a_n| \left(|z|^n - \beta \frac{|z|^n - 1}{|z| - 1} \right)$$

$$= |a_n| \left(|z|^n - \frac{\beta |z|^n}{|z| - 1} + \frac{\beta}{|z| - 1} \right)$$

$$\geq |a_n| \left(|z|^n - \frac{\beta |z|^n}{|z| - 1} \right), \qquad \text{provided} \quad |z| > 1.$$

Hence,

$$|P(z)| \geq |a_n||z|^n\left(1 - \frac{\beta}{|z| - 1}\right), \qquad \text{provided} \quad |z| > 1.$$

Now if $|z| > 2\beta + 1$ then $\beta/(|z| - 1) < \frac{1}{2}$ and we have

$$|P(z)| \geq |a_n| \cdot |z|^n \cdot \tfrac{1}{2} \qquad \text{for} \quad |z| > 2\beta + 1,$$

Thus we can achieve $|P(z)| \geq |a_0|$ by choosing z so that

$$|z|^n > \frac{2|a_0|}{|a_n|} \quad \text{and} \quad |z| > 2\beta + 1,$$

which means the disk K can have any radius $> 2\beta + 1$ whose nth power exceeds $2|a_0|/|a_n|$.

Exercise 2. That such a radius exists is due to the unboundedness of $\{x^n\}$ when $x > 1$. Proof? (Hint: Use Bernoulli's inequality, $(1 + \alpha)^n \geq 1 + n\alpha$ when $\alpha \geq -1$ and $n \in \mathbb{N}$.)

§3. Sequences and Series of Reals. The Number e

If the numbers $\{x_n\}$ in a sequence are generated as the successive sums of the numbers from another sequence, i.e.,

$$x_n = a_1 + a_2 + \cdots + a_n, \qquad n = 1, 2, 3, \ldots,$$

then $\{x_n\}$ is called an **infinite series** and is denoted by the symbol $\sum_{i=1}^{\infty} a_i$.

The motivation for such terminology is the recognition that it would be possible and useful to assign values to certain indicated sums of infinitely many numbers. For example, the symbolic long division of 1 by $1 - x$:

$$\begin{array}{r|l} & 1 + x + x^2 + \cdots \\ \hline 1 - x & 1 \\ & \underline{1 - x} \\ & \quad x \\ & \quad \underline{x - x^2} \\ & \qquad x^2 \\ & \qquad \vdots \end{array}$$

suggests that the formula

$$\frac{1}{1 - x} = 1 + x + x^2 + \cdots + x^n + \cdots = \sum_{i=0}^{\infty} x^i$$

can be made valid, at least for some values of x, by appropriately defining the right-hand side.

The numbers $a_1, a_2, a_3, \ldots$ are called the **terms** of the series $\sum_{i=1}^{\infty} a_i$ and the numbers $x_1, x_2, x_3, \ldots$ are called the **partial sums** of the series.

The series is said to be **convergent** if the sequence of partial sums has a limit, otherwise, it is **divergent**. If $\sum_{i=1}^{\infty} a_i$ is convergent and S denotes the limit of the partial sums, $S = \lim x_n$, then S is called the **sum** of the series $\sum_{i=1}^{\infty} a_i$. This is the most natural and straightforward way to assign values to certain series, as their sums. The symbol $\sum_{i=1}^{\infty} a_i$ for a series is also used to denote the sum in case the series is convergent, so that

$$S = \sum_{i=1}^{\infty} a_i.$$

Thus, the formula above,

$$\frac{1}{1-x} = \sum_{i=0}^{\infty} x^i$$

can be shown to be valid if $|x| < 1$.

Of course, *any* sequence $\{x_n\}$ is also a series; just take the terms to be the differences

$$a_1 = x_1 \quad \text{and} \quad a_i = x_i - x_{i-1}, \qquad i = 2, 3, 4, \ldots,$$

to obtain the series $\sum_{i=1}^{\infty} a_i$ which is the same as $\{x_n\}$. This means that the study of series in general coincides logically with the study of sequences; any statement about the one subject has an equivalent reformulation as a statement about the other. The distinctions that are made reflect the different motivations and interpretations that arise in the two contexts:

Sequences: successive approximations of a number. Series: defining "sums" of infinitely many terms.

EXERCISE 1. Reformulate, in terms of the sequence of terms $\{a_i\}$, the following properties of the sequence $\{x_n\}$ of partial sums:

$\{x_n\}$ is increasing,

$\{x_n\}$ is strictly increasing.

EXERCISE 2. If the terms $\{a_i\}$ are all nonnegative, what condition on $\{x_n\}$ expresses the convergence of the series?

EXERCISE 3. Prove that

$$\sum_{i=1}^{\infty} a_i \text{ converges} \quad \text{implies} \quad \lim_{i\to\infty} a_i = 0.$$

Show that the converse is false using the terms $\{1/i\}$.

EXERCISE 4. Translate the theorem on limits of sums into a theorem about the three series

$$\sum_{i=1}^{\infty} a_i, \quad \sum_{i=1}^{\infty} b_i \quad \text{and} \quad \sum_{i=1}^{\infty} (a_i + b_i).$$

In the rest of this section, we shall pursue the study of sequences, only occasionally using series interpretations. Our formal treatment of series is found in Chapter 6.

The facts about the interaction between convergence and the operations of arithmetic, established in the previous section, must be supplemented, in the case of real numbers, by consideration of the order relation. The most immediate observation is that order is preserved in taking limits, but only in the wide sense.

Proposition. *If $x_n \leq y_n$ for $n = 1, 2, 3, \ldots, x_n \to x$ and $y_n \to y$, then $x \leq y$. It can happen that $x = y$ even though $x_n < y_n$ for all n.*

Proof. We show $x \leq y$ in the form $x < y + \varepsilon$ for every $\varepsilon > 0$, or

$$x - \frac{\varepsilon}{2} < y + \frac{\varepsilon}{2} \qquad \text{for every} \quad \varepsilon > 0.$$

There are indices N_1 and N_2 giving, for $\varepsilon > 0$, $n \geq N_1$ implies $x - \varepsilon/2 < x_n$ and $n \geq N_2$ implies $y_n < y + \varepsilon/2$. For $n \geq \max\{N_1, N_2\}$ we have the result:

$$x - \frac{\varepsilon}{2} < x_n \leq y_n \leq y + \frac{\varepsilon}{2}.$$

EXERCISE 5. Show the second assertion in the proposition by giving an example.

EXERCISE 6. The reformulation of $x \leq y$ in the, apparently more complicated, form $x < y + \varepsilon$ for every $\varepsilon > 0$, is suited to arguments in which x, y, or both, are limits, as is seen in this proof. Write an argument that $x < y + \varepsilon$ for every $\varepsilon > 0$ implies $x \leq y$, based on the observation that $x > y$ means there is *an amount* (positive!, namely $x - y$) by which x exceeds y.

EXERCISE 7. Prove the proposition by contradiction, by supposing $x > y$ and taking $\varepsilon = (x - y)/2$ in the definitions of $x_n \to x$ and $y_n \to y$.

An additional observation is the so-called *squeeze*.

Proposition. *If $x_n \leq y_n \leq z_n$ for $n = 1, 2, 3, \ldots$ and $\{x_n\}$ and $\{z_n\}$ converge to the same limit l then $y_n \to l$ as well.*

EXERCISE 8. Proof? Hint: Use $x_n - l \leq y_n - l \leq z_n - l$ to show, given $\varepsilon > 0$, that $-\varepsilon < y_n - l < \varepsilon$ for all sufficiently large n.

Techniques are needed for deciding the convergence or divergence of specific sequences, and, if possible, finding the limit. The squeeze is useful for this, as is the fact that bounded monotone sequences converge, and any result on the continuity of functions. Some ingenuity is often needed to apply these ideas. We consider some examples next, the more important results being cited as propositions.

EXAMPLE. Intuition suggests that the sequence

$$y_n = \frac{n^2 - 2}{n^3 + 1}$$

has limit 0, since the effects of the -2 and the $+1$ are comparatively negligible when n is very large; y_n is "like" $n^2/n^3 = 1/n \to 0$. This intuition is justified by a squeeze:

$$0 \leq y_n \leq \frac{1}{n} \qquad \text{for} \quad n \geq 2$$

that results when the -2 and $+1$ are suppressed.

The same sequence can be rewritten, dividing numerator and denominator by the highest power present, n^3:

$$y_n = \frac{1/n - 2/n^3}{1 + 1/n^3}.$$

This shows $\{y_n\}$ as an arithmetic combination of $\{1/n\}$ and constant sequences, yielding

$$\lim y_n = \frac{0 - 2 \cdot 0}{1 + 0} = 0.$$

Alternatively, this form of y_n shows it to be the value of $(x - 2x^3)/(1 + x^3)$ for $x = 1/n$. The continuity of this function at 0, i.e., $\lim 1/n$, gives the result $\lim y_n = f(0) = 0$.

EXERCISE 9. Guess the limits.

(a) $\left\{\dfrac{2n(n-4)}{3n^2 + 7}\right\}$. (b) $\sqrt{n+1} - \sqrt{n}$.

(c) $\left\{\dfrac{\cos n\pi}{n}\right\}$. (d) $\left\{\dfrac{2^n}{n^n}\right\}$.

Prove your answers by using a squeeze.

EXERCISE 10. Suppose $x_n \to 0$ and $\{y_n\}$ is a bounded sequence. Show $x_n y_n \to 0$.

Proposition. *For* $p > 0$, $1/n^p \to 0$.

For integer values of p, this follows from the continuity of x^p at 0.

EXERCISE 11. Base a proof for general $p > 0$ on a squeeze, using

$$0 < p < q \quad \text{implies} \quad \left(\frac{1}{n}\right)^q \leq \left(\frac{1}{n}\right)^p.$$

EXERCISE 12. Discuss the convergence of $\{\sin n/n^p\}$ for $p > 0$.

Proposition. *For $|a| < 1$, $a^n \to 0$.*

Proof. Writing $|a| = 1/(1 + y)$, with $y > 0$, which is possible since $|a| < 1$, allows the use of the Binomial Theorem to compute $|a|^n$, then to establish an inequality simply by dropping positive terms from a sum:

$$0 \leq |a|^n = \frac{1}{(1+y)^n} = \frac{1}{1 + ny + \dfrac{n(n-1)}{2!}y^2 + \cdots + y^n} < \frac{1}{ny}.$$

Thus $|a^n| \to 0$, since $1/ny \to 0$. Since $-|a^n| \leq a^n \leq |a^n|$, another squeeze gives $a^n \to 0$.

EXERCISE 13. Modify the above argument to show that if $|a| < 1$ then $na^n \to 0$.

EXERCISE 14. Consider $\{n^2 a^n\}$, $\{n^3 a^n\}$, ... State and prove a generalization of the proposition.

We shall establish the familiar facts about the existence and properties of nth roots of real numbers in the next chapter. Assuming them for now, we can consider further interesting sequences.

EXERCISE 15. Suppose $a > 1$. Put $a^{1/n} = 1 + y_n$ with $y_n > 0$ and use the binomial expansion to see that $y_n < (a - 1)/n$ for all n. Then complete the proof of the following.

Proposition. *If $0 < a$ then $a^{1/n} \to 1$.*

The plot thickens if we replace a by n, to consider the interesting sequence $\{n^{1/n}\}$. In an effort to gain some insight, one is led to the calculation of the first few terms, which reveals that

$$1 < \sqrt{2} < \sqrt[3]{3} > \sqrt[4]{4} > \sqrt[5]{5} > \sqrt[6]{6},$$

i.e.,

$$1 < 1.414\ldots < 1.442\ldots > 1.414\ldots > 1.379\ldots > 1.348\ldots.$$

The sequence appears to be decreasing after the third term:

$$(n+1)^{1/(n+1)} < n^{1/n} \qquad \text{if} \quad n \geq 3.$$

The convergence would follow from this since the sequence is bounded below. This can be rewritten, taking the $n(n + 1)$ power, as

$$(n+1)^n < n^{n+1}, \qquad n \geq 3,$$

$$\left(\frac{n+1}{n}\right)^n < n, \qquad n \geq 3,$$

or

$$\left(1 + \frac{1}{n}\right)^n < n \quad \text{if} \quad n \geq 3.$$

Examining $(1 + 1/n)^n$ by the Binomial Theorem, in case $n \geq 3$:

$$\left(1+\frac{1}{n}\right)^n = 1 + n\frac{1}{n} + \frac{1}{2!}n(n-1)\cdot\frac{1}{n^2} + \frac{1}{3!}n(n-1)(n-2)\frac{1}{n^3} + \cdots + \frac{n!}{n!}\frac{1}{n^n}$$

$$= 1 + 1 + \frac{1}{2!}\frac{(n-1)}{n} + \frac{1}{3!}\frac{(n-1)(n-2)}{n^2} + \cdots + \frac{1}{n!}\frac{(n-1)\ldots 3\cdot 2}{n^{n-1}}$$

$$= 1 + 1 + \frac{1}{2!}\left(1-\frac{1}{n}\right) + \frac{1}{3!}\left(1-\frac{1}{n}\right)\left(1-\frac{2}{n}\right) + \cdots$$

$$+ \frac{1}{n!}\left(1-\frac{1}{n}\right)\left(1-\frac{2}{n}\right)\ldots\left(1-\frac{n-1}{n}\right)$$

$$< 1 + 1 + \frac{1}{2!} + \frac{1}{3!} + \cdots + \frac{1}{n!} \quad \text{(each parenthesis is } < 1\text{)}$$

$$< 1 + 1 + \frac{1}{2} + \frac{1}{2^2} + \cdots + \frac{1}{2^{n-1}} \quad \text{(reducing each larger factor to a 2)}$$

$$= 1 + \frac{1 - (\frac{1}{2})^n}{1 - \frac{1}{2}} \quad \text{(finite geometric sum)}$$

$$= 3 - (\tfrac{1}{2})^{n-1}.$$

Hence $(1 + 1/n)^n < 3$ if $n \geq 3$, so the sequence $\{n^{1/n}\}$ is decreasing when $n \geq 3$.

Since 1 is a lower bound of $\{n^{1/n}\}$,

$$n^{1/n} = 1 + x_n \quad \text{with} \quad x_n > 0 \quad \text{for} \quad n = 2, 3, 4, \ldots,$$

and the Binomial Theorem (again!) yields

$$n = (1 + x_n)^n = 1 + nx_n + \frac{n(n-1)}{2!}x_n^2 + \cdots + x_n^n$$

$$> \frac{n(n-1)}{2!}x_n^2.$$

Thus

$$0 < x_n < \sqrt{\frac{2}{n-1}}, \qquad n \geq 2,$$

a squeeze which shows that $x_n \to 0$, hence the

Proposition. $n^{1/n} \to 1$.

In the course of the above calculations we encountered two other interesting sequences: $\{(1 + 1/n)^n\}$ and $\{1 + 1 + 1/2! + 1/3! + \cdots + 1/n!\}$, the sequence of partial sums of $\sum_{i=0}^{\infty} 1/k!$.[1] We saw that

$$\left(1 + \frac{1}{n}\right)^n < 1 + 1 + \frac{1}{2!} + \cdots + \frac{1}{n!} < 3, \qquad n \geq 3.$$

Since the sequence of partial sums is clearly increasing, we infer its convergence to a limit ≤ 3, or

$$\sum_{k=0}^{\infty} \frac{1}{k!} \leq 3.$$

Is the sequence $\{(1 + 1/n)^n\}$ also increasing? We saw that

$$\left(1 + \frac{1}{n}\right)^n = 1 + 1 + \frac{1}{2!}\left(1 - \frac{1}{n}\right) + \frac{1}{3!}\left(1 - \frac{1}{n}\right)\left(1 - \frac{2}{n}\right) + \cdots$$
$$+ \frac{1}{n!}\left(1 - \frac{1}{n}\right)\left(1 - \frac{2}{n}\right)\cdots\left(1 - \frac{n-1}{n}\right),$$

hence also that (putting $n + 1$ for n)

$$\left(1 + \frac{1}{n+1}\right)^{n+1}$$
$$= 1 + 1 + \frac{1}{2!}\left(1 - \frac{1}{n+1}\right) + \frac{1}{3!}\left(1 - \frac{1}{n+1}\right)\left(1 - \frac{2}{n+1}\right) + \cdots$$
$$+ \frac{1}{n!}\left(1 - \frac{1}{n+1}\right)\left(1 - \frac{2}{n+1}\right)\cdots\left(1 - \frac{n-1}{n+1}\right)$$
$$+ \frac{1}{(n+1)!}\left(1 - \frac{1}{n+1}\right)\cdots\left(1 - \frac{n}{n+1}\right)$$

Clearly, the second expression is term-by-term greater than or equal to the first except it contains an additional term; the sequence is increasing.

Consequently, $\lim(1 + 1/n)^n$ exists. Moreover

$$\lim\left(1 + \frac{1}{n}\right)^n \leq \sum_{k=0}^{\infty} \frac{1}{k!}.$$

[1] $0! = 1$ is by definition.

The reverse inequality also holds. To see this, we fix an $m \leq n$ and drop terms from the expression for $(1 + 1/n)^n$ above to get

$$\left(1 + \frac{1}{n}\right)^n \geq 1 + 1 + \frac{1}{n!}\left(1 - \frac{1}{n}\right) + \frac{1}{3!}\left(1 - \frac{1}{n}\right)\left(1 - \frac{2}{n}\right) + \cdots$$
$$+ \frac{1}{m!}\left(1 - \frac{1}{n}\right)\left(1 - \frac{2}{n}\right)\ldots\left(1 - \frac{m-1}{n}\right).$$

Letting $n \to \infty$ gives

$$\lim_{n\to\infty}\left(1 + \frac{1}{n}\right)^n \geq 1 + 1 + \frac{1}{2!} + \frac{1}{3!} + \cdots + \frac{1}{m!} \qquad \text{for all } m.$$

Hence

$$\lim_{n\to\infty}\left(1 + \frac{1}{n}\right)^n \geq \lim_{m\to\infty}\left(1 + 1 + \frac{1}{2!} + \cdots + \frac{1}{m!}\right) = \sum_{k=0}^{\infty} \frac{1}{k!}.$$

Definition. The common value of these two limits is denoted e;

$$e = \lim_n \left(1 + \frac{1}{n}\right)^n = \sum_{k=0}^{\infty} \frac{1}{k!}.$$

Remark. The significance of e, as that base for an exponential function e^x which is preferred because e^x is its own derivative, will be shown in Chapter 4.

That the values of the limits of the two sequences we have just studied could not have been easily guessed is explained by the

Proposition. *e is irrational.*

Proof. If e were rational, $e = p/q$ with $p, q \in \mathbb{N}$, then $m!\,e$ would be in $\mathbb{N}$ as soon as $m \geq q$. Thus it is enough to show that

$$\text{for all sufficiently large } m \in \mathbb{N}, \qquad m!\,e \notin \mathbb{N}.$$

We express $m!\,e$ in a form that separates out a part that is clearly a whole number:

$$m!\,e = m!\left(1 + \frac{1}{2!} + \cdots + \frac{1}{m!}\right) + m! \lim_{n\to\infty}\left[\frac{1}{(m+1)!} + \cdots + \frac{1}{n!}\right].$$

It is enough to show that the other part is a fraction:

$$m! \lim_{n\to\infty}\left[\frac{1}{(m+1)!} + \cdots + \frac{1}{n!}\right] < 1,$$

when m is large. In the form

$$\frac{1}{(m+1)} \lim_{n\to\infty} \left[1 + \frac{1}{(m-2)} + \cdots + \frac{1}{n(n-1)\ldots(m+2)} \right] < 1$$

this is seen by showing that these limits are bounded in their dependence on m. But when $n > m$

$$1 + \frac{1}{(m+2)} + \cdots + \frac{1}{n(n-1)\ldots(m+2)} < 1 + \frac{1}{2} + \cdots$$
$$+ \frac{1}{(n-m)(n-m-1)\ldots 2}$$

(each factor in each denominator has been reduced by m), so

$$1 + \frac{1}{m+2} + \cdots + \frac{1}{n(n-1)\ldots(m+2)} < 1 + \frac{1}{2!} + \cdots + \frac{1}{(n-m)!}.$$

We have seen that 2 is an upper bound for these last numbers, hence also for the sequence of limits.

When no good guess for the value of $\lim x_n$ presents itself, as in these examples, one seeks, as a compromise, the *existence* of the limit (without its value), usually via boundedness and monotonicity. Knowing a limit exists, one can find decimal approximations to its value from the terms of the sequence.

EXERCISE 16. Prove that $\sum_{k=1}^{\infty} 1/k^2$ converges by showing that

$$1 + \frac{1}{2^2} + \frac{1}{3^2} + \cdots + \frac{1}{n^2} \leq 2 - \frac{1}{n}, \qquad n = 1, 2, 3, \ldots,$$

using induction.

Some sequences are too difficult to analyze for convergence, or in fact diverge. Nevertheless, something useful can be said.

First, the simplest divergent sequences of reals are those that "diverge to infinity" in the sense of the

Definition. $x_n \to \infty$ means, for every $B > 0$ there is an index N such that

$$n \geq N \quad \text{implies} \quad x_n > B.$$

$x_n \to -\infty$ means $-x_n \to \infty$.

EXERCISE 17. For $x_n > 0$, $n = 1, 2, 3, \ldots$,

$$x_n \to \infty \quad \text{if and only if} \quad \frac{1}{x_n} \to 0.$$

Definition. A sequence is **oscillating** if it does not converge and does not diverge to either ∞ or $-\infty$.

Extreme examples of oscillating sequences are provided by $\{(-1)^n\}$ on one hand, and $\{r_n\}$, where $n \leftrightarrow r_n$ is an enumeration of $\mathbb{Q}$, on the other.

The notion of a neighborhood of a point is helpful in the discussion of oscillating sequences.

Definition. A **neighborhood** of $x_0 \in \mathbb{R}$ is any set $\mathcal{U} \subset \mathbb{R}$ that contains some interval $(x_0 - \varepsilon, x_0 + \varepsilon)$ with $\varepsilon > 0$.

Remark. The intervals $(x_0 - \varepsilon, x_0 + \varepsilon)$ are the crucial neighborhoods of x_0; it is a matter of convenience to extend the term to embrace all sets containing such an interval.

In the language of neighborhoods, the convergence, $x_n \to x_0$, is formulated as

For every neighborhood $\mathcal{U}$ of x_0 there is an N such that $n \geq N$ gives $x_n \in \mathcal{U}$.

The range $n \geq N$ is a special kind of infinite set of indices. By admitting *any* infinite set of indices, we obtain the notion of a cluster point of $\{x_n\}$:

Definition. A point x_0 is a **cluster point** of a sequence $\{x_n\}$ if, for every neighborhood $\mathcal{U}$ of x_0 there are infinitely many indices n for which $x_n \in \mathcal{U}$.

EXAMPLES. 1 and -1 are the cluster points of $\{(-1)^n\}$. Any real number x_0 is a cluster point of $\{r_n\}$ if $n \leftrightarrow r_n$ is an enumeration of $\mathbb{Q}$. (Why?)

Clearly, if $x_n \to x_0$ then x_0 is the only cluster point of $\{x_n\}$, and if $x_n \to \pm\infty$ there are no cluster points of $\{x_n\}$. Any number x_0 that occurs as the value x_{n_k} of the sequence for infinitely many indices $\{n_k\}$ is a cluster point of $\{x_n\}$.

A characterization of cluster points in terms of convergence is given by the

Proposition. *x_0 is a cluster point of $\{x_n\}$ if and only if x_0 is the limit of a subsequence of $\{x_n\}$.*

Proof. If x_0 is the limit of a subsequence $\{x_{n_k}\}$ then, by the definition of subsequence, $\{n_k\}$ is an infinite set of indices. Any neighborhood of x_0 contains x_{n_k} for all k sufficiently large, i.e., for infinitely many indices, showing x_0 is a cluster point of $\{x_n\}$.

Conversely, if x_0 is a cluster point of $\{x_n\}$ and $\mathcal{U}_k = (x_0 - 1/k, x_0 + 1/k)$, we can choose a subsequence converging to x_0 by the following induction:

Fix $x_{n_1} \in \mathcal{U}_1$.

Having chosen $x_{n_k} \in \mathscr{U}_k$, there are indices $l > n_k$ for which $x_l \in \mathscr{U}_{k+1}$ (why?) and any such l can be taken as n_{k+1}.

That $x_{n_k} \to x_0$ is clear.

EXERCISE 18. Show that if $\{x_n\}$ is bounded then the set of all cluster points of $\{x_n\}$ is nonempty and bounded.

Applying this exercise and the completeness of $\mathbb{R}$, we can make the

Definition. *If* Γ *is the set of cluster points of the bounded sequence* $\{x_n\}$, $\inf \Gamma$ *and* $\sup \Gamma$ *are called, respectively, the* **limit inferior** *of* $\{x_n\}$ *and the* **limit superior** *of* $\{x_n\}$.

There are two common notations:

$$\inf \Gamma = \underline{\lim}\, x_n \quad \text{or} \quad \lim\inf x_n,$$

$$\sup \Gamma = \overline{\lim}\, x_n \quad \text{or} \quad \lim\sup x_n.$$

Clearly,

$$\underline{\lim}\, x_n \le \overline{\lim}\, x_n$$

for any bounded sequence, and $\{x_n\}$ converges exactly when equality holds.

EXERCISE 19. Prove that for any bounded sequence $\{x_n\}$, both $\underline{\lim}\, x_n$ and $\overline{\lim}\, x_n$ are themselves cluster points of $\{x_n\}$. Thus $\underline{\lim}\, x_n$ and $\overline{\lim}\, x_n$ can be described as, respectively, the smallest and the largest cluster point of $\{x_n\}$.

It is sometimes convenient to adapt the notation $\overline{\lim}\, x_n = \infty$ ($\underline{\lim}\, x_n = -\infty$) to express that $\{x_n\}$ is not bounded above (below).

EXERCISE 20. $\overline{\lim}\, x_n \le \sup x_n$ and strict inequality can hold.

EXERCISE 21. Find $\underline{\lim}\, x_n$ and $\overline{\lim}\, x_n$.

(a) $\left\{\sin \dfrac{n\pi}{3}\right\}$. (c) $\left\{(-1)^n + \dfrac{1}{n}\right\}$.

(b) $\left\{\left[\dfrac{n}{3}\right] - \dfrac{n}{3}\right\}$. (d) $\{2^{(-1)^n}\}$.

The usefulness of these concepts can be illustrated by two observations.

1. Recall the Ratio Test for the (absolute) convergence of a series $\sum_{n=1}^{\infty} a_n$: that $\lim |a_{n+1}/a_n| < 1$. The existence of this limit need not be shown since the proof of the test actually shows that $\overline{\lim}\, |a_{n+1}/a_n| < 1$ insures convergence.
2. If f has compact domain, the conclusion that f is bounded above

and assumes its maximum value follows from the condition, called **upper semicontinuity**:

$$\text{If } x_0 \in \operatorname{dom} f \quad \text{and} \quad x_n \to x_0 \text{ in } \operatorname{dom} f, \quad \text{then} \quad f(x_0) \geq \overline{\lim}\, f(x_n).$$

EXERCISE 22. Proof?

We shall not need this generalization of the theorem on maximum values.

EXERCISE 23. Generalize the theorem on minimum values via an appropriate definition of lower semicontinuity.

For a bounded sequence $\{x_n\}$, consider the "tail-ends" $\{x_k, x_{k+1}, \dots\}$ for each $k = 1, 2, 3, \dots$, and their infima, $\inf_{n\geq k} x_n$, and suprema, $\sup_{n\geq k} x_n$. Since $\{x_k, x_{k+1}, \dots\} \supset \{x_{k+1}, x_{k+2}, \dots\}$,

$$\inf_{n\geq k} x_n \leq \inf_{n\geq k+1} x_n \quad \text{and} \quad \sup_{n\geq k} x_n \geq \sup_{n\geq k+1} x_n$$

(adding a "candidate" x_k can only, at most, lower the inf and raise the sup).

The monotone sequences $\{\inf_{n\geq k} x_n\}$ and $\{\sup_{n\geq k} x_n\}$ are bounded, hence convergent. In fact, their limits are

$$\lim_{k\to\infty} \inf_{n\geq k} x_n = \underline{\lim}\, x_n$$

and

$$\lim_{k\to\infty} \sup_{n\geq k} x_n = \overline{\lim}\, x_n.$$

EXERCISE 24. Prove one of these formulas.

EXERCISE 25. For any $\varepsilon > 0$ there is a K such that

$$k \geq K \quad \text{gives} \quad x_k < \overline{\lim}\, x_n + \varepsilon,$$

where $\{x_n\}$ is any bounded sequence.

EXERCISE 26. Show that

$$\overline{\lim}(x_n + y_n) \leq \overline{\lim}\, x_n + \overline{\lim}\, y_n$$

and give an example of the strict inequality.

EXERCISE 27. For a sequence $\{x_n\}$, consider the sequence of successive averages

$$\sigma_1 = x_1, \quad \sigma_2 = \frac{x_1 + x_2}{2}, \dots, \quad \sigma_n = \frac{x_1 + x_2 + \cdots + x_n}{n}, \dots.$$

It is intuitively clear that the numbers σ_n would be more "bunched together" than the numbers x_n, so we conjecture

$$\underline{\lim}\, x_n \leq \underline{\lim}\, \sigma_n \leq \overline{\lim}\, \sigma_n \leq \overline{\lim}\, x_n.$$

In particular

$$x_n \to a \quad \text{implies} \quad \sigma_n \to a.$$

Prove this last statement directly by a (careful!) $\varepsilon - N$ argument.

EXERCISE 28. Show that $\sigma_n \to \frac{1}{2}$ when $\{x_n\}$ is 1, 0, 1, 0, 1, 0, ..., a divergent sequence. Thus lim σ_n generalizes the notion of "limit" to a broader class of sequences $\{x_n\}$ than just the convergent ones. This hints at the existence of other senses in which something useful might be said about divergent sequences (see Hardy [15].)

§4. Sets of Reals. Limits of Functions

A number x_0 that belongs to a set $A \subset \mathbb{R}$ can be approximated by members of A (to within ε, for any $\varepsilon > 0$) in the trivial way of choosing x_0 itself as the approximation. To distinguish those numbers that can be approximated by members of A in a nontrivial way, we introduce some terms.

Definition. A **punctured neighborhood** of x_0 is any set that includes, for some $\varepsilon > 0$, the set $\{x: 0 < |x - x_0| < \varepsilon\}$.

Thus, the deletion of x_0 itself from a neighborhood of x_0 yields a punctured neighborhood of x_0.

Definition. A point x_0 of A is an **isolated point of** A if some punctured neighborhood of x_0 is disjoint from A.

Definition. A **limit point** of A is any point x_0 such that every punctured neighborhood of x_0 contains points of A.

The isolated points of A are the points that can be approximated by A only in the trivial way, whereas, the limit points of A are those points, whether or not members of A, that can be approximated by A in a nontrivial way. All points of A are either isolated points of A or limit points of A. Some limit points of A may fail to belong to A.

EXAMPLES. 1. [0, 1) has as limit points all members of [0, 1]. There are no isolated points.

2. $[0, 1] \cup \{2\}$ has isolated point 2.

3. $\{1, \frac{1}{2}, \frac{1}{3}, \dots\}$ has 0 as limit point and all points of the set are isolated points.

4. Consider any sequence $\{x_n\}$ and its value set $A = \{x: x = x_n \text{ for some } n \in \mathbb{N}\}$. Only a cluster point of $\{x_n\}$ can be a limit point of A, but a cluster point of $\{x_n\}$ can be an isolated point of A: one that is named infinitely often

in the sequence. Thus, for $\{(-1)^n\}$, the two cluster points -1 and 1 make up the value set and are both isolated points of that set.

EXERCISE 1. If x_0 is a limit point of A then every neighborhood of x_0 contains infinitely many points of A.

Proposition. *A point x_0 is the limit of a sequence of points in A if and only if either $x_0 \in A$ or x_0 is a limit point of A.*

Proof. If $x_0 \in A$, the sequence $x_0, x_0, x_0, \ldots$ converges to x_0. In case x_0 is a limit point of A, we define a sequence in A converging to x_0 by the induction:

$$x_1 = \text{any point of } (x_0 - 1, x_0 + 1) \cap A$$

and, having chosen $x_n \in (x_0 - 1/n, x_0 + 1/n) \cap A$, any point in $(x_0 - 1/(n + 1), x_0 + 1/(n + 1)) \cap A$ may be chosen as x_{n+1}. Of course, it is the hypothesis that insures the existence of points in these sets.

Conversely, suppose $x_0 = \lim x_n$, where each $x_n \in A$. We must show that if $x_0 \notin A$ then necessarily x_0 is a limit point of A. But any neighborhood of x_0 contains some x_N. The punctured neighborhood intersects A since $x_N \in A$ and $x_0 \notin A$. Thus x_0 is a limit point of A.

Since a closed set was defined to be one that includes the limit of every convergent sequence it contains, we have the

Corollary. *A set A is closed if and only if every limit point of A belongs to A.*

Proposition. 1. *A union of finitely many closed sets is a closed set.*

2. *A union of countably many closed sets may fail to be a closed set.*
3. *The intersection of any family of closed sets is closed.*

EXERCISE 2. Prove 1.

EXERCISE 3. Give an example showing 2.

EXERCISE 4. Prove 3. (If you choose to introduce notation for the family of sets, avoid integer notation that suggests a spurious countability of the family. A common notation is to consider any set I for the "indices," the family of sets $\{A_\alpha: \alpha \in I\}$ and the intersection $\bigcap_{\alpha \in I} A_\alpha$.)

EXERCISE 5. The Cantor set is closed.

Definition. *The* **closure** $\bar{A}$ *of a set A is the union of A and the set of all limit points of A.*

Clearly, $\bar{A}$ is contained in any closed set that contains A. It is not clear that $\bar{A}$ is itself a closed set because the limit points of A could themselves have limit points; they are limit points of $\bar{A}$ and only limit points of A are adjoined to A in forming $\bar{A}$. However, it is easy to see that a limit point of $\bar{A}$ is necessarily a limit point of A, hence the

Proposition. *For any set A, $\bar{A}$ is closed.*

EXERCISE 6. Proof?

In fact, $\bar{A}$ is seen to be the smallest closed set containing A; the intersection of all closed sets containing A.

Recall that a set D is said to be dense if there is a member of D between every two reals, which means that every neighborhood of every point contains points of D. Thus,

Proposition. *A set D is dense if and only if $\bar{D} = \mathbb{R}$, i.e., every real number is the limit of some sequence of members of D.*

Thus, in particular, every real number is the limit of a sequence of rationals, or of irrationals.

EXERCISE 7. Generalize the idea of dense set by defining, for any set $A \subset \mathbb{R}$, "D is dense in A." Specify three sets that are dense in $[0, 1]$. Three more!

We wish next to define limits of functions, $l = \lim_{x \to a} f(x)$ or $f(x) \to l$ as $x \to a$. The definition is to express the idea that l is approximated by the function values $f(x)$ for x near a, independent of the value *at* a, or even whether $a \in \operatorname{dom} f$. Clearly, such a definition should only consider points a that are limit points of dom f.

Definition. *If a is a limit point of* dom f, *we say* ***l* is the limit of *f* at *a*** *if, for every $\varepsilon > 0$ there is a δ such that*

$$x \in \operatorname{dom} f \quad \textit{and} \quad 0 < |x - a| < \delta \quad \textit{implies} \quad |f(x) - l| < \varepsilon.$$

Remarks. 1. The requirement that a be a limit point of dom f insures the existence of points $x \in \operatorname{dom} f$ satisfying $0 < |x - a| < \delta$ for any $\delta > 0$, hence that the implication imposes a nonvacuous condition on f and l.

2. If $l \neq l'$ then, when $\varepsilon = |l - l'|/2$, it is impossible to have both $|f(x) - l| < \varepsilon$ and $|f(x) - l'| < \varepsilon$. Thus at most one number l can satisfy the definition; the limit is unique, if it exists.

3. If dom f is an interval and a is one of the endpoints, this definition yields the usual one-sided limit at a.

4. The definition excludes consideration of the existence or value of f at a through the use of the punctured neighborhood, i.e., $0 < |x - a| < \delta$.

Proposition. $\lim_{x \to a} f(x) = l$ *if and only if* $\lim f(x_n) = l$ *for every sequence* $\{x_n\} \subset \operatorname{dom} f - \{a\}$ *converging to a.*

Proof. Suppose $\lim_{x \to a} f(x) = l$ and $\{x_n\}$ is such a sequence. For $\varepsilon > 0$, choose δ so that

$$x \in \operatorname{dom} f \quad \text{and} \quad 0 < |x - a| < \delta \quad \text{implies} \quad |f(x) - l| < \varepsilon.$$

Since $x_n \to a$, there is an N such that

$$n \geq N \quad \text{gives} \quad |x_n - a| < \delta.$$

Hence, since $0 < |x_n - a|$ by assumption,

$$n \geq N \quad \text{implies} \quad |f(x_n) - l| < \varepsilon.$$

This shows that $\lim f(x_n) = l$.

Conversely, suppose $\lim_{x \to a} f(x) = l$ to be false. Then there is an $\varepsilon_0 > 0$ such that $|f(x) - l| \geq \varepsilon_0$ for some $x \in \operatorname{dom} f$ with $0 < |x - a| < \delta$, no matter what δ may be. Applying this successively for $\delta = 1, \frac{1}{2}, \frac{1}{3}, \ldots$, we get points x_1, $x_2, x_3, \ldots$ in dom f satisfying, for each n,

$$0 < |x_n - a| < \frac{1}{n} \quad \text{and} \quad |f(x_n) - l| \geq \varepsilon_0.$$

This contradicts the hypothesis. Hence $\lim_{x \to a} f(x) = l$ must hold under that hypothesis, concluding the proof.

Theorem. *If a is a limit point of* $\operatorname{dom} f \cap \operatorname{dom} g$, $\lim_{x \to a} f(x) = l$ *and* $\lim_{x \to a} g(x) = m$ *then*

1. $\lim\limits_{x \to a} (f(x) \pm g(x)) = l \pm m$;
2. $\lim\limits_{x \to a} f(x) \cdot g(x) = l \cdot m$; *and*
3. $\lim\limits_{x \to a} \dfrac{f(x)}{g(x)} = \dfrac{l}{m}$ *provided* $m \neq 0$.

The conclusions include the existence of the limits as well as the indicated values.

The proof can be achieved via the preceding proposition, reducing each assertion to the corresponding fact about sequences. In the same way, we obtain the

Proposition. *If a is both a point and a limit point of* dom f, *then*

$$f \text{ is continuous at } a \text{ if and only if } \lim_{x \to a} f(x) = f(a).$$

In view of this, the continuity at a has the $\varepsilon - \delta$ formulation:

f* is continuous at *a if and only if for $\varepsilon > 0$ there is a δ such that

$$x \in \operatorname{dom} f \quad \text{and} \quad |x - a| < \delta \quad \text{implies} \quad |f(x) - f(a)| < \varepsilon.$$

Since it is important for the student to deal with $\varepsilon - \delta$ formulations, a direct proof of the theorem, partly left as exercises, is discussed next.

EXERCISE 8. Prove 1.

***Proof of* 2.** Fix $\varepsilon > 0$. We have

$$\begin{aligned} |f(x)g(x) - l \cdot m| &\le |f(x)g(x) - lg(x)| + |lg(x) - lm| \\ &= |g(x)|\,|f(x) - l| + |l|\,|g(x) - m| \end{aligned}$$

and we wish to replace $|g(x)|$ by a constant before proceeding.

The hypothesis $\lim_{x \to a} g(x) = m$ gives a δ_1 such that $x \in \operatorname{dom} g$ and $0 < |x - a| < \delta_1$ gives $|g(x) - m| < 1$, hence that $|g(x)| < |m| + 1$. Thus,

$$x \in \operatorname{dom} f \cap \operatorname{dom} g \quad \text{and} \quad 0 < |x - a| < \delta_1$$

gives

$$|f(x)g(x) - lm| \le (|m| + 1)\,|f(x) - l| + |l|\,|g(x) - m|.$$

Now we choose δ_2 so that

$$x \in \operatorname{dom} f \quad \text{and} \quad 0 < |x - a| < \delta_2 \quad \text{implies} \quad |f(x) - l| < \frac{\varepsilon}{2(|m| + 1)}.$$

If $l = 0$, choose $\delta_3 = \delta_2$. If $l \neq 0$, choose δ_3 so that

$$x \in \operatorname{dom} g \quad \text{and} \quad 0 < |x - a| < \delta_3 \quad \text{implies} \quad |g(x) - m| < \frac{\varepsilon}{2|l|}.$$

The choice $\delta = \min\{\delta_1, \delta_2, \delta_3\}$ gives

$$x \in \operatorname{dom} f \cap \operatorname{dom} g \quad \text{and} \quad 0 < |x - a| < \delta \quad \text{implies} \quad |f(x)g(x) - lm| < \varepsilon.$$

The proof of 3 makes use of a preliminary result that has other uses, the

Persistence of Sign Lemma. *If* $\lim_{x \to a} f(x) = l$ *and* $l > 0$ *then there is a punctured neighborhood* $\mathcal{U}$ *of* a *such that*

$$x \in \mathcal{U} \cap \operatorname{dom} f \quad \textit{implies} \quad f(x) > 0.$$

Similarly for $l < 0$, *i.e., the sign of the limit persists throughout some punctured neighborhood of* a.

Proof. Take $\varepsilon = l/2$ in the definition of $\lim_{x \to a} f(x) = l$. There is a δ such that

$$x \in \operatorname{dom} f \quad \text{and} \quad 0 < |x - a| < \delta \quad \text{gives} \quad |f(x) - l| < \frac{l}{2},$$

or $-l/2 < f(x) - l < l/2$, which includes $f(x) > l/2 > 0$. Thus $\mathscr{U} = \{x: 0 < |x - a| < \delta\}$ gives the conclusion.

This lemma insures that the quotient $f(x)/g(x)$ is defined for all x in $\operatorname{dom} f \cap \operatorname{dom} g \cap \mathscr{U}$ for some neighborhood $\mathscr{U}$ of a, when $m \neq 0$ as in 3.

EXERCISE 9. Prove 3.

The usefulness of the fact that continuity is preserved when functions are combined by arithmetic is greatly augmented by the fact that it is preserved under combination by composition.

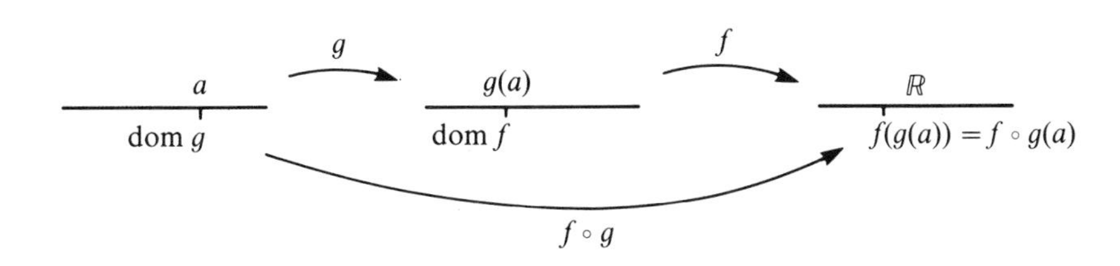

Theorem. *If a is a point and a limit point of* $\operatorname{dom} f \circ g$ *then $f \circ g$ is continuous at a if g is continuous at a and f is continuous at $g(a)$.*

Proof. Fix $\varepsilon > 0$. The continuity of f at $g(a)$ provides a δ_1 such that

$$y \in \operatorname{dom} f \quad \text{and} \quad |y - g(a)| < \delta_1 \quad \text{implies} \quad |f(y) - f(g(a))| < \varepsilon.$$

The continuity of g at a provides a δ such that

$$x \in \operatorname{dom} g \quad \text{and} \quad |x - a| < \delta \quad \text{implies} \quad |g(x) - g(a)| < \delta_1.$$

It follows that

$$x \in \operatorname{dom} f \circ g \quad \text{and} \quad |x - a| < \delta \quad \text{implies} \quad |f \circ g(x) - f \circ g(a)| < \varepsilon.$$

Since any composite of rational functions is again a rational function, the fruits of this result appear only after the introduction of some nonrational functions.

EXAMPLE. Given that every $x \geq 0$ has a unique positive square root $\sqrt{x}$, we prove the continuity of the function so defined. Since $x - a = (\sqrt{x} - \sqrt{a})(\sqrt{x} + \sqrt{a})$ we have, for $a > 0$,

$$|\sqrt{x} - \sqrt{a}| = \frac{|x - a|}{\sqrt{x} + \sqrt{a}} \leq \frac{1}{\sqrt{a}}|x - a|.$$

Hence, for $\varepsilon > 0$, the choice $\delta = \sqrt{a\varepsilon}$ yields

$$x \geq 0 \quad \text{and} \quad |x - a| < \delta \quad \text{implies} \quad |\sqrt{x} - \sqrt{a}| < \varepsilon.$$

This proves the continuity at a if $a > 0$. The continuity at $a = 0$ uses $\delta = \varepsilon^2$.

EXERCISE 10. Give the details of the $a = 0$ case.

EXERCISE 11. For what values of x is the given function continuous? Explain.

(a) $\sqrt{\dfrac{x^2 - 1}{x^2 + 1}}$. (c) $\dfrac{1}{1 + \sqrt{1 + \sqrt{x}}}$.

(b) $x^{3/2} - 2\sqrt{x} + 4$. (d) $\sqrt[4]{x}$.

EXERCISE 12. Prove the continuity of $\sqrt[3]{x}$, assuming it is defined on $\mathbb{R}$.

EXERCISE 13. If P is a polynomial, discuss the continuity of $P(\sqrt{x})$ and $\sqrt{P(x)}$.

EXERCISE 14. Use $|x| = \sqrt{x^2}$, $x \in \mathbb{R}$, to formulate and prove a proposition about the continuity of $|f(x)|$.

CHAPTER 3

Continuous Functions

By pursuing the theme of building new classes of functions, we are led to some general considerations about continuity in this chapter. One outcome of this approach is a sketch of one way (there are several) of providing rigorous definitions of the so-called elementary functions and establishing their familiar properties. Perhaps the more important outcome is the appearance of certain fundamental concepts and theorems of Analysis and some clarification of exactly what variety of functions possess the property of continuity.

§1. Implicit Functions. $\sqrt[n]{x}$. The Intermediate-Value Theorem

A rational function is an *explicit* function in that the value $f(x)$ results from an explicit list of computations involving x and given constants, e.g., as in

$$f(x) = \frac{8x^3 - 7x + 2}{x^4 - 3x^2 + 9};$$

cube x, multiply by 8, etc.

By contrast, the function $\sqrt{x}$ results from solving the equation $y^2 = x$, being given x. This is an example of a function that is *defined implicitly*, the domain consisting of values of x for which a solution exists.

The study of functions that are defined implicitly is one of the basic issues of Analysis and its applications. It is frequently the case in both theoretical and applied contexts that an important quantity y is known to

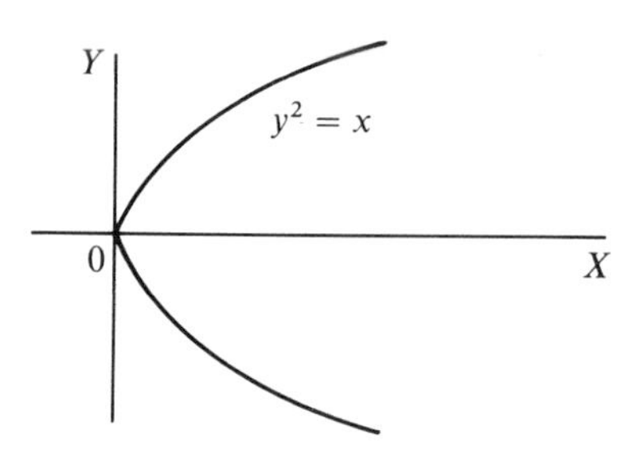

satisfy an equation involving x, $F(x, y) = 0$, for some function of two variables F, while an explicit formula for y in terms of x is not readily available. Thus in an example like

$$x^3y^5 + xy^4 - (3x^2 - 1)y^3 + (2x^3 - 4)y - 7x = 0,$$

an explicit presentation of y as a function of x would have to include steps for solving a fifth-degree equation.

The general implicit function problem is to infer information about the implicitly defined function, its existence and properties, from properties of F.

The implicit function problem includes as a special case the general question of solving equations: given a function g, find those x for which the equation $g(y) = x$ has a solution y, and describe the dependence of y on x. This is just the case $F(x, y) = g(y) - x$.

Consider the simplest nontrivial case, $g(y) = y^2$, which leads to the function $\sqrt{x}$. The pairs of real numbers (x, y) satisfying $y^2 = x$ constitute the parabola shown, which makes it clear that for each $x > 0$ there are two values y that solve $y^2 = x$; one positive value, which we denote $\sqrt{x}$, and its negative $-\sqrt{x}$.

If we make different choices between $\sqrt{x}$ and $-\sqrt{x}$ for different values of x in some set, say, dom $f \subset [0, \infty)$, we obtain a discontinuous function f that is defined implicitly by $y^2 = x$. The graph of one such function is shown; there are obviously infinitely many of them.

The two functions

$$f(x) = \sqrt{x} \qquad \text{for all} \quad x \geq 0$$

and

$$g(x) = -\sqrt{x} \qquad \text{for all} \quad x \geq 0$$

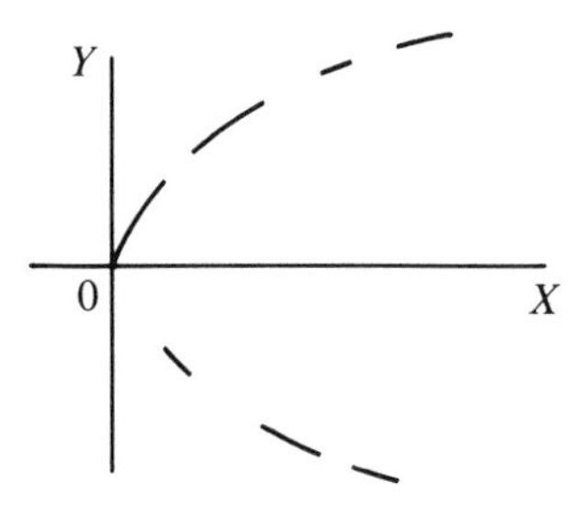

are distinguished as the only functions defined implicitly by $y^2 = x$ that are defined *and continuous* for all $x \geq 0$.

This simple example shows that one cannot expect to infer from nice properties of F that the functions defined implicitly by $F(x, y) = 0$ have nice properties; it is a question, rather, of the *existence* of nice implicitly defined functions; say, continuous with "maximal" domain.

Remark. The problem is of most interest when "nice" means continuously differentiable, the case that leads to the very important Implicit Function Theorem. See Chapter 4, §5.

EXERCISE 1. The function

$$f(x) = \begin{cases} \sqrt{x}, x \text{ rational}, & x \geq 0, \\ -\sqrt{x}, x \text{ irrational}, & x \geq 0, \end{cases}$$

is defined implicitly by $y^2 = x$. Where is f continuous?

We consider the task of defining $\sqrt[n]{x}$, for $n \in \mathbb{N}$, via the equation $y^n = x$. (Since we are concerned here with real-valued functions of real variables, we shall ignore our earlier discussion of the Fundamental Theorem of Algebra.)

First, since $y^n \geq 0$ for n even, the maximal domain one can expect for $\sqrt[n]{x}$ is: $\mathbb{R}$ when n is odd, and $[0, \infty)$ when n is even.

For the existence of solutions of $y^n = x$, we observe that it is obvious for the perfect nth power values of x

$$\ldots, (-3)^n, (-2)^n, (-1)^n, 0, 1^n, 2^n, 3^n, \ldots$$

and that every x in the prospective domain is intermediate to two of these values;

$$(p-1)^n \leq x \leq p^n, \qquad \text{for some} \quad p \in \mathbb{Z}.$$

This puts forward the question whether a continuous function (e.g., y^n) assumes every value intermediate to two of its values.

We have the important

Intermediate-Value Theorem. *If f is continuous on $[a, b]$ and γ is any value between $f(a)$ and $f(b)$ then there exists at least one $c \in [a, b]$ such that $f(c) = \gamma$.*

Proof. Replacing f by $-f$ if necessary, we may suppose $f(a) \leq f(b)$. Thus $f(a) \leq \gamma \leq f(b)$ and only the case of strict inequalities requires further discussion.

Taking $f(a) < \gamma < f(b)$, the set

$$S = \{x \in [a, b] : f(x) < \gamma\}$$

is nonempty (contains a) and bounded above (by b), so completeness provides a reasonable prospect for c, namely $c = \sup S$. We show $f(c) = \gamma$ by eliminating the alternatives, $f(c) < \gamma$ and $f(c) > \gamma$, using the continuity at c.

If $f(c) < \gamma$ then corresponding to $\varepsilon = \gamma - f(c) > 0$ there is a δ such that $x \in [a, b]$ and $|x - c| < \delta$ implies $|f(x) - f(c)| < \varepsilon$. In particular,

$$x \in [a, b] \quad \text{and} \quad c \le x < c + \delta \quad \text{implies} \quad f(x) < f(c) + \varepsilon = \gamma,$$

hence

$$[a, b] \cap [c, c + \delta) \subset S.$$

This contradicts the definition of c as $\sup S$ if there are in fact any elements in $[a, b] \cap [c, c + \delta)$, that is, if $c < b$. That $c < b$ is true follows from $f(b) > \gamma$ and the continuity of f at b by a similar argument: For $\varepsilon = f(b) - \gamma > 0$ there is a δ giving

$$x \in (b - \delta, b] \quad \text{implies} \quad f(b) - \varepsilon < f(x) < f(b) + \varepsilon.$$

In particular, $\gamma = f(b) - \varepsilon < f(x)$, i.e., $(b - \delta, b]$ contains no points of S; $c \le b - \delta$.

The argument that eliminates $f(c) > \gamma$ is completely similar.

EXERCISE 2. Supply the details.

EXERCISE 3. Give examples (draw graphs) showing that the continuity at the endpoints a and b must be assumed, to draw the conclusion.

EXERCISE 4. Give another proof of the Intermediate-Value Theorem by using the Nested Intervals Theorem as follows: If $f(a) < \gamma < f(b)$ and the midpoint $(a + b)/2$ does not give f the value γ, put

$$a_1 = \frac{a+b}{2} \quad \text{and} \quad b_1 = b \qquad \text{if} \quad f\left(\frac{a+b}{2}\right) < \gamma,$$

$$a_1 = a \qquad \text{and} \quad b_1 = \frac{a+b}{2} \quad \text{if} \quad f\left(\frac{a+b}{2}\right) > \gamma.$$

By induction, define a nest $\{[a_n, b_n]\}$ and use the continuity at $c = \bigcap_{n=1}^{\infty} [a_n, b_n]$ to show $f(c) = \gamma$, assuming no midpoint gives f the value γ.

We shall see some of the uses of this theorem in what follows. Some of its interesting consequences make good exercises.

EXERCISE 5. If f is continuous on $[a, b]$, then the range of f is a closed interval.

EXERCISE 6. If f is continuous and carries $[a, b]$ into $[a, b]$ then f must have a **fixed point**; $f(c) = c$ for some $c \in [a, b]$. Hint: Consider $f(x) - x$.

EXERCISE 7. Every real polynomial of odd degree has a real root.

The existence of solutions of $y^n - x = 0$, for all x when n is odd, and for

$x \geq 0$ when n is even, follows from the theorem. To define $\sqrt[n]{x}$, one seeks a choice of solution that yields continuity on the maximal domain.

EXERCISE 8. Show that the solution of $y^n - x = 0$ is unique if n is odd, and, if n is even and $x > 0$, there is a unique positive solution.

EXERCISE 9. Choosing the unique solution to define $\sqrt[n]{x}$, the continuity at $x_0 \in \operatorname{dom} \sqrt[n]{x}$ can be shown by using the identity

$$a^n - b^n = (a - b)(a^{n-1} + a^{n-2}b + a^{n-3}b^2 + \cdots + ab^{n-2} + b^{n-1})$$

with $a = \sqrt[n]{x}$ and $b = \sqrt[n]{x_0}$ to prove an inequality. Carry this out in the case $n = 3$.

§2. Inverse Functions. x^r for $r \in \mathbb{Q}$

We shall pursue the definition and continuity of $\sqrt[n]{x}$ in a different context: The function $f(x) = \sqrt[n]{x}$ is to be inverse to the function $g(x) = x^n$ in the sense that $f(g(x)) = x$ and $g(f(x)) = x$ for any x making these compositions meaningful.

We consider the question of continuous inverses of continuous functions in general, beginning with the

Definition. If f, with domain $\operatorname{dom} f$ and range $\operatorname{ran} f$, is such that there is a function g: $\operatorname{ran} f \to \operatorname{dom} f$ for which

$$g \circ f \text{ is the identity on } \operatorname{dom} f \quad (\text{i.e., } g(f(x)) = x,\ x \in \operatorname{dom} f),$$

and

$$f \circ g \text{ is the identity on } \operatorname{ran} f \ (\text{i.e., } f(g(x)) = x,\ x \in \operatorname{ran} f),$$

we say f is **invertible** and g is the **inverse of f**.

The usage, *the* inverse, is justified since there is obviously only one possibility for a function g of the required kind; one could say that the issue is whether looking at f: $\operatorname{dom} f \to \operatorname{ran} f$ "backwards," $\operatorname{ran} f \to \operatorname{dom} f$, one sees a function, a single-valued correspondence.

This observation makes clear the

Proposition. *f is invertible if and only if f is a one–one correspondence between* dom f *and* ran f.

The notation used for the inverse function of f is f^{-1}, despite the need to beware that it does *not* mean the -1 power, which is written as a reciprocal to avoid any confusion.

EXERCISE 1. Show that if f is invertible, so is f^{-1} and $(f^{-1})^{-1} = f$. Write an equation $F(x, y) = 0$ that defines f^{-1} implicitly.

EXERCISE 2. Every restriction of an invertible function is invertible. What is the domain of $(f|_X)^{-1}$?

The invertibility of $f: \mathbb{R} \to \mathbb{R}$ is obvious if f is strictly monotonic as in the

Definition. f is **strictly increasing** if $x < y$ in dom f implies $f(x) < f(y)$.

f is **strictly decreasing** if $x < y$ in dom f implies $f(x) > f(y)$.
f is **strictly monotonic** if it has one of these properties.

(Replacing $<$ everywhere by $\leq$ here gives the definitions of the (wide sense) terms: **increasing, decreasing,** and **monotonic** function.)

One observes at once that there are invertible functions that are not strictly monotonic, e.g., piecewise strictly monotonic functions, as in the figure. Observing the need to introduce discontinuities to create such examples, we conjecture the

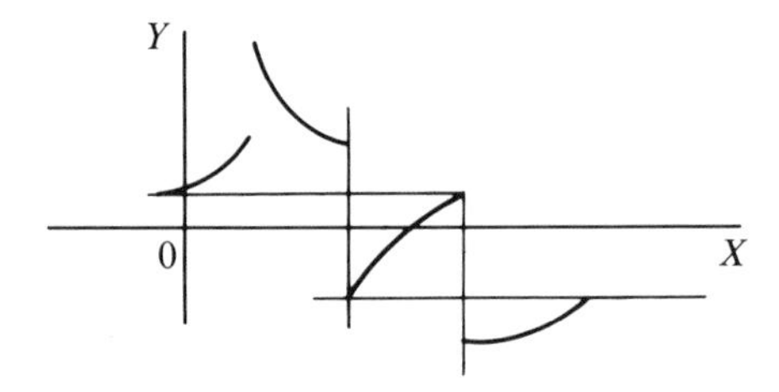

Theorem. *If f is continuous and invertible on a closed interval $[a, b]$ then f is strictly monotonic and f^{-1} is continuous.*

Proof. Since f is one–one, either $f(a) < f(b)$ or $f(a) > f(b)$. We show f is strictly increasing in the case $f(a) < f(b)$, the other case being similar.

Supposing $a < x < x' < b$, we must show

$$f(a) < f(x) < f(x') < f(b).$$

First, $f(a) < f(x)$, since otherwise $f(x) \leq f(a) < f(b)$, which makes $f(a)$ an intermediate value of f on $[x, b]$, and the Intermediate-Value Theorem yields a point in $[x, b]$ (that is, different from a) where the value $f(a)$ occurs, contrary to the one–one property of f.

The same argument shows that $f(x) < f(b)$. Similarly, $f(a) < f(x') < f(b)$ is established, leaving $f(x) < f(x')$ to be shown.

Now suppose $f(x') \leq f(x)$. Then $f(x')$ is an intermediate value of f on

$[a, x]$. Since $[a, x]$ does not contain x', a contradiction to the one–one property follows once more.

To show f^{-1} is continuous, let $y_0 \in \operatorname{ran} f$ and consider any $\{y_n\} \subset \operatorname{ran} f$ converging to y_0. With the notation

$$x_n = f^{-1}(y_n), \qquad n = 0, 1, 2, \ldots,$$

we must show $x_n \to x_0$. Since $\underline{\lim}\, x_n$ and $\overline{\lim}\, x_n$ are limits of subsequences of $\{x_n\}$, we obtain the result, namely that $\underline{\lim}\, x_n = \overline{\lim}\, x_n = x_0$, by showing that every convergent subsequence of $\{x_n\}$ must have limit x_0. If $\{x_{n_k}\}$ is a convergent subsequence, the continuity of f at $\lim x_{n_k}$ gives

$$f(\lim x_{n_k}) = \lim f(x_{n_k}) = \lim y_{n_k} = y_0 = f(x_0).$$

But f is one–one. Hence $f(\lim x_{n_k}) = f(x_0)$ yields $\lim x_{n_k} = x_0$, which we wished to show.

Remark. The question of the existence of an inverse for a given function f is just the question of the existence of a unique solution of the equation $f(x) = y$ for each $y \in \operatorname{ran} f$. This, in turn, is the implicit function problem with $F(x, y) = f(x) - y$, i.e., using the equation $f(x) - y = 0$ to define x as a function of y implicitly on ran f.

Applying the theorem to the problem in its "solve the equation $f(x) = y$" formulation gives:

If f is strictly monotonic and continuous on $[a, b]$, then the equation $f(x) = y$ has a unique solution, $y = f^{-1}(x)$, which is defined on $f([a, b])$ and depends continuously on y, the "data" of the equation.

The theorem applies when f is any one of the functions $f(y) = y^n$, $n \in \mathbb{N}$, on any interval $[0, A]$, and, in case n is odd, on any interval $[-A, A]$, with $A > 0$. There follows the existence of nth root functions defined and continuous on the images of these intervals. But those images fill out the sets $[0, \infty)$, for n even, and $(-\infty, \infty)$ for n odd, and the inverses coincide in overlapping images. In this way, we obtain well-defined continuous functions $\sqrt[n]{x}$, for $n \in \mathbb{N}$, on the maximal domains.

By forming composites,

$$x^{m/n} = (\sqrt[n]{x})^m, \qquad m, n \in \mathbb{N},$$

we complete the definitions of the continuous functions x^r for all rational powers r.

Exercise 3. Describe the domain of x^r, $r \in \mathbb{Q}$, using the lowest-terms expression for r.

Exercise 4. Show that $(\sqrt[n]{x})^m = \sqrt[n]{x^m}$ for any $m, n \in \mathbb{N}$ and any x for which both sides are defined.

§3. Continuous Extension. Uniform Continuity. The Exponential and Logarithm

The functions $f(x) = x^y$, $x > 0$, have been defined for rational values of y. The introduction of such functions for irrational values of y calls for the definition of an arbitrary real power of a fixed positive number ($3^{\sqrt{2}}$ and 2^π, for example). Thus the issue is in fact that of defining *exponential functions* a^x, $a > 0$, $x \in \mathbb{R}$.

One requirement on the definitions of such functions will be the preservation of the laws of exponents, as properties of the resulting functions. In particular, if we define a^x on $\mathbb{R}$ for $a > 1$, the law

$$\left(\frac{1}{a}\right)^x = a^{-x}$$

then provides a meaning for a^x when $0 < a < 1$. Of course, $1^x = 1$ for all x is the only choice for $a = 1$. Suppose, therefore, that $a > 1$.

Now a^r, $r \in \mathbb{Q}$, is already defined; the issue is that of *extending* this function to the closure of its domain. The obvious way to attempt the extension is by continuity, i.e., by

$$a^x = \lim a^{r_n}, \qquad \text{where} \quad \{r_n\} \subset \mathbb{Q} \text{ and } r_n \to x.$$

This raises the general question of the possibility of extending a given function to the closure of its domain so as to obtain a continuous extension. Let us consider this question as a basis for further discussion of exponential functions.

If f is to have a continuous extension then f itself must be continuous. Is the continuity of f sufficient to insure the existence of the extension? Some examples show that the answer is no, and suggest how to modify the assumption.

EXERCISE 1. $f(x) = 1/x$, $x \neq 0$, is continuous, but cannot be assigned a value at $x = 0$ that yields a function continuous on $\mathbb{R}$.

EXERCISE 2. $f(x) = \tan x$ on $[0, \pi/2)$ cannot be extended continuously to $[0, \pi/2]$.

Both of these examples result from the fact that $f(x_n) \to \infty$ for sequences $\{x_n\}$ that converge to the missing limit point of dom f. An example of a different kind is

EXERCISE 3. $f(x) = \sin 1/x$, $x \in (0, 1/\pi]$. Whatever value is considered for $f(0)$, some sequence with limit 0 gives values that don't approach that $f(0)$.

The common feature of all three examples can be described by saying that the graph becomes "arbitrarily steep" near the limit point of dom f that is missing from dom f. Let us examine this phenomenon in terms of the $\varepsilon - \delta$ expression of the continuity of the functions.

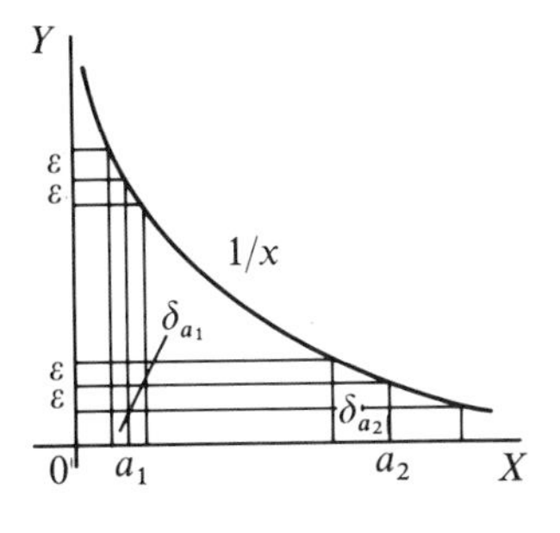

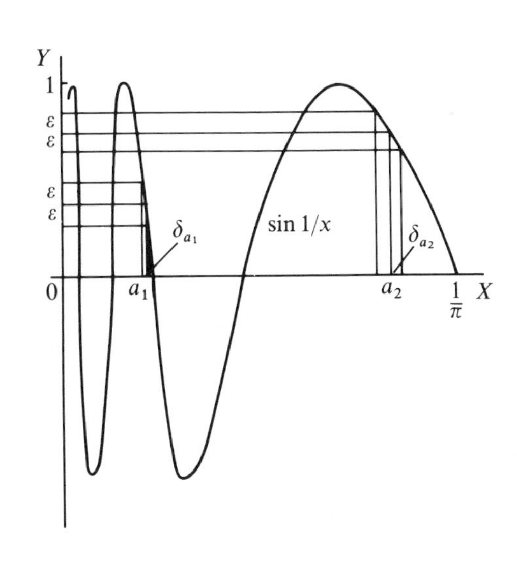

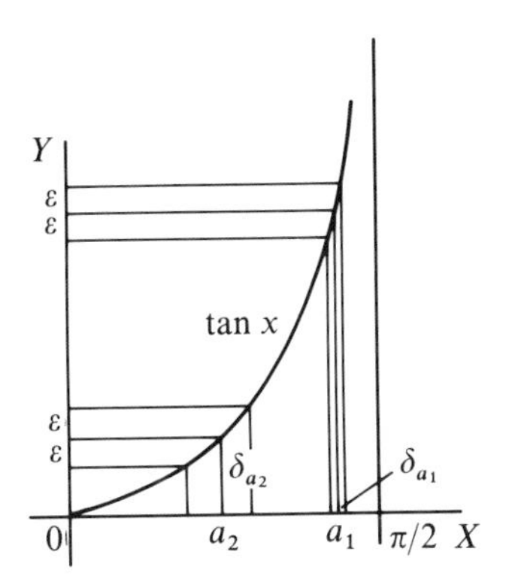

The figures show how, once $\varepsilon > 0$ is fixed, the size of the δ_a that insures

$$x \in (a - \delta_a, a + \delta_a) \quad \text{implies} \quad |f(x) - f(a)| < \varepsilon$$

depends critically on the location of a; the closer a is to the missing limit point, the smaller δ_a must be, and *no single δ works at every $a \in$* dom f.

The difficulty is removed by supposing f to be uniformly continuous as in the

Definition. If $S \subset$ dom f, we say **f is continuous uniformly on S**, if for each $\varepsilon > 0$ there is a δ such that

$$x, x' \in S \quad \text{and} \quad |x - x'| < \delta \quad \text{implies} \quad |f(x) - f(x')| < \varepsilon.$$

When f is continuous uniformly on dom f we say simply that f **is uniformly continuous**.

Remark. In rough terms, this says that $|f(x) - f(x')| < \varepsilon$ holds for any pair of nearby points in S, the meaning of nearby, i.e., δ, being *independent of the location* of the pair in S. The property includes the point-by-point continuity in S, but requires it to hold in a special way.

The functions in the examples are not uniformly continuous, even though they are continuous—and, indeed, even though that continuity is uniform on various proper subsets of their domains; it is not uniform on the entire domain.

EXERCISE 4. Prove that $1/x$ is continuous uniformly on $[a, \infty)$ for any $a > 0$.

We conjecture that the uniform continuity of f suffices for the existence of the continuous extension $\bar{f}$ of f to domain $\overline{\text{dom } f}$, the closure of dom f.

The value to be assigned to $\bar{f}$ at a limit point $\bar{x}$ of dom f is dictated by the desired continuity of $\bar{f}$, namely $\bar{f}(\bar{x}) = \lim f(x_n)$, where $\{x_n\}$ is any sequence in dom f that converges to $\bar{x}$. This obliges us to show that $\lim f(x_n)$ exists, i.e., that $\{f(x_n)\}$ is a Cauchy sequence, and that the limit is independent of the choice of the sequence $\{x_n\}$.

The key observation is the

Proposition. *If f is continuous uniformly on S and $\{x_n\}$ is a Cauchy sequence in S, then $\{f(x_n)\}$ is a Cauchy sequence.*

EXERCISE 5. Proof?

Theorem. *If f is continuous uniformly on S there is a unique continuous extension of f to an $\bar{f}$ on $\bar{S}$.*

Proof. Let $\bar{x}$ be any point or limit point of S. Let $\{x_n\}$ be a sequence in S converging to $\bar{x}$. The proposition and completeness insure the existence of $\lim f(x_n)$. This value is independent of the choice of the sequence $\{x_n\}$ converging to $\bar{x}$ since if $\{x_n^*\}$ is another such sequence, the sequence $x_1, x_1^*, x_2, x_2^*, \ldots,$ converges, and so, therefore, does $f(x_1), f(x_1^*), f(x_2), f(x_2^*), \ldots,$ which forces its two subsequences $\{f(x_n)\}$ and $\{f(x_n^*)\}$ to have the same limit.

It follows that $\bar{f}$ on $\bar{S}$ is well defined by putting

$$\bar{f}(\bar{x}) = \lim f(x_n) \qquad \text{for any} \quad \{x_n\} \subset S \text{ with } x_n \to \bar{x},$$

for every $\bar{x} \in \bar{S}$. The continuity of f insures that $\bar{f}$ is an extension of f. The uniqueness of the extension is clear, by continuity also.

To show $\bar{f}$ is continuous, fix $\varepsilon > 0$ and then choose δ so that

$$x, y \in S \quad \text{and} \quad |x - y| < \delta \quad \text{implies} \quad |f(x) - f(y)| < \frac{\varepsilon}{3}.$$

We shall show that

$$\bar{x}, \bar{y} \in \bar{S} \quad \text{and} \quad |\bar{x} - \bar{y}| < \frac{\delta}{2} \quad \text{implies} \quad |\bar{f}(\bar{x}) - \bar{f}(\bar{y})| < \varepsilon$$

by introducing sequences $\{x_n\} \subset S$ and $\{y_n\} \subset S$ with $x_n \to \bar{x}$ and $y_n \to \bar{y}$. For any sufficiently large index n, the following hold:

$$|x_n - y_n| < \delta \quad \left(\text{since } |\bar{x} - \bar{y}| < \frac{\delta}{2} \text{ and } x_n \to x, \bar{y}_n \to \bar{y}\right),$$

$$|\bar{f}(\bar{x}) - f(x_n)| < \frac{\varepsilon}{3} \quad \text{and} \quad |\bar{f}(\bar{y}) - f(y_n)| < \frac{\varepsilon}{3}.$$

Now, the first of these insures that $|f(x_n) - f(y_n)| < \varepsilon/3$, hence for any such index n, we have the estimate we seek:

$$|\bar{f}(\bar{x}) - \bar{f}(\bar{y})| \leq |\bar{f}(\bar{x}) - f(x_n)| + |f(x_n) - f(y_n)| + |\bar{f}(\bar{y}) - f(y_n)| < \varepsilon.$$

(This shows in fact that $\bar{f}$ is continuous uniformly on $\bar{S}$.)

Returning to the consideration of the function a^r, $r \in \mathbb{Q}$, we show next that it is continuous uniformly on $\mathbb{Q} \cap [-B, B]$ for any $B > 0$. The theorem then gives the continuous extension a^x on $[-B, B]$. These extensions certainly agree on the common parts of their domains, so that a well-defined a^x on $\mathbb{R}$ results.

Denote a pair of points of $\mathbb{Q} \cap [-B, B]$ by x and $x + h$. Then the examination of the uniformity we claim begins with the estimate

$$|a^{x+h} - a^x| = a^x|a^h - 1| \leq a^B|a^h - 1| < \varepsilon \qquad \text{if} \quad |a^h - 1| < \frac{\varepsilon}{a^B}.$$

EXERCISE 6. We used $a^x \leq a^B$ if $x \leq B$, i.e., that a^r is increasing on $\mathbb{Q}$. Prove it.

That $|a^h - 1| < \varepsilon/a^B$ for h sufficiently small is easily shown on the basis of the fact that $a^{1/n} \to 1$, and this completes the proof that a^x is continuous uniformly on $\mathbb{Q} \cap [-B, B]$.

EXERCISE 7. Prove $a^{1/n} \to 1$ for $a \to 0$. Hint: $0 < a^{1/n} - 1 < \varepsilon$ means $(1 + \varepsilon)^n > a$. Show $(1 + \varepsilon)^n \to \infty$.

EXERCISE 8. Show that a^x, $a > 1$, is a strictly increasing function on $\mathbb{R}$.

EXERCISE 9. Show that, for $a > 1$,

$$B_n \to \infty \qquad \text{implies} \quad a^{B_n} \to \infty,$$

and

$$B_n \to -\infty \quad \text{implies} \quad a^{B_n} \to 0.$$

As a continuous, strictly increasing function, a^x on $[-B, B]$, $a > 1$, has a continuous inverse, which is strictly increasing on $[a^{-B}, a^B]$, for any $B > 0$. These inverse functions obviously agree on the common part of their domains. The existence and continuity on $(0, \infty)$ of the inverse function $\log_a x$ of a^x results from these observations, in the case $a > 1$.

The laws of exponents, such as

$$a^x \cdot a^y = a^{x+y},$$

which govern computations with the exponential functions, hold for rational exponents by the very definition of rational powers. Whether they continue to hold for arbitrary real exponents is seen to be the question of the persistence of an identity under continuous extension. By viewing the equality as a pair of inequalities ($\leq$ and $\geq$), the question is settled by the following general lemma.

Definition. A set A is **dense in a set** B if $\bar{A} \supset B$.

Lemma. *If f and g are continuous on B and if*

$$f(x) \leq g(x) \qquad \textit{for all } x \textit{ in a dense subset of } B,$$

then

$$f(x) \leq g(x) \qquad \textit{for all} \quad x \in B.$$

EXERCISE 10. Prove the lemma.

EXERCISE 11. Prove that strict inequality can be lost in the continuous extension.

EXERCISE 12. Prove that, for $a > 0$,

$$a^x \cdot a^y = a^{x+y}, \qquad \text{all} \quad x, y \in \mathbb{R}.$$

EXERCISE 13. Prove that a continuous function from $\mathbb{R}$ which satisfies ⊛ $f(x) \cdot f(y) = f(x + y)$ for all $x, y \in \mathbb{R}$ and $f(1) = a > 0$ can only be the function a^x. Thus a^x can be viewed as the unique solution of the *functional equation* ⊛ that satisfies $f(1) = a$.

The laws of logarithms are simply translations of the laws of exponents, using the definition of $\log_a x$, as the inverse of a^x.

EXERCISE 14. Prove that $\log_a x \cdot y = \log_a x + \log_a y$, for any positive x and y.

Remark. In most situations, the choice of base $a > 1$ is at one's disposal and can be made on the grounds of convenience. This indicates the choice e whenever calculus is to be used. The simple notation $\log x$ is usually used for the function $\log_e x$, although $\ln x$ is also used.

§4. The Elementary Functions

The **elementary functions** consist of 1, x, $\sin x$ and a^x, together with all of the functions that can be generated via arithmetic combinations and compositions of these, or that can be defined implicitly by equations $F(x, y) = 0$ in which F is so generated.

Of course, this includes all rational functions and the functions $\sqrt[n]{x}$ for $n \in \mathbb{N}$, defined implicitly by the equations $y^n - x = 0$. Consider, more generally, the implicit function problem in which $F(x, y)$ is a polynomial in y whose coefficients are polynomials in x:

$$p_n(x)y^n + p_{n-1}(x)y^{n-1} + \cdots + p_1(x)y + p_0(x) = 0, \qquad n \geq 1, \quad p_n(x) \not\equiv 0.$$

If $y = f(x)$ solves this equation and f is continuous then f is called an **algebraic function**. Functions involving radicals, such as $1/(1 + \sqrt{x})$ and $\sqrt[3]{2 + \sqrt{x}}$ are examples. For example, if $y = 1/(1 + \sqrt{x})$, then $y + \sqrt{xy} = 1$, $\sqrt{xy} = 1 - y$, $xy^2 = (1 - y)^2$, and so $(1 - x)y^2 - 2y + 1 = 0$.

EXERCISE 1. Show that $\sqrt[3]{2 + \sqrt{x}}$ is algebraic.

A detailed analysis of general algebraic functions is a complex task, well beyond the scope of this book. (At the outset, one recognizes the need for the Fundamental Theorem of Algebra in such an analysis; the subject properly belongs to the study of complex-valued functions of complex variables.) Despite their complexity, real-valued algebraic functions are among the elementary functions.

EXERCISE 2. Is the composite of two algebraic functions algebraic?

In the interests of completeness, we provide a sketch of one way of defining the trigonometric functions.

The definition of arclength for an arc of a circle has been discussed in Exercise 11, p. 46. If the unit circle in the X–Y plane is considered, and the point $(1, 0)$ is taken as reference point, with counterclockwise as the positive direction, then there is a well-determined point $P(\theta)$ on the unit circle, assigned to every real number θ: the point at arclength θ from $(1, 0)$.

The $X-Y$ coordinates of $P(\theta)$ are, by definition, $\cos\theta$ and $\sin\theta$, respectively.

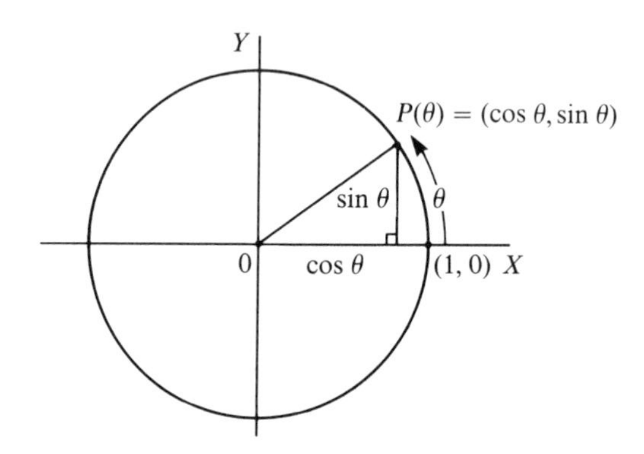

Remark. This sequence of instructions, leading from θ to, say, $\sin\theta$ gives an example of an explicit function of a different kind from the rational functions.

Some things that are clear from the geometry are

$$0 < \sin\theta < \theta, \qquad \text{if} \quad 0 < \theta < \frac{\pi}{2},$$

$$\cos\theta = \sqrt{1 - \sin^2\theta}, \qquad \text{if} \quad 0 < \theta < \frac{\pi}{2}$$

and

$$\sin(-\theta) = -\sin\theta.$$

The continuity at $\theta = 0$ of both $\sin\theta$ and $\cos\theta$ follows directly from these facts.

EXERCISE 3. Give the details.

The identity (proved as in trigonometry courses)

$$\sin(\theta_0 + h) = \sin\theta_0 \cos h + \cos\theta_0 \sin h$$

then shows, by letting $h \to 0$, that the function $\sin\theta$ is continuous at all θ_0.

EXERCISE 4. Why is $\cos\theta$ continuous?

The remaining trigonometric functions are defined as the usual ratios. They are clearly continuous.

It is evident from the geometry that $\sin\theta$ is strictly increasing on $[-\pi/2, \pi/2]$, hence that the inverse function arcsine is defined and continuous on $[-1, 1]$. Since it is defined implicitly by the equation $\sin y - x = 0$, $\arcsin x$ is an elementary function.

The same observation applies to other inverse trigonometric functions and to the logarithm functions; they are all elementary functions.

The great surge of discoveries in the mathematical sciences following the introduction of the formal manipulations of the calculus naturally focused attention on the idea of function-as-formula. The exponential function served the need to express growth phenomena which no algebraic expression could yield, and the trigonometric functions encompassed periodic phenomena; one could expect, perhaps, that all future needs would be met similarly. The idea of elementary function is probably a relic of such attitudes. Later critical analyses of the function concept showed the importance of the abstract interpretation, as a many-to-one correspondence, divorced from the issue of how it is, or can be, presented via formulas. Thus the elementary functions can be taken in modern terms as merely a description of the class of functions to which the manipulative techniques of calculus apply. It is an extraordinarily rich class of functions, so it is understandable that it may have been seen to be adequate for usual needs ... in spite of certain gaps (an antiderivative for e^{x^2}, for example).

We have made continuity central to the above discussion of the elementary functions since it is our focus of interest. Of course the elementary functions are all differentiable and this gives their continuity at once. A point of view which focuses on the differentiability would introduce the functions as solutions of problems in differential equations, e.g., e^x as the unique solution of

$$\begin{cases} y' = y, \\ y(0) = 1. \end{cases}$$

A general existence and uniqueness theorem for solutions of differential equations provides the basis for such a treatment. See Simmons [25] for a discussion of this point of view.

Another way to introduce the elementary functions is by power series, e.g., e^x is defined as the function given by the series, which converges for all x, $\sum_{n=0}^{\infty} x^n/n!$

Exercise 10 of the previous section hints at a "functional equation" point of view to the definition of the basic elementary functions.

It should be remarked that the natural extensions of the meanings of e^x, $\sin x$, and $\cos x$ to complex variables yield the identity

$$e^{i\theta} = \cos\theta + i\sin\theta, \qquad \theta \text{ real.}$$

Thus, in the complex setting the trigonometric functions are simple combinations of exponential functions, so that 1, x, and e^x are the basic elementary functions.

Finally, many calculus texts make the definition

$$\log x = \int_1^x \frac{1}{t}\, dt, \qquad x > 0,$$

the basis for the careful discussion of the logarithm (e^x is then defined as the inverse). This definition is elegant in the way that the properties of $\log x$ are easily derived, but it is awkward in that it draws upon so much theoretical background (integration, Fundamental Theorem of Calculus) to treat a relatively primitive question—essentially that of defining x^y.

§5. Uniformity. The Heine–Borel Theorem

The theorem on continuous extension exposes the need for a general result identifying sets S on which the continuity of a function is uniform. We shall see other uses for such a result in later chapters. The path to the appropriate theorem is best revealed in terms of a reformulation of the idea of continuity, which is of interest for its own sake as well.

Suppose $f\colon \mathbb{R} \to \mathbb{R}$ has domain $\operatorname{dom} f$ and range, or value set, $\operatorname{ran} f$ and let T be any set. Whether T is contained in $\operatorname{ran} f$ or not, we can consider the subset of $\operatorname{dom} f$ consisting of the points whose images belong to T, or "map into" T, denoted $f^{-1}(T)$ and called the **inverse image** of T under f;

$$f^{-1}(T) = \{x \in \operatorname{dom} f\colon f(x) \in T\}.$$

(The use of the overworked -1 notation is standard and should not be taken to imply the use, or existence, of an inverse function.)

EXERCISE 1. What statement in terms of inverse images expresses the existence of the inverse function?

EXERCISE 2. Show that the operation of taking inverse image is interchangeable with the set theoretic operations:

1. $$f^{-1}(T \cup T') = f^{-1}(T) \cup f^{-1}(T').$$
2. $$f^{-1}(T \cap T') = f^{-1}(T) \cap f^{-1}(T').$$
3. $$f^{-1}(\mathbb{R} - T) = \operatorname{dom} f - f^{-1}(T).$$

Taking the case $\operatorname{dom} f = \mathbb{R}$ at first, we translate the continuity property into different terminology as follows: The statement

$$|x - x_0| < \delta \quad \text{implies} \quad |f(x) - f(x_0)| < \varepsilon$$

says that $(x_0 - \delta, x_0 + \delta)$ maps under f into $(f(x_0) - \varepsilon, f(x_0) + \varepsilon)$, which we denote $T_\varepsilon(x_0)$, to rewrite the statement as

$$f^{-1}(T_\varepsilon(x_0)) \supset (x_0 - \delta, x_0 + \delta).$$

Thus, the continuity of f at x_0 can be restated as

$$\text{for every } \varepsilon > 0,\ f^{-1}(T_\varepsilon(x_0)) \text{ is a neighborhood of } x_0.$$

Now, if $\mathcal{U}$ is any neighborhood of $f(x_0)$ then $\mathcal{U}$ contains some $T_\varepsilon(x_0)$ and so

$$f^{-1}(\mathcal{U}) \supset f^{-1}(T_\varepsilon(x_0)).$$

The continuity of f at x_0 can therefore be further restated;

$$\text{for every neighborhood } \mathcal{U} \text{ of } f(x_0);\ f^{-1}(\mathcal{U}) \text{ is a neighborhood of } x_0.$$

In order to reformulate the continuity of f (i.e., on its domain $\mathbb{R}$) we make the

Definition. *A set that is a neighborhood of each of its points is called an* **open set**.

The reformulation of continuity is then the statement

$$f^{-1}(G) \text{ is open for every open set } G.$$

EXERCISE 3. Write the details of how this statement is equivalent to the preceding formulation.

We must modify the restatement to cover a function f with *any* domain. First, a few facts about open sets are noted.

EXAMPLES. Any open interval is an open set. $[0, 1)$ contains no $(-\delta, \delta)$, $\delta > 0$, hence is not open. The empty set ϕ is open, by logical convention. $\mathbb{R}$ is open.

Proposition. 1. *The union of any collection of open sets is open.*
2. *An intersection of finitely many open sets is open.*
3. *A set is open if and only if its complement is closed.*

***Proof of* 2.** If $G_1, G_2, \ldots, G_n$ are open sets and $x \in G_1 \cap G_2 \cap \cdots \cap G_n$, we must show there is a $\delta > 0$ such that $(x - \delta, x + \delta) \subset G_1 \cap G_2 \cap \cdots \cap G_n$.

But for each $i = 1, 2, \ldots, n$ there is a $\delta_i > 0$ such that $(x - \delta_i, x + \delta_i) \subset G_i$. Clearly, $\delta = \min\{\delta_1, \ldots, \delta_n\}$ works.

Since infinitely many positive numbers can fail to have a positive infimum, the proof does not cover a general infinite intersection of open sets.

EXERCISE 4. Give examples of infinite families of open sets whose intersection is, respectively is not, open.

EXERCISE 5. Prove 1. (Select your notation to avoid impugning countability to the collection.)

EXERCISE 6. Prove 3.

Remark. Sets in general are neither open nor closed, ϕ and $\mathbb{R}$ are both; the terms suggest a dichotomy that is false and must be guarded against.

Definition. If D is any set, a set S is **open relative to D** if S is the intersection of D and an open set.

EXAMPLE. $[0, \frac{1}{2})$ is not open, but it is open relative to $[0, 1]$, as the intersection of $[0, 1]$ and, say, $(-\frac{1}{2}, \frac{1}{2})$.

Clearly, if D itself is open then the sets open relative to D are open. Generally, as the example shows, "relatively open" sets need not be open.

Obviously, S is open relative to D if and only if, for each $x \in S$, there is a $\delta_x > 0$ such that

$$(x - \delta_x, x + \delta_x) \cap D \subset S.$$

The characterization of continuity in general takes the form of the

Proposition. *f is continuous if and only if $f^{-1}(G)$ is open relative to* dom f *for every open set G.*

Proof. Let f be continuous and consider any open set G. If no $x_0 \in \operatorname{dom} f$ maps into G then $f^{-1}(G)$ is empty, hence open relative to $\operatorname{dom} f$. There remains those G for which $f^{-1}(G) \neq \phi$. If $x_0 \in f^{-1}(G)$ then $f(x_0) \in G$, so there is an $\varepsilon > 0$ giving $(f(x_0) - \varepsilon, f(x_0) + \varepsilon) \subset G$. The continuity at x_0 gives $\delta > 0$ such that

$$x \in \operatorname{dom} f \quad \text{and} \quad |x - x_0| < \delta \quad \text{implies} \quad |f(x) - f(x_0)| < \varepsilon,$$

hence $\operatorname{dom} f \cap (x_0 - \delta, x_0 + \delta) \subset f^{-1}(G)$. This shows that $f^{-1}(G)$ is open relative to $\operatorname{dom} f$.

Conversely, suppose $f^{-1}(G)$ is open relative to $\operatorname{dom} f$ for every open set G. Fix $x_0 \in \operatorname{dom} f$ and $\varepsilon > 0$. Since $G = (f(x_0) - \varepsilon, f(x_0) + \varepsilon)$ is open, $f^{-1}(G)$ is open relative to $\operatorname{dom} f$. Hence there is a $\delta > 0$ such that

$$\operatorname{dom} f \cap (x_0 - \delta, x_0 + \delta) \subset f^{-1}(G),$$

that is,

$$x \in \operatorname{dom} f \quad \text{and} \quad |x - x_0| < \delta \quad \text{implies} \quad |f(x) - f(x_0)| < \varepsilon,$$

the continuity of f at x_0.

Remark. If $f: \mathbb{R} \to \mathbb{R}$ is continuous then the members of the following two collections of special sets are all open relative to dom f:

$$f^{-1}((a, \infty)) = \{x: f(x) > a\} \qquad \text{for} \quad a \in \mathbb{R},$$

and

$$f^{-1}((-\infty, b)) = \{x: f(x) < b\} \qquad \text{for} \quad b \in \mathbb{R}.$$

The converse holds as well since any open set in $\mathbb{R}$ is a union of open intervals (indeed of the form $(x - \delta, x + \delta)$ as x ranges over the set) and the general open interval (a, b) is the intersection of two sets from these collections.

EXERCISE 7. Give the detailed arguments.

EXERCISE 8. Define "S is closed relative to D" for sets S and D and characterize the continuity of f in terms of the closure relative to dom f of each member of a pair of families of sets.

EXERCISE 9. Suppose $G \subset \mathbb{R}$ is open and $x \in G$. Let

$$L_x = \{l \in \mathbb{R}: (l, x] \subset G\},$$

$$U_x = \{u \in \mathbb{R}: [x, u) \subset G\},$$

and put

$$a_x = \begin{cases} \inf L_x & \text{if } L_x \text{ is bounded below,} \\ -\infty & \text{otherwise,} \end{cases}$$

$$b_x = \begin{cases} \sup U_x & \text{if } U_x \text{ is bounded above,} \\ \infty & \text{otherwise,} \end{cases}$$

so that (a_x, b_x) is the maximal open subinterval of G containing x. Use the density of $\mathbb{Q}$ to show the

Proposition. *Every open set in $\mathbb{R}$ is the union of countably many open intervals.*

Now suppose f is continuous and $S \subset \text{dom } f$. We investigate the uniformity of the continuity on S.

For $\varepsilon > 0$, the continuity gives values δ_x at each x for which

$$(x - 2\delta_x, x + 2\delta_x) \subset f^{-1}((f(x) - \varepsilon, f(x) + \varepsilon)).$$

Thus any two points x' and x'' lying in the same one of these open intervals $(x - 2\delta_x, x + 2\delta_x)$ (and in dom f) give $|f(x') - f(x'')| < \varepsilon$; we establish uniformity by ensuring that *every* pair of sufficiently close points in S are in a *common* $(x - 2\delta_x, x + 2\delta_x)$ interval.

For technical reasons, we halve these intervals and consider the family

$$g_\varepsilon = \{(x - \delta_x, x + \delta_x): x \in S\}$$

of open sets, whose union contains S ... the "open covering" g_ε of S.

Suppose S is already covered by a finite subcollection $\{(x_i - \delta_{x_i}, x_i + \delta_{x_i}): i = 1, 2, \ldots, n\}$ (g_ε "has a finite subcover"). Then, putting $\delta = \min_{1 \le i \le n} \delta_{x_i}$, we have

1. $\delta > 0$ because of the finiteness.
2. If x' and $x'' \in S$ satisfy $|x' - x''| < \delta$ then we shall see that they lie in the same interval:

$$x', x'' \in (x_{i_0} - 2\delta_{x_{i_0}}, x_{i_0} + 2\delta_{x_{i_0}}) \qquad \text{for some } i_0,$$

which insures, as remarked above, that

$$|f(x') - f(x'')| < \varepsilon.$$

This holds because, if i_0 denotes the index for which $x' \in (x_{i_0} - \delta_{x_{i_0}}, x_{i_0} + \delta_{x_{i_0}})$, we have

$$|x'' - x_{i_0}| \le |x'' - x'| + |x' - x_{i_0}| < \delta + \delta_{x_{i_0}} \le 2\delta_{x_{i_0}},$$

hence x' and x'' both belong to $(x_{i_0} - 2\delta_{x_{i_0}}, x_{i_0} + 2\delta_{x_{i_0}})$ (and, incidentally, the reason for the halving is revealed).

We make the definitions

Definition. An **open covering** of a set S is any family, or collection, of open sets whose union contains S, i.e., each $x \in S$ is in at least one set in the family.

Definition. A set S has the **Heine–Borel property** if every open covering of S contains a finite subcovering, i.e., a finite subfamily that still covers S.

Applying this to the g_ε coverings of S for all $\varepsilon > 0$, we conclude that the continuity of f is uniform on any subset $S \subset \operatorname{dom} f$ that has the Heine–Borel property.

EXAMPLES. 1. $\{(1/n, 1 - (1/n)): n \in \mathbb{N}\}$ is an open covering of $(0, 1)$ that has no finite subcovering. Thus $(0, 1)$ does not have the Heine–Borel property.

2. $\{(n - 1, n + 1): n \in \mathbb{Z}\}$ is an open covering of $\mathbb{R}$ that has no finite subcovering; $\mathbb{R}$ does not have the Heine–Borel property.

EXERCISE 10. Does $[0, 1)$ have the Heine–Borel property? Give an opinion on what sets do.

It is clear that finite sets have the Heine–Borel property, but other positive assertions are not at all obvious—which makes it pleasant to have the

Heine–Borel Theorem. *S has the Heine–Borel property, if and only if S is compact.*

Proof. Consider a noncompact set S; some sequence $\{x_n\} \subset S$ has no subsequence converging to a point of S. For each $x \in S$ there is an open interval $(x - \delta_x, x + \delta_x)$, $\delta_x > 0$, that excludes all x_n from some index N_x on. These intervals form an open covering of S that has no finite subcovering, for if it did there would be a largest index N among the numbers N_x associated with the finite subcovering, so that $x_{N+1}, x_{N+2}, \ldots$ are all excluded from all sets of the subcovering, hence from S, which is absurd.

This shows that only a compact set can have the Heine–Borel property.

For the converse, suppose S is compact and some open covering $\mathscr{G} = \{G_x: x \in S\}$ has no finite subcovering. We are led to a contradiction as follows. Since S is bounded, there is a bounded closed interval I_1 containing S. Since it requires infinitely many sets in $\mathscr{G}$ to cover S, the same must be true of the part of S in at least one of the closed halves of I_1. Let I_2 denote such a closed half of I_1. Repeat the argument to define, inductively, a nest $\{I_n\}$ of closed intervals such that for every $n = 1, 2, 3, \ldots$, $S \cap I_n$ cannot be covered by finitely many members of $\mathscr{G}$.

The point $x_0 = \bigcap_{n=1}^{\infty} I_n$ (completeness!) is obviously a limit point of S, hence a member of S since S is closed. Choose a $G_{x_0} \in \mathscr{G}$ that contains x_0 and a $\delta > 0$ such that $(x_0 - \delta, x_0 + \delta) \subset G_{x_0}$ (G_{x_0} is open).

As soon as length $I_n < \delta/2$ we find $I_n \cap S$ covered by a single member of g, contrary to the definition of I_n.

With this result we can claim the

Theorem. *If f is continuous and $S \subset$ dom f is compact, the continuity is uniform on S.*

EXERCISE 11. Write out an elegant proof based on the earlier discussion.

EXERCISE 12. Give an example of a uniformly continuous function whose domain is not compact.

Proposition. *The continuous image of a compact set is compact.*

Proof. Let $S \subset$ dom f be compact and consider its image $f(S) = \{y: y = f(x)$ for some $x \in S\}$. Let $\mathscr{G}$ be an open covering of $f(S)$.

The inverse images of the members of $\mathscr{G}$ are open relative to dom f since f is continuous. For each $G \in \mathscr{G}$ choose an open set H such that $H \cap \text{dom } f = f^{-1}(G)$. The family $\mathscr{H}$ of all such H is an open covering of S. Choose a finite subcovering $\{H_1, H_2, \ldots, H_n\}$ of S. Now for $i = 1, 2, \ldots, n$

$$H_i \cap S \subset H_i \cap \text{dom } f = f^{-1}(G_i),$$

where G_i denotes the member of $\mathscr{G}$ that gives this last equality. It follows that

$$S = \bigcup_{i=1}^{n} H_i \cap S \subset \bigcup_{i=1}^{n} f^{-1}(G_i) = f^{-1}\left(\bigcup_{i=1}^{n} G_i\right)$$

and hence

$$f(S) \subset \bigcup_{i=1}^{n} G_i.$$

The compactness of $f(S)$ follows.

EXERCISE 13. Identify $f(S)$ if f is continuous and $S = [a, b] \subset \operatorname{dom} f$.

§6. Uniform Convergence. A Nowhere Differentiable Continuous Function

The taking of limits provides a means to construct new functions and to express functions in terms of simpler functions.

Definition. *If f_n, $n = 1, 2, 3, \ldots$, are defined on a set S and $\{f_n(x)\}$ converges for each $x \in S$, the function*

$$f(x) = \lim f_n(x), \qquad x \in S,$$

is the **pointwise limit** *on S of $\{f_n\}$. The functions f_n are said to* **converge pointwise** *to f on S.*

An obvious question is whether assuming each f_n continuous it follows that f is continuous. Here are two counterexamples.

EXAMPLE 1. Let $f_n(x) = x^n$ for $x \in [0, 1]$, $n = 1, 2, 3, \ldots$. For $x \in [0, 1]$, $\lim f_n(x)$ exists and is

$$f(x) = \begin{cases} 0, & 0 \le x < 1, \\ 1, & x = 1, \end{cases}$$

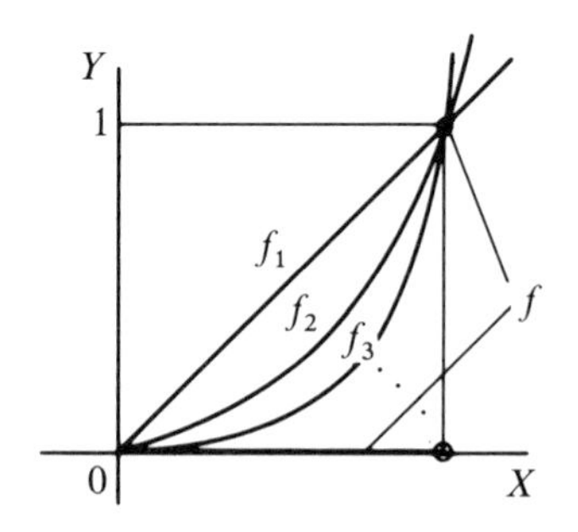

which is discontinuous.

EXAMPLE 2. Let

$$f_n(x) = \begin{cases} 0, & 0 \le x < 1 - 1/n, \\ n(x-1) + 1, & 1 - 1/n \le x \le 1 + 1/n \text{ (linear!)}, \\ 2, & 1 + 1/n < x \le 2. \end{cases}$$

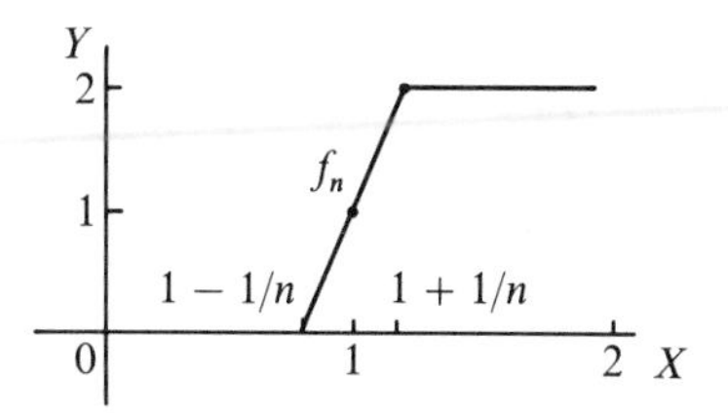

Clearly, if $x \in [0, 1)$ there is an index beyond which all these functions are 0 at x. Similarly, if $x \in (1, 2]$ then $f_n(x) = 2$ for all n sufficiently large. Hence the sequence of continuous functions $\{f_n\}$ converges pointwise on $[0, 2]$ to the discontinuous function

$$f(x) = \begin{cases} 0, & 0 \leq x < 1, \\ 1, & x = 1, \\ 2, & 1 < x \leq 2. \end{cases}$$

In both of these examples, for $\varepsilon > 0$ fixed, the index N at which $|f_n(x) - f(x)| < \varepsilon$ begins to hold depends critically on x; the closer an $x \neq 1$ is to the discontinuity point $x = 1$, the larger the index N must be, and *no single index works for all* x.

Definition. The convergence of f_n to f is **uniform on a set** S if for each $\varepsilon > 0$ there is an N such that

$$x \in S \quad \text{and} \quad n \geq N \quad \text{implies} \quad |f_n(x) - f(x)| < \varepsilon.$$

Note: This says that the *same* index N "works" at *all* points of S, for a given $\varepsilon > 0$.

The convergence is not uniform on the common domains in the two examples. A reasonable conjecture is that continuity carries over from the terms of a sequence to the limit function f on any set S on which the convergence is uniform. A proof would involve an estimate, for $x_0 \in S$, of $|f(x) - f(x_0)|$, obviously, in view of what is given, to be made in terms of $|f(x) - f_n(x)|$ and $|f_n(x_0) - f(x_0)|$.

Now,

$$|f(x) - f(x_0)| \leq |f(x) - f_n(x)| + |f_n(x) - f_n(x_0)| + |f_n(x_0) - f(x_0)|$$

and each term on the right is to be made small, say $< \varepsilon/3$, when $\varepsilon > 0$ is fixed.

Notice that

1. If n were fixed, the continuity at x_0 of that f_n would provide a restriction on x that makes the middle term small.

2. If x were fixed, the convergence at x and x_0 would provide a restriction on n ($n \geq N$) that makes the first and last terms small.
3. The apparent paradox, in which each of the variables, n and x, needs to be fixed before the other can be appropriately restricted, is bypassed when the choice of N is independent of x, as in uniformity.

This points out why we have the important

Theorem. *If f_n, $n = 1, 2, 3, \ldots$, are continuous on S and converge to f uniformly on S then f is continuous on S.*

Proof. Fix $\varepsilon > 0$ and $x_0 \in S$. Choose N so that $|f_n(x) - f(x)| < \varepsilon/3$ for $n \geq N$, independent of $x \in S$. The continuity of f_N at x_0 gives δ such that

$$x \in S \quad \text{and} \quad |x - x_0| < \delta \quad \text{implies} \quad |f_N(x) - f_N(x_0)| < \varepsilon/3.$$

Then $x \in S$ and $|x - x_0| < \delta$ gives

$$|f(x) - f(x_0)| \leq |f(x) - f_N(x)| + |f_N(x) - f_N(x_0)| + |f_N(x_0) - f(x_0)| < \varepsilon,$$

establishing the continuity of f at x_0.

EXERCISE 1. The convergence doesn't have to be uniform for the limit function to be continuous: the functions

$$f_n(x) = \frac{nx}{1 + n^2x^2}, \qquad x \in \mathbb{R},$$

are continuous and converge to $f(x) = 0$, for $x \in \mathbb{R}$. Show that the convergence is not uniform. Hint: $f_n(1/n) = \frac{1}{2}$.

A graphical interpretation of uniform convergence is available. The graphs of $f + \varepsilon$ and $f - \varepsilon$, for $\varepsilon > 0$, form a band centered about the graph of f. The collection, $\mathscr{U}_\varepsilon(f)$, of all functions on S whose graphs lie in this band is the "uniform ε neighborhood" of f, and the assertion $f_n \to f$ uniformly on S can be stated:

$$\varepsilon > 0 \text{ implies there is an } N \text{ such that } f_n \in \mathscr{U}_\varepsilon(f) \text{ for all } n \geq N.$$

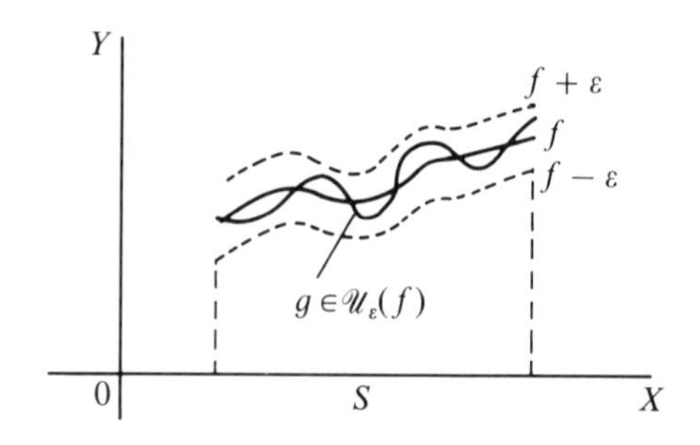

For f and g bounded on S the statement $g \in \mathscr{U}_\varepsilon(f)$ can be written

$$\sup_{x \in S} |f(x) - g(x)| < \varepsilon.$$

The quantity

$$\|f - g\| = \sup_{x \in S} |f(x) - g(x)|$$

generalizes the idea of "distance between points" to the setting in which "points" are bounded functions on S.

EXERCISE 2. Prove the triangle inequality

$$\|f - g\| \leq \|f - h\| + \|h - g\|.$$

The convergence of the "points" f_n to the "point" f in the sense of distance, $\|f_n - f\| \to 0$, is just the uniform convergence on S. The Cauchy condition for uniform convergence is best cast in the same notation.

Definition. The sequence $\{f_n\}$ of bounded functions on S is **uniformly Cauchy on S** if, for each $\varepsilon > 0$, there is an N such that

$$n \geq N \quad \text{and} \quad k \in \mathbb{N} \quad \text{implies} \quad \|f_{n+k} - f_n\| < \varepsilon.$$

EXERCISE 3. A sequence of functions bounded on f can only converge uniformly on S to a bounded function.

The thing that is not obvious is that a uniformly Cauchy sequence of bounded functions on S has a bounded function on S to serve as its uniform limit on S. (Compare with the completeness property.)

Theorem. *If $\{f_n\}$ is a uniformly Cauchy sequence of functions bounded on S, there is a bounded function f on S such that $f_n \to f$ uniformly on S.*

Proof. For $x \in S$ fixed, we have

$$|f_{n+k}(x) - f_n(x)| \leq \|f_{n+k} - f_n\|$$

which shows that $\{f_n(x)\}$ is a Cauchy sequence of real numbers. By the completeness of $\mathbb{R}$, we can define a function f on S as the pointwise limit

$$f(x) = \lim f_n(x), \qquad x \in S.$$

We show $f_n \to f$ uniformly on S. Let $\varepsilon > 0$ and choose N so that

$$n \geq N \quad \text{and} \quad k \in \mathbb{N} \quad \text{implies} \quad \|f_{n+k} - f_n\| < \frac{\varepsilon}{2}.$$

For every $x \in S$,

$$|f_{n+k}(x) - f_n(x)| \le \|f_{n+k} - f_n\| < \frac{\varepsilon}{2}.$$

Letting $k \to \infty$,

$$|f(x) - f_n(x)| \le \frac{\varepsilon}{2} \quad \text{for} \quad n \ge N \quad \text{and} \quad x \in S.$$

Hence, taking $\sup_{x \in S}$ gives $\|f - f_n\| \le \varepsilon/2$, hence

$$\|f - f_n\| < \varepsilon \qquad \text{for} \quad n \ge N,$$

the desired result. (The boundedness of f is implicit in the above argument, or results from Exercise 3.)

If the functions f_n are continuous on S the limit function on S is also continuous, in view of the previous theorem.

New kinds of continuous functions can be generated as the uniform limits of continuous functions. Here is a well-known example of a continuous function that is not differentiable anywhere!

The continuous function $|x|$ is not differentiable at $x = 0$ (the graph has a "corner" there). Let $u_1(x) = |x|$ on $[-\frac{1}{2}, \frac{1}{2}]$ and extend u_1 to $\mathbb{R}$ by periodicity with period 1, i.e., $u_1(x + 1) = u_1(x)$ for all $x \in \mathbb{R}$. Clearly, u_1 is continuous and has no derivative at $0, \pm\frac{1}{2}, \pm 1, \pm\frac{3}{2}, \ldots$.

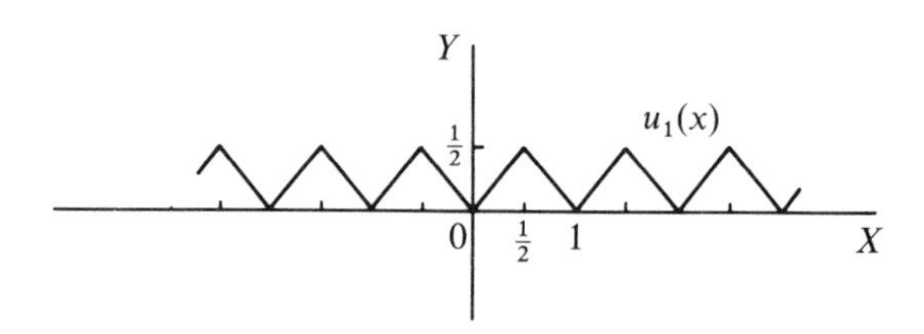

The function $u_2(x) = |x|$ on $[-\frac{1}{4}, \frac{1}{4}]$ and periodic with period $\frac{1}{2}$, $u_2(x + \frac{1}{2}) = u_2(x)$ for all $x \in \mathbb{R}$, is continuous and has no derivative at the points $0, \pm\frac{1}{4}, \pm\frac{1}{2}, \pm\frac{3}{4}, \ldots$. In general, let

$$u_n(x) = |x| \quad \text{on} \quad \left[-\frac{1}{2^n}, \frac{1}{2^n}\right] \quad \text{and} \quad u_n\left(x + \frac{1}{2^{n-1}}\right) = u_n(x), \qquad x \in \mathbb{R}.$$

EXERCISE 4. Show $u_n \to 0$ uniformly on $\mathbb{R}$. (The example is not $\lim u_n$; 0 is differentiable.)

Letting

$$v_m(x) = u_1(x) + u_2(x) + \cdots + u_m(x), \qquad x \in \mathbb{R},$$

we obtain continuous functions that fail to be differentiable at many points. The example is $v = \lim v_n$. In infinite series notation

$$v(x) = \sum_{n=1}^{\infty} u_n(x).$$

The existence and continuity of v follows from our theory if we show $\{v_m\}$ to be uniformly Cauchy on $\mathbb{R}$. But

$$|v_{m+k}(x) - v_m(x)| = \left| \sum_{n=m+1}^{m+k} u_n(x) \right| \leq \sum_{n=m+1}^{m+k} |u_n(x)|$$

$$\leq \sum_{n=m+1}^{m+k} \frac{1}{2^n} = \frac{1}{2^m}\left(\frac{1}{2} + \frac{1}{4} + \cdots + \frac{1}{2^k}\right)$$

$$< \frac{1}{2^m} \to 0 \quad \text{as} \quad m \to \infty, \text{independent of } k \text{ and } x.$$

The nondifferentiability of v at any $a \in \mathbb{R}$ results from the existence of a sequence of points $\{a + h_n\}$, with $h_n \to 0$, such that the difference quotients

$$\left\{ \frac{f(a + h_n) - f(a)}{h_n} \right\}$$

do not converge. The existence of such points is seen by examining the graphs of u_n, its predecessor u_{n-1} and its successors in the sequence.

Each h_n will be either $1/2^n$ or $-1/2^n$, depending on how a is situated relative to the segment of the graph of u_{n-1} above a. One of the numbers $a \pm 1/2^n$ (both, if a is at a vertex of the graph) is under the same segment of the graph of u_{n-1} that a is under. Choose $h_n = \pm 1/2^n$ accordingly (in the figure, $h_n = 1/2^n$.)

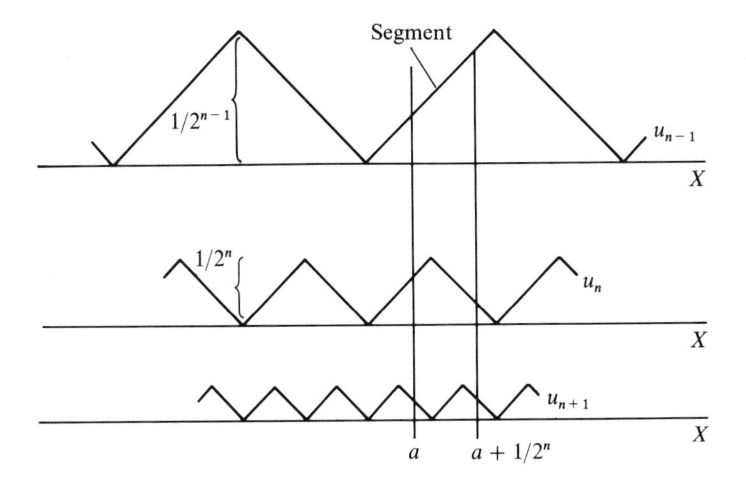

For all $k \leq n-1$, the points a and $a+h_n$ lie below the same segment of the graph of u_k, and those segments all have the same slope, either $+1$ or -1 ($+1$ in the figure). Thus, the difference quotient, using $a+h_n$ and a, of $v_{n-1} = u_1 + \cdots + u_{n-1}$ is either $n-1$ or $-(n-1)$.

The functions $u_{n+1}, u_{n+2}, \ldots$ are all periodic of period $1/2^n$, so that $u_{n+p}(a+h_n) - u_{n+p}(a) = 0$ for all $p \in \mathbb{N}$; these terms contribute nothing to the difference quotient of v.

The term u_n of v contributes an amount between ± 1 to the difference quotient of v. Thus either

$$-(n-1)-1 \leq \frac{v(a+h_n)-v(a)}{h_n} \leq -(n-1)+1$$

or

$$(n-1)-1 \leq \frac{v(a+h_n)-v(a)}{h_n} \leq (n-1)+1$$

according to the situation of a vis-à-vis the graph of u_{n-1}. In any case, the sequence of difference quotients has no limit.

§7. The Weierstrass Approximation Theorem

The second use for uniform convergence is to express functions in terms of simpler functions (as uniform limits) or, put otherwise, to approximate functions by simpler functions, within an error that holds uniformly over a set.

The simplest functions from the point of view of arithmetic computation are the polynomials. It is a remarkable fact that every continuous function (even v of the previous example!) on an interval $[a, b]$ is the uniform limit there of a sequence of polynomials.

EXERCISE 1. Show that the result for $[a, b]$ holds if and only if it holds for $[0, 1]$, by changing variables from x to $t = a + (b-a)x$.

The proof due to S. Bernstein provides formulas for polynomials in terms of the function to be approximated, rather than just proving their existence.

Weierstrass Approximation Theorem. *Let f be continuous on* $[0, 1]$ *and put*

$$B_n(f)(x) = \sum_{k=0}^{n} f\left(\frac{k}{n}\right)\binom{n}{k} x^k(1-x)^{n-k}, \qquad x \in [0, 1],$$

where

$$\binom{n}{k} = \frac{n!}{k!\,(n-k)!}.$$

Then $B_n(f) \to f$ *uniformly on* $[0, 1]$.

Remark. $B_n(f)$, the **nth Bernstein polynomial of f**, has an interpretation in terms of probability. Suppose x is the probability of success in a single trial of a random event, so that $(1 - x)$ is the probability of failure (e.g., success means heads in the flip of a biased coin). In n trials of the event, which are independent in the sense that the outcome of one trial has no effect on the outcomes of the others, the probability of exactly k successes is immediately seen to be

$$\binom{n}{k} x^k (1 - x)^{n-k}.$$

(Note that the total probability should be 1,

$$\sum_{k=0}^{n} \binom{n}{k} x^k (1 - x)^{n-k} = 1,$$

and this is true since the left-hand side is just the binomial expansion of $[x + (1 - x)]^n$.)

Now, if the "payoff" for exactly k successes is expressed as the value of f at k/n, then $B_n(f)$ totals the payoffs, weighted by their probabilities; gives what is called the expected payoff from the n trials.

Note: If we multiply the last equation by $f(x)$ we get an expression for $f(x)$ that is compatible in form with that for $B_n(f)(x)$;

$$f(x) = \sum_{k=0}^{n} f(x) \binom{n}{k} x^k (1 - x)^{n-k}.$$

Proof of the Theorem. We have

$$\begin{aligned} |B_n(f)(x) - f(x)| &= \left| \sum_{k=0}^{n} \left(f\left(\frac{k}{n}\right) - f(x) \right) \binom{n}{k} x^k (1 - x)^{n-k} \right| \\ &\le \sum_{k=0}^{n} \left| f\left(\frac{k}{n}\right) - f(x) \right| \binom{n}{k} x^k (1 - x)^{n-k}. \end{aligned}$$

We seek estimates, given $\varepsilon > 0$, showing this is $< \varepsilon$ for all $x \in [0, 1]$, when n is sufficiently large.

The continuity of f on $[0, 1]$ is uniform, so there is a δ such that $|y - x| < \delta$ gives $|f(y) - f(x)| < \varepsilon/2$. Certain indices k gives points k/n that are suitable for y here, and we separate the sum over such indices from the rest of the terms:

$$\begin{aligned} |B_n(f)(x) - f(x)| &\leq \sum_{|k/n - x| < \delta} \left| f\left(\frac{k}{n}\right) - f(x) \right| \binom{n}{k} x^k (1-x)^{n-k} \\ &\quad + \sum_{|k/n - x| \geq \delta} \left| f\left(\frac{k}{n}\right) - f(x) \right| \binom{n}{k} x^k (1-x)^{n-k} \\ &< \frac{\varepsilon}{2} \sum_{|k/n - x| < \delta} \binom{n}{k} x^k (1-x)^{n-k} \\ &\quad + \sum_{|k/n - x| \geq \delta} \left| f\left(\frac{k}{n}\right) - f(x) \right| \binom{n}{k} x^k (1-x)^{n-k} \end{aligned}$$

Extending the first sum over all indices k increases it to 1, so

$$|B_n(f)(x) - f(x)| < \frac{\varepsilon}{2} + \sum_{|k/n - x| \geq \delta} \left| f\left(\frac{k}{n}\right) - f(x) \right| \binom{n}{k} x^k (1-x)^{n-k}.$$

In estimating the remaining sum, the estimate for $|f(k/n) - f(x)|$ can be crude; it must be the rest that insures the sum is small. Since f is continuous and $[0, 1]$ is compact there is an M such that $|f(x)| \leq M$, for $x \in [0, 1]$, so $|f(k/n) - f(x)| \leq 2M$ and it remains to show that

$$\sum_{|k/n - x| \geq \delta} \binom{n}{k} x^k (1-x)^{n-k} < \frac{\varepsilon}{4M} \qquad \text{for} \quad n \geq \text{some } N, \quad \text{all } x \in [0, 1].$$

When $|k/n - x| \geq \delta$, $(k - nx)^2 \geq n^2\delta^2$, or $(k - nx)^2/n^2\delta^2 \geq 1$ and the sum is at most

$$\sum_{|k/n - x| \geq \delta} \frac{(k - nx)^2}{n^2\delta^2} \binom{n}{k} x^k (1-x)^{n-k}.$$

We extend this sum over all indices $k = 0, 1, \ldots, n$, which can only increase it, and face the computational task of showing that

$$\sum_{k=0}^{n} (k - nx)^2 \binom{n}{k} x^k (1-x)^{n-k} < \frac{n^2\delta^2\varepsilon}{4M}, \qquad x \in [0, 1],$$

for n sufficiently large.

Computation will show that this sum equals $nx(1 - x)$, so the result follows from the boundedness of $x(1 - x)$ on $[0, 1]$ (and the n^2 on the right, of course).

Here are the computations:

(a)

$$\begin{aligned} \sum_{k=0}^{n} (k - nx)^2 \binom{n}{k} x^k (1-x)^{n-k} &= \sum_{k=0}^{n} k^2 \binom{n}{k} x^k (1-x)^{n-k} \\ &\quad - 2nx \sum_{k=0}^{n} k \binom{n}{k} x^k (1-x)^{n-k} + n^2x^2. \end{aligned}$$

(b)

$$\sum_{k=0}^{n} k\binom{n}{k}x^k(1-x)^{n-k} = \sum_{k=1}^{n} k\binom{n}{k}x^k(1-x)^{n-k}$$
$$= \sum_{k=1}^{n} n\binom{n-1}{k-1}x^k(1-x)^{n-k},$$

since, for $k = 1, 2, \ldots, n$,

$$k\binom{n}{k} = k\frac{n!}{k!\,(n-k)!} = \frac{n(n-1)!}{(k-1)!\,(n-k)!} = n\frac{(n-1)!}{(k-1)!\,[(n-1)-(k-1)]!}$$
$$= n\binom{n-1}{k-1}.$$

Changing the summation index to $l = k - 1$,

$$\sum_{k=0}^{n} k\binom{n}{k}x^k(1-x)^{n-k} = \sum_{l=0}^{n-1} n\binom{n-1}{l}x^{l+1}(1-x)^{(n-1)-l}$$
$$= nx\sum_{l=0}^{n-1}\binom{n-1}{l}x^l(1-x)^{(n-1)-l} = nx$$

(c)

$$\sum_{k=0}^{n} k^2\binom{n}{k}x^k(1-x)^{n-k} = \sum_{k=0}^{n} k(k-1)\binom{n}{k}x^k(1-x)^{n-k}$$
$$+ \sum_{k=0}^{n} k\binom{n}{k}x^k(1-x)^{n-k}$$
$$= \sum_{k=2}^{n} k(k-1)\binom{n}{k}x^k(1-x)^{n-k} + nx$$
$$= \sum_{k=2}^{n} n(n-1)\binom{n-2}{k-2}x^k(1-x)^{n-k} + nx$$
$$= n(n-1)\sum_{l=0}^{n-2}\binom{n-2}{l}x^{l+2}(1-x)^{(n-2)-l} + nx$$
$$(l = k-2),$$
$$= n(n-1)x^2\sum_{l=0}^{n-2}\binom{n-2}{l}x^l(1-x)^{(n-2)-l} + nx$$
$$= n(n-1)x^2 + nx$$

Substituting (b) and (c) in (a) gives the result:

$$\sum_{k=0}^{n}(k-nx)^2\binom{n}{k}x^k(1-x)^{n-k} = nx(1-x).$$

Summary of the Main Properties of Continuous Functions

If f is continuous and dom f is compact, then:

1. f is bounded.
2. min f and max f exist.
3. The continuity is uniform.
4. ran f is compact.

If, in addition, dom $f = [a, b]$,

5. f has the intermediate-value property.
6. ran $f = [\min f, \max f]$.
7. If f is one–one, it is strictly monotonic and has continuous inverse.

Finally,

8. The limit of a uniformly convergent sequence of continuous functions is continuous.

Appendix. A Space-Filling Continuous Curve

To conclude this chapter, we give an example of a continuous curve in the plane which fills an area; a so-called space-filling curve, or Peano curve. Certain details are left as exercises.

Familiar curves can all be represented parametrically, each in many ways. The notion of "curve" itself is best formulated in full generality by considering all mappings of intervals into the plane and then identifying those that only differ in the sense that one results from the other by a "change of parameter." For our purposes, it is enough to consider specific mappings of intervals into the plane and accept the fact that each such mapping determines a curve.

Exercise 1. Define *continuous curve*, i.e., continuity for a mapping $\phi: I \to X-Y$ plane, where I is an interval in $\mathbb{R}$.

However the definition is formalized, it will express the idea that two points on the curve can be made closer in the plane than $\varepsilon > 0$ by choosing their parameter values closer than some δ, for any $\varepsilon > 0$.

The example we propose is a mapping ϕ of $[0, 1]$ into a fixed triangle (closed, i.e., including interior and boundary) in the plane. The value of ϕ at x is described as follows.

1. Let $x = 0.b_1 b_2 b_3 \ldots$ be the terminating binary expansion of x.
2. Consider one vertex of the given triangle T_0 and the perpendicular to the

opposite side, generating two closed subtriangles: the left one, looking along the perpendicular from the vertex, and the right one. Let T_1 = the left one if $b_1 = 0$ and the right one if $b_1 = 1$.
3. Proceed inductively to select a sequence $\{T_n\}$ of closed subtriangles according to the sequence of digits and the rule $0 \leftrightarrow$ left, $1 \leftrightarrow$ right. The case $x = 0.100100011\ldots$ is illustrated, in part.
4. The point $\phi(x)$ is the point in the plane common to the "nested triangles" so determined by x.

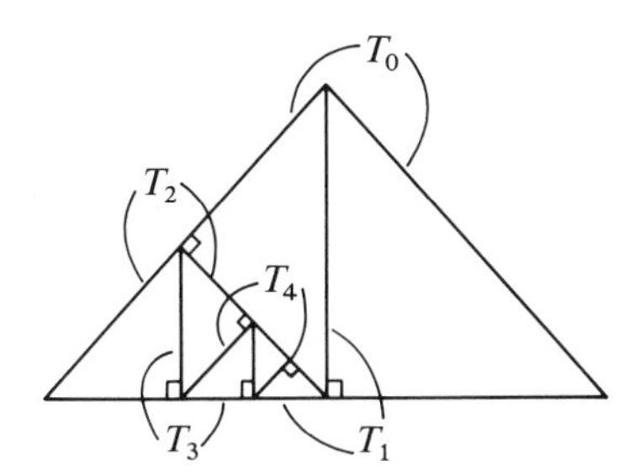

Clearly, the existence of such a point calls for a "Nested Closed Triangles Theorem" in the plane. We digress to outline some exercises that cover this need.

EXERCISE 2. Define: $\{P_n\}$ is a Cauchy sequence of points in the plane.

EXERCISE 3. Use rectangular coordinates, the X–Y plane, and rectangular neighborhoods, to show that every Cauchy sequence in the plane converges.

Definition. A **bounded set** in the plane is a set of points that is contained in some disk. The **diameter** of a bounded set S is

$$\operatorname{diam} S = \sup\{\sqrt{(x - x')^2 + (y - y')^2}\colon (x, y) \text{ and } (x', y') \in S\}.$$

EXERCISE 4. Define limit point and closed set in the plane and show that a closed triangle is a closed set.

EXERCISE 5. Prove the

Nested Closed Sets Theorem. *If $\{S_n\}$ is a sequence of closed bounded sets in the plane satisfying*

$$S_{n+1} \subset S_n, \qquad n = 1, 2, 3, \ldots,$$

and

$$\operatorname{diam} S_n \to 0,$$

there exists a unique point common to all S_n.

Having defined $\phi(x)$, the continuity of ϕ is intuitively clear in the form "close points in [0, 1] have close images" since "close" points have the same initial binary digits, hence their images are in the same "small" triangle.

EXERCISE 6. Give the details.

Finally, we observe that ϕ maps [0, 1] *onto* T_0. This follows from the observation that for any $P \in T_0$ there is a well-determined sequence of subtriangles $T_1(P)$, $T_2(P)$, $T_3(P)$, ... to which P belongs;

$$P = \bigcap_{n=1}^{\infty} T_n(P).$$

One simply chooses the left or right closed subtriangle that contains P inductively, choosing the left one in case P is on an altitude, to correspond to the terminating choice of binary expansions. Writing 0 for left and 1 for right yields the binary expansion of the point $x \in [0, 1]$ such that $\phi(x) = P$.

EXERCISE 7. Find the numbers x_0 and x_1 that map onto the other two vertices of T_0.

Part II

Foundations of Calculus

The words calculus and calculation obviously have the same root. A calculus is an organized body of techniques of calculation and **The Calculus** refers to the one that concerns differentiation, integration, and infinite series. It is not our purpose to review its techniques; only to establish its foundations in the real number system and the theory of limits, and to indicate a few extensions and connections.

CHAPTER 4

Differentiation

Science begins with observation. At first, the observations may be only crude and tentative: more of one thing seems always to be accompanied by more of another (direct variation), or, perhaps, by less of the other (inverse variation). Once a science has established ways of measuring its central attributes (measuring temperature was a major achievement of early physical science) the primary observations become expressible as formulas, relating quantities. That is, attributes become quantities via measurement, and the science, or part of it, emerges from the qualitative stage to a quantitative stage, amenable to mathematical descriptions. The fundamental laws of the science are expressed by formulas.

Almost invariably, these fundamental laws take the form of proportionality statements; the joint variation becomes direct proportionality

$$y - y_0 = m(x - x_0),$$

where x_0 and y_0 are appropriate reference levels for the measurement of the respective attributes, and the constant of proportionality m incorporates the choices of units for the two measurements. The choices of x_0, y_0, and m are dictated by considerations within the science itself (scale, instrumentation, accepted standards and practices, etc.) and the mathematics, in order to be applicable to a variety of situations, must *remain abstract.*

Specific examples of this practice in the development of a science are abundant. To mention a few: price varies directly with demand; the length of a metal rod varies directly with its temperature; the rate at which a heated object cools varies directly with the difference between its temperature and that of the surrounding medium; the acceleration of a moving object varies directly with the force applied to it. Basic science abounds with formulas like $F = ma$, $e = mc^2$, $E = IR$, and $s = vt$. (Some such formulas embody major advances in understanding—force was long thought proportional to velocity, and the notion that energy and mass might be convertible, one to the other, was revolutionary—while others are just definitions within the science: velocity = distance per unit time; $v = s/t$.)

The constant appearance of direct proportionality in the sciences reveals the central importance in mathematical modeling of the linear functions:

$$y = y_0 + m(x - x_0).$$

Even inverse proportionalities: the atmospheric pressure is inversely proportional to the altitude, and more complicated fundamental laws: the force of attraction between two masses is inversely proportional to the square of the distance between them, assert proportionality and so can be reduced to the composition of a function, like $1/x$ or $1/x^2$, followed by a linear function.

A second consideration accompanies each assertion of a fundamental law: the range of values of the variables over which it is claimed to have validity. Obviously, the assertion about atmospheric pressure is not valid at altitudes where there is no atmosphere. The fundamental law, recognizing that the phenomenon it deals with is nonlinear, only asserts that a linear approximation adequately describes the phenomenon over a certain range of the variables. For example, *Hooke's Law* asserts that stretching a spring requires a force that is proportional to the elongation produced. The constant of proportionality k, called the spring constant, embodies not only the choices of units, but the properties of the individ-

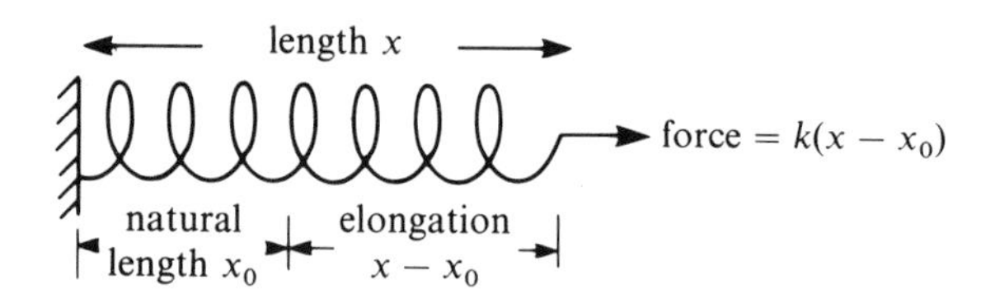

ual spring—such as its material and dimensions. Since the natural length x_0 is the length at which the force is zero, Hooke's Law is the linear "model" of the spring:

$$\text{force} = k(x - x_0),$$

for x in some range.

The overall phenomenon is clearly nonlinear; at some elongation, the material will lose its elasticity and break down by stretching large amounts in response to small forces. The figure shows how an actual plot of force versus length might look.

The fundamental law is, mathematically, a linear approximation through a point (x_0, y_0), here $y_0 = 0$, to the actual force function $f(x)$, which is "valid" for some range of x. The abstract formulation of this notion of "valid" is that, for an acceptable maximum error $\varepsilon > 0$, there is some neighborhood of x_0 throughout which the error $|f(x) - k(x - x_0)|$ is less

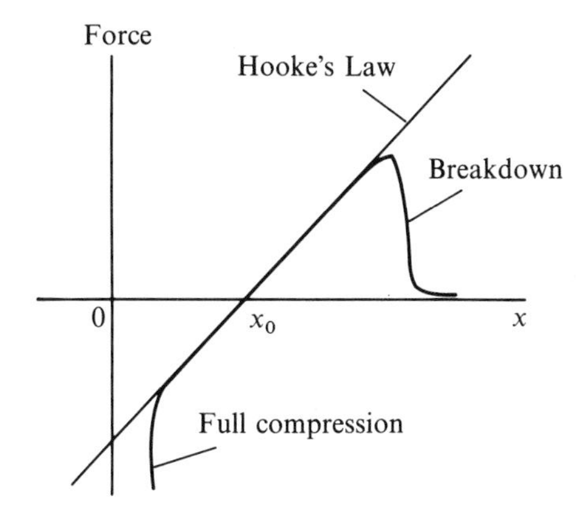

than ε. Recall that the abstract statement has by its very nature abandoned any framework within which to judge that some amount of error ε is "satisfactory" or that a range of x, or neighborhood of x_0, is "adequate"; these judgments belong to the science, not to the mathematics being used to model the science. For a suitably abstract statement we must use—"for *every* $\varepsilon > 0$, *there is* a neighborhood of x_0." This is called **local** approximation of f at x_0 (or at the point (x_0, y_0) on the graph of f).

We are led by these considerations to the conclusion that the very basis of the mathematization of science is in large part the *local linear approximation* of functions, which is precisely the subject matter of differentiation.

§1. Differential and Derivative. Tangent Line

Consider a function f and a point a that is **interior** to dom f in the sense that some interval about a, $(a - \delta, a + \delta)$, $\delta > 0$, is contained in dom f.

For the purpose of discussing local approximation near a, it is natural to choose $(a, f(a))$ as the origin in a new coordinate system and introduce, as new variables h and k, the "increments" in x and y

$$h = x - a,$$
$$k = y - f(a).$$

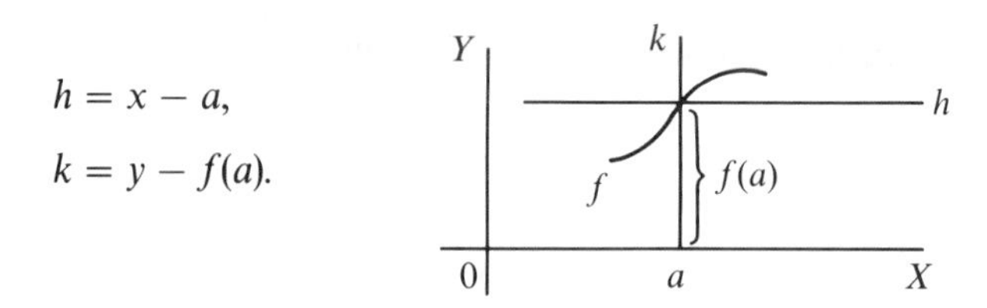

The linear functions through $(a, f(a))$ among which we seek an approximator to f are expressed mh, $m \in \mathbb{R}$, in these coordinates. For each h, the error when mh is chosen as approximator is

$$e(h) = [f(a+h) - f(a)] - mh. \tag{$*$}$$

The local approximation at a of f by mh is simply stated as

$$\lim_{h \to 0} e(h) = 0.$$

Observe that this holds precisely when $[f(a+h) - f(a)] \to 0$, i.e., when f is continuous at a, *regardless* of the value of m; *every* nonvertical line through $(a, f(a))$ approximates f locally at a. To single out a line that approximates better than the others, we impose the stronger requirement that "$e(h)$ be small compared to h," in the sense

$$\lim_{h \to 0} \frac{e(h)}{h} = 0.$$

We see from the formula $(*)$ that division of $e(h)$ by h^α for *any* $\alpha \geq 1$ has the effect of leaving the m term present in the limit as $h \to 0$, resulting in a condition on m.

If $\lim_{h \to 0} e(h)/h^\alpha = 0$, it is said that $e(h)$ **vanishes of order** α as $h \to 0$, written $e(h) = o(h^\alpha)$. In this terminology, the above stronger requirement is that $e(h) = o(h)$; $e(h)$ vanishes of first order.

Clearly, when $1 \leq \alpha \leq \beta$,

$$e(h) = o(h^\beta) \quad \text{implies} \quad e(h) = o(h^\alpha),$$

so that our requirement $e(h) = o(h)$ is the mildest one of this kind that invokes a restriction on m.

The requirement has a geometric interpretation. The line $k = mh$ is the central axis of the two-dimensional "cones" consisting of the regions bounded by the lines $k = (m \pm \varepsilon)h$, for every $\varepsilon > 0$. If the graph of f remains in each such "cone" over some range $(a - \delta, a + \delta)$ of x, the δ depending on ε, intuition suggests that the line $k = mh$ be called **tangent** to the graph of f at $(a, f(a))$. This says that for every $\varepsilon > 0$, there is a δ such that

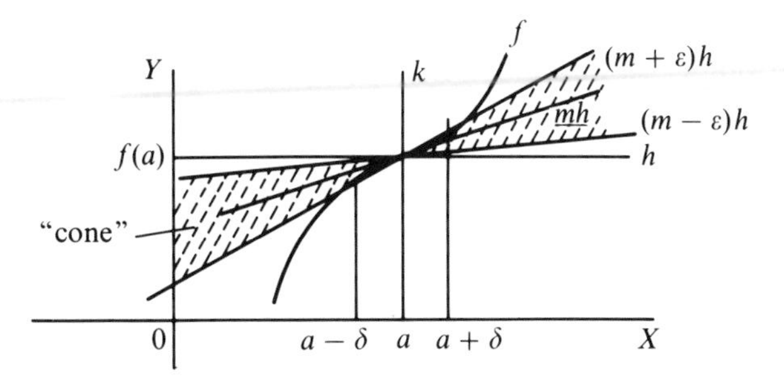

$$|h| < \delta \quad \text{implies} \quad |e(h)| < \varepsilon |h|,$$

which means

$$\lim_{h \to 0} \frac{e(h)}{h} = 0 \quad \text{or} \quad e(h) = o(h).$$

Spelled out in X–Y notation, the condition is

$$\lim_{x \to a} \frac{[f(x) - f(a)] - m(x - a)}{x - a} = 0$$

or

$$\lim_{x \to a} \frac{f(x) - f(a)}{x - a} = m.$$

The second version highlights the **difference quotient** $[f(x) - f(a)]/(x - a)$ (defined on a punctured neighborhood of a) and its limit m, also denoted $f'(a)$, the **derivative of** f **at** a. The existence of the limit is expressed by saying f **is differentiable at** a.

The first version of the statement stresses the idea of local linear approximation and the best linear approximator $m(x - a)$, or $f'(a)(x - a)$, called the **differential of** f **at** a.

The graph of the differential is the tangent line. Its equation in X–Y coordinates is

$$y = f(a) + f'(a)\,(x - a),$$

and in h–k (or "local") coordinates, it is

$$k = f'(a)\,h.$$

(Changing notation for the local coordinates by putting dx for h and dy for k, this gives

$$dy = f'(a)\;dx,$$

which conforms with the Leibniz notation dy/dx for the derivative.)

Remark. The restriction that a be an interior point of dom f can be relaxed, to simply require that a be a point and a limit point of dom f. In particular, if dom f is a closed interval $[a, b]$, this leads to definitions of one-sided derivatives at a (derivative from the right) and at b (derivative from the left). We shall not split these particular hairs in our brief treatment; the main interest is in derivatives at interior points.

The observation we have already made is of sufficient importance to be called a

Theorem. *If f is differentiable at a then f is continuous at a.*

EXERCISE 1. Base a proof on the product rule for limits and the formulation of continuity at a embodied in the statement $\lim_{x \to a} [f(x) - f(a)] = 0$.

We have seen that there exist continuous functions that are nowhere differentiable, using the idea that a "corner" in the graph, as with $|x|$ at $x = 0$, preserves continuity but represents nondifferentiability. Here are some additional examples.

EXAMPLE. A function whose graph has a vertical tangent at a point is continuous and nondifferentiable there. A simple example is $y = \sqrt[3]{x}$ at $x = 0$. The difference quotients $\sqrt[3]{x}/x$ have no limit as $x \to 0$.

EXAMPLE. A continuous function can have difference quotients that oscillate wildly, as with the following function at the point $x = 0$:

$$f(x) = \begin{cases} x \sin(1/x), & x \neq 0, \\ 0, & x = 0. \end{cases}$$

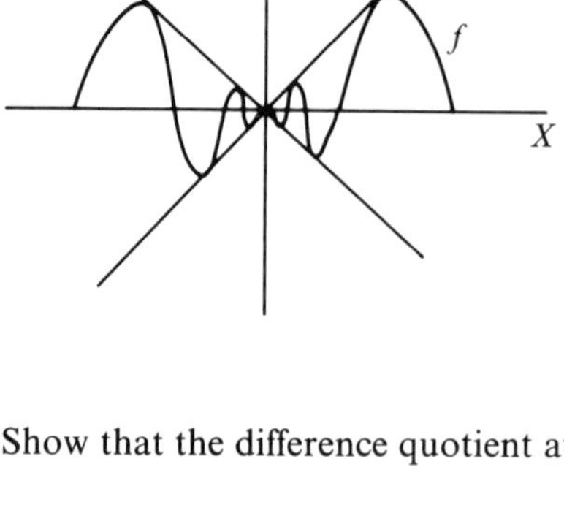

EXERCISE 2. Why is this f continuous? Show that the difference quotient at 0 is $\sin(1/x)$.

A function can be differentiable and the derivative function fail to be continuous, as is shown in the

EXAMPLE. For

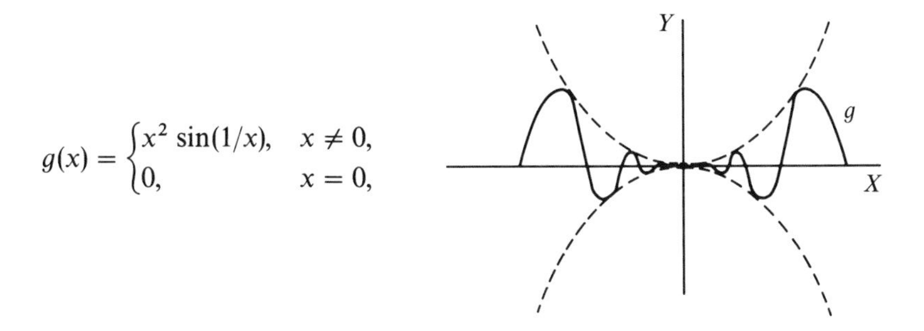

$$g(x) = \begin{cases} x^2 \sin(1/x), & x \neq 0, \\ 0, & x = 0, \end{cases}$$

the difference quotient at 0 is $x \sin(1/x)$, which $\to 0$ as $x \to 0$. Thus $g'(0) = 0$. For any $x \neq 0$,

$$g'(x) = 2x \sin\frac{1}{x} - \cos\frac{1}{x}.$$

Thus, g is differentiable everywhere. But g' is not continuous at 0 because $\cos(1/x)$ has no limit as $x \to 0$ while $2x \sin(1/x)$ does.

A function f with open domain (every point is an interior point) is called **continuously differentiable** if f' exists and is continuous at all points of dom f.

If S is an open set, we introduce notations for certain classes of functions on S: the collection of all functions

continuous on S is denoted $C(S)$,
differentiable on S is denoted $D(S)$,
continuously differentiable on S is denoted $C^1(S)$.

We have seen that

$$C(S) \supsetneqq D(S) \supsetneqq C^1(S).$$

The **second derivative** f'' of f is the derivative of f'. That is, f'' has as domain all interior points a of dom f' at which the limit

$$\lim_{x \to a} \frac{f'(x) - f'(a)}{x - a}$$

exists, and this limit is the value $f''(a)$. The **nth derivative** $f^{(n)}$ is defined inductively in the same way for all $n \in \mathbb{N}$.

EXERCISE 3. Define classes $D^n(S)$ and $C^n(S)$ for $n \in \mathbb{N}$ and S open. Show that $C \supsetneqq D^1 \supsetneqq C^1 \supsetneqq D^2 \supsetneqq C^2 \supsetneqq \cdots$, using the functions

$$g_n(x) = \begin{cases} x^n \sin(1/x), & x \neq 0, \\ 0, & x = 0. \end{cases}$$

Exercise 4. To what classes does the function

$$h_n(x) = \begin{cases} x^n, & x \geq 0, \\ 0, & x < 0, \end{cases}$$

belong?

The class of all functions that have derivatives of all orders (necessarily all continuous) on S is denoted $C^\infty(S)$.

Exercise 5. Show that $\lim_{x\to 0}(1/x^k)e^{-1/x^2} = 0$, for $k = 1, 2, 3, \ldots$. (Hint: Apply L'Hôpital's rule.)

Exercise 6. Show that $f \in C^\infty(\mathbb{R})$, where

$$f(x) = \begin{cases} e^{-1/x^2}, & x > 0 \\ 0, & x \leq 0. \end{cases}$$

(Hint: Show that, on $(0, \infty)$, $f^{(n)}$ is a finite sum of constant multiples of the functions in Exercise 5.)

§2. The Foundations of Differentiation

A significant portion of a course in Calculus is devoted to the techniques whereby one writes, from inspection, the derivative of a given elementary function. These techniques are based on three things:

1. The formulas

$$\frac{dx^n}{dx} = nx^{n-1}, \quad n \in \mathbb{N}, \qquad \frac{d\sin x}{dx} = \cos x \quad \text{and} \quad \frac{de^x}{dx} = e^x.$$

2. The sum, difference, product, and quotient rules.
3. The chain rule.

Exercise 1. Use induction and the product rule to prove $dx^n/dx = nx^{n-1}$, $n \in \mathbb{N}$.

Exercise 2. Prove $d\sin x/dx = \cos x$ on the assumption that

$$\lim_{h\to 0} \frac{\sin h}{h} = 1 \quad \text{and} \quad \lim_{h\to 0} \frac{\cos h - 1}{h} = 0.$$

Exercise 3. Prove

$$\lim_{h\to 0} \frac{\cos h - 1}{h} = 0$$

on the assumption that $\lim_{h\to 0} \dfrac{\sin h}{h} = 1$.

EXERCISE 4. Show that $\cos h < \dfrac{\sin h}{h} < 1$ for $|h|$ small, by comparing areas in the figure. Infer that

$$\lim_{h\to 0} \frac{\sin h}{h} = 1.$$

EXERCISE 5. Show that $de^x/dx = e^x$ on the assumption that

$$\lim_{h\to 0} \frac{e^h - 1}{h} = 1.$$

EXERCISE 6. Show that

$$\lim_{h\to 0} \frac{e^h - 1}{h} = 1$$

follows if the one-sided limit

$$\lim_{\substack{h\to 0\\ h>0}} \frac{e^h - 1}{h} = 1$$

is proved.

Since e^h is defined as the continuous extension of e^r, $r \in \mathbb{Q}$, it is enough to show the one-sided limit holds for h restricted to $\mathbb{Q}$. We have, for $p, q \in \mathbb{N}$,

$$e^{p/q} = \left[\lim_n \left(1 + \frac{1}{n}\right)^n\right]^{p/q} = \left[\lim_m \left(1 + \frac{1}{mq}\right)^{mq}\right]^{p/q}$$

(restricting n to the subsequence of all multiples of q). By the continuity at e of $x^{p/q}$,

$$e^{p/q} = \lim_m \left(1 + \frac{1}{mq}\right)^{mp}.$$

Expanding the terms by the Binomial Theorem, we find

$$\frac{e^{p/q} - 1}{p/q} = \lim_m \left[1 + \frac{(mp-1)}{2!}\cdot\frac{1}{mq} + \frac{(mp-1)(mp-2)}{3!}\frac{1}{(mq)^2} + \cdots \right.$$
$$\left. + \frac{(mp-1)(mp-2)\cdots(mp-[mp-1])}{(mp)!}\cdot\frac{1}{(mq)^{mp-1}}\right].$$

Hence, subtracting 1 and factoring out p/q,

$$\frac{e^{p/q} - 1}{p/q} - 1 = \frac{p}{q}\lim_m \left[\frac{1}{2!}\left(1 - \frac{1}{mp}\right) + \frac{1}{3!}\left(1 - \frac{1}{mp}\right)\left(1 - \frac{2}{mp}\right)\frac{p}{q} + \cdots \right.$$
$$\left. + \frac{1}{(mp)!}\left(1 - \frac{1}{mp}\right)\left(1 - \frac{2}{mp}\right)\cdots\left(1 - \frac{mp-1}{mp}\right)\left(\frac{p}{q}\right)^{mp-2}\right].$$

The desired conclusion follows from the fact that the second factors here, $\lim_m [\ldots]$, are bounded functions of p/q.

EXERCISE 7. Show that, for $p/q < 1$, the second factor, $\lim_m [\ldots]$ is less than

$$\lim_m \left[\frac{1}{2!} + \frac{1}{3!} + \cdots + \frac{1}{(mp)!}\right] = e - 2.$$

This argument shows, as well, that e is the choice of base for which the exponential function is its own derivative.

Concerning the combinations of functions by the operations of arithmetic, we have the

Theorem. *If f and g are differentiable at a then $f \pm g$ and $f \cdot g$ are differentiable at a and*

$$(f \pm g)'(a) = f'(a) \pm g'(a),$$

$$(f \cdot g)'(a) = f(a) \cdot g'(a) + f'(a) \cdot g(a).$$

If, in addition, $g(a) \neq 0$ then f/g is differentiable at a and

$$\left(\frac{f}{g}\right)'(a) = \frac{g(a) \cdot f'(a) - f(a) \cdot g'(a)}{g(a)^2}.$$

Notice that if a is an interior point of dom f and of dom g then a is an interior point of dom $f \cap$ dom g, so that the differentiability at a of $f \pm g$ and $f \cdot g$ can be considered. In the quotient case, the differentiability of g at a insures its continuity there, which insures the persistence of the sign of $g(a) \neq 0$ throughout some neighborhood of a. Thus a is an interior point of dom f/g under the hypotheses.

Beyond this, each proof consists of expressing the difference quotient of the combined function in terms of the difference quotients of f and g: the theorem on limits of combined functions then yields the result.

Here are the details in the quotient case:

$$\begin{aligned}
&\frac{(f/g)(x) - (f/g)(a)}{x - a} \\
&\quad = \frac{1}{x - a}\left[\frac{f(x)}{g(x)} - \frac{f(a)}{g(a)}\right] \\
&\quad = \frac{1}{x - a}\left[\frac{f(x)g(a) - f(a)g(x)}{g(x)g(a)}\right] \\
&\quad = \frac{1}{(x - a)g(x)g(a)}[f(x)g(a) - f(a)g(a) + f(a)g(a) - f(a)g(x)] \\
&\quad = \frac{1}{g(x)g(a)}\left[g(a)\frac{f(x) - f(a)}{x - a} - f(a)\frac{g(x) - g(a)}{x - a}\right].
\end{aligned}$$

We know $g(x) \to g(a)$ as $x \to a$ since g is differentiable, hence continuous, at a. Taking the limit as $x \to a$, the quotient rule is obtained.

Exercise 8. Supply the details in the product case.

Exercise 9. Show that every rational function is differentiable and that the derivative is again a rational function. Infer that every rational function f is in $C^\infty(\operatorname{dom} f)$.

Consider the combination of two functions by composition, in the same spirit. Suppose g maps a neighborhood of a into dom f, so that $f \circ g$ is defined on a neighborhood of a. Suppose $g'(a)$ and $f'(g(a))$ exist and consider the difference quotient for $f \circ g$ at a:

$$\frac{f \circ g(x) - f \circ g(a)}{x - a} = \frac{f(g(x)) - f(g(a))}{x - a}.$$

The obvious way to express this in terms of difference quotients of f and g is to multiply and divide (if possible) by $g(x) - g(a)$:

$$\frac{f \circ g(x) - f \circ g(a)}{x - a} = \frac{f(g(x)) - f(g(a))}{g(x) - g(a)} \cdot \frac{g(x) - g(a)}{x - a}.$$

Now, g is continuous at a, hence $x \to a$ gives $g(x) \to g(a)$, suggesting the result

$$(f \circ g)'(a) = f'(g(a)) \cdot g'(a).$$

The difficulty, that division by zero; $g(x) - g(a) = 0$, can occur for x arbitrarily near a, is overcome by the observation that at such points both the difference quotient for $f \circ g$ on the left and that of g on the right are zero. Thus the undefined first factor on the right can be replaced by *anything* for such values of x, and the equality that results holds for all x. Choosing $f'(g(a))$ as the replacement value, i.e.,

$$\frac{f \circ g(x) - f \circ g(a)}{x - a} = \begin{cases} \dfrac{f(g(x)) - f(g(a))}{g(x) - g(a)} \cdot \dfrac{g(x) - g(a)}{x - a} & \textit{if } g(x) \neq g(a) \\[2ex] f'(g(a)) \cdot \dfrac{g(x) - g(a)}{x - a} & \text{if } g(x) = g(a), \end{cases}$$

the limit as $x \to a$ gives the

Chain Rule.

$$(f \circ g)'(a) = f'(g(a)) \cdot g'(a).$$

(The formula states a theorem implicitly, in the sense that the conclusion, "$(f \circ g)'(a)$ exists and is given as in the formula," follows from the least hypotheses that make the symbols meaningful.)

A corollary of the chain rule is the standard result on inverse functions, the

Proposition. *Suppose a is an interior point of* dom f, *f is invertible and $f(a)$ is an interior point of* dom f^{-1}. *If $f'(a) \neq 0$ then f^{-1} is differentiable at $f(a)$ and*

$$(f^{-1})'(f(a)) = \frac{1}{f'(a)}.$$

Proof. Under the hypotheses, the graph of f has a nonhorizontal tangent at $(a, f(a))$, which means that the graph of f^{-1} at the point $(f(a), a) = (f(a), f^{-1}(f(a)))$ has a nonvertical tangent. Thus f^{-1} is differentiable at $f(a)$ and hence the chain rule applies to the composite $f^{-1} \circ f$ at a. Applying it, in order to differentiate both sides of the identity $f^{-1}(f(x)) = x$, at a, we get

$$(f^{-1})'(f(a)) \cdot f'(a) = 1.$$

EXERCISE 10. Derive the formulas for the derivatives of $\cos x$ and $\tan x$, using trigonometric identities.

EXERCISE 11. Let $y = \sin^{-1} x$. Show that $y' = 1/\cos y$ and express y' in terms of x.

EXERCISE 12. Let $y = \log x$, $x > 0$. Show $y' = 1/x$.

EXERCISE 13. For $x > 0$, $x^r = e^{r \log x}$ for $r \in \mathbb{R}$ (why?). Show that $dx^r/dx = rx^{r-1}$, $x > 0$, for any $r \in \mathbb{R}$.

These results provide the means to write the derivative of any explicitly defined elementary function. Since the result is a function of the same kind, derivatives of any order can be found.

If a function is defined implicitly, say by $F(x, y) = 0$, i.e., $F(x, f(x)) = 0$ for $x \in$ dom f (see §5 below for details), the differentiation of both sides of this identity leads to an equation for $f'(x)$, expressing it in terms of x and $y = f(x)$. Once partial derivatives and the chain rule for several variables have been introduced, the general form of this formula can be stated:

If $y = f(x)$ is defined by $F(x, y) = 0$, then

$$F_x + F_y \cdot y' = 0 \quad \text{and so} \quad f'(x) = -\frac{F_x(x, f(x))}{F_y(x, f(x))} = -\frac{F_x(x, y)}{F_y(x, y)}.$$

This so-called implicit differentiation completes the techniques of the differential calculus; *any elementary function can be easily differentiated.*

EXERCISE 14. Compute the second and third derivatives of a product fg. Guess, and prove inductively, a formula for the nth derivative of a product.

EXERCISE 15. Give a formula for the derivative of the product of n functions.

EXERCISE 16. Find the equation of the tangent line to the curve $x^{2/3} + y^{2/3} = 1$ at $(1/2\sqrt{2}, 1/2\sqrt{2})$.

Remark. The manipulation of the differential symbols—thinking of dy as representing an "infinitesimal" increment in the quantity y—provides a way to formulate conjectures, some of which can be verified rigorously. For example, suppose $y = f(x)$ and an "increment" dx starting at x yields

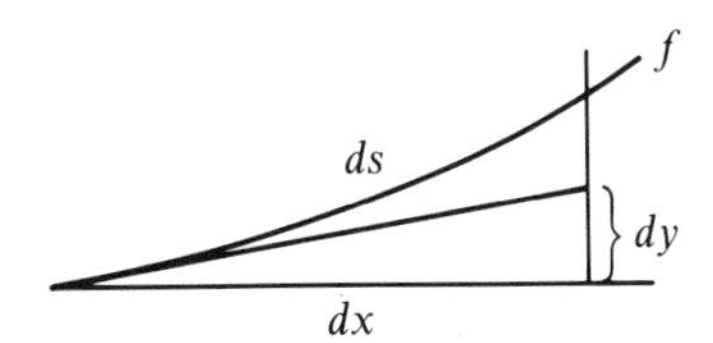

the "increment" dy, as in the figure. Then the graph of f and the segments dx and dy "approximately" form a right triangle with "hypotenuse" an "infinitesimal" increment ds in the arclength of the graph. Thus a formula involving the arclength suggests itself via the pythagorean theorem;

$$ds^2 = dx^2 + dy^2,$$

i.e., the rate of change (at x) of the arclength accumulating along the graph with respect to x is $ds/dx = \sqrt{1 + (y')^2}$. In view of the Fundamental Theorem of Calculus, this amounts to saying that the integral $\int_a^b \sqrt{1 + [f'(x)]^2}\, dx$ yields the length of the portion of the graph of f over the interval $[a, b]$. (This formula is established rigorously on the basis of a careful definition of arclength in Chapter 5.)

As an additional example, consider the "curvature" of the graph of f at x. The intuitive idea of curvature is the amount of direction change in the tangent line due to an "increment" in arclength, ds. Let $\theta(x)$ denote the

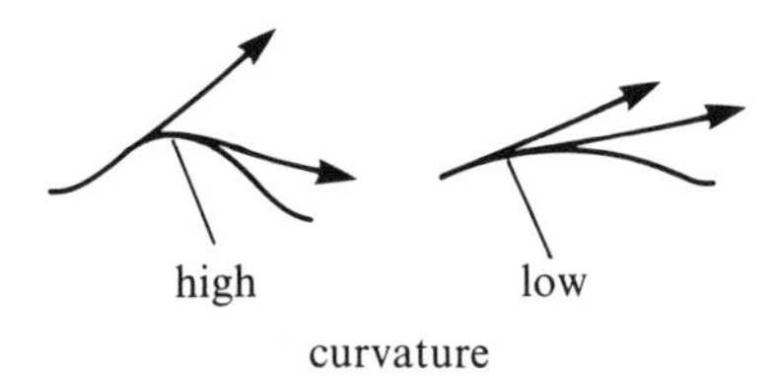

angle from the horizontal (positive x) direction to the tangent at x (direction that gives $-\pi/2 < \theta < \pi/2$). Then the curvature is $|(d\theta/ds)(x)|$ according to this intuition.

Now, since $y' = \tan\theta =$ slope of tangent,

$$\theta = \tan^{-1} y'.$$

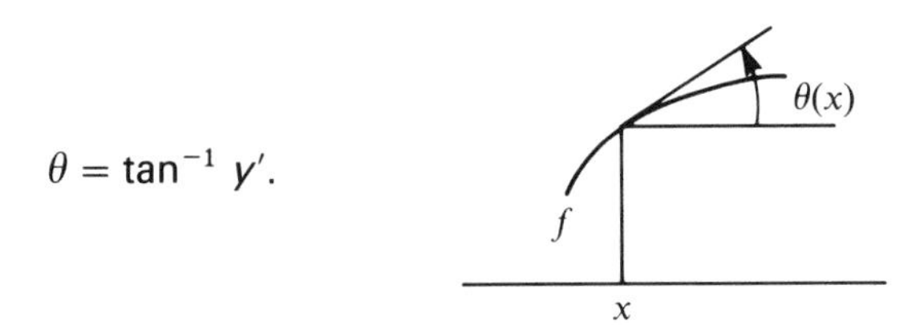

This suggests introducing x and y' as intermediate variables, to express $d\theta/ds$ in terms of derivatives with respect to x.

The chain rule in differential notation is

$$\frac{dy}{dx} = \frac{dy}{du}\cdot\frac{du}{dx}$$

where u is an intermediate variable between x and y, i.e., y is a function of u and u is a function of x. Applying this manipulation of differentials, we obtain a formula which can be validated in the context of a careful discussion of curves:

$$\begin{aligned} \text{curvature} = \left|\frac{d\theta}{ds}\right| &= \left|\frac{d\theta}{dy'}\cdot\frac{dy'}{dx}\cdot\frac{dx}{ds}\right| \\ &= \frac{1}{1+(y')^2}\cdot|y''|\cdot\frac{1}{\sqrt{1+(y')^2}} \\ &= \frac{|y''|}{[1+(y')^2]^{3/2}}. \end{aligned}$$

EXERCISE 17. Find the curvature of the graph of $\sin(1/x)$ at the points $2/\pi$ and $2/3\pi$.

§3. Curve Sketching. The Mean-Value Theorem

The Extreme-Value Theorem establishes the existence of extreme values of functions but does not help in finding the points where they occur. If we sketch some continuous graphs over an interval we notice that the extremes of the function are included among certain points that stand out—the "relative extreme" points, such as $A, B, \ldots, G$ in the figure. Moreover, these points seem always to be one of three kinds:

endpoints of the interval, A and G; "corners" in the graph, A, C, F, and G; points where the tangent line is horizontal, B, D, and E.

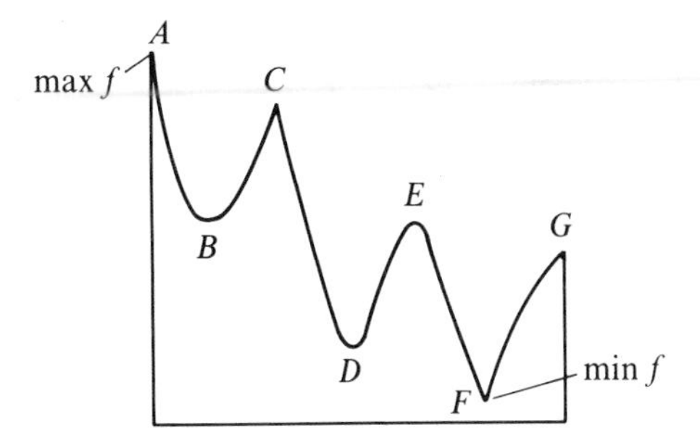

We make definitions and prove a theorem which make these observations precise.

Definition. A point, $a \in \operatorname{dom} f$ is a **relative maximum of** f if there is a neighborhood $\mathscr{U}$ of a such that

$$x \in \mathscr{U} \cap \operatorname{dom} f \quad \text{implies} \quad f(x) \leq f(a).$$

Reversing the inequality yields the definition of a **relative minimum of** f. The term **relative extreme** embraces both.

Clearly, $\max f$ and $\min f$, if they exist, occur at relative extremes of f (take $\mathscr{U} = \mathbb{R}$); locating all relative extremes is a first step toward locating the points where $\min f$ and $\max f$ occur (and is itself interesting).

The possibilities conjectured above will be recognized in the

Relative Extreme Theorem. *If a is a relative extreme of f that is interior to* $\operatorname{dom} f$ *then either f' is undefined at a or* $f'(a) = 0$.

Proof. We shall show that at a point a interior to $\operatorname{dom} f$ where f' exists, the alternatives $f'(a) > 0$ and $f'(a) < 0$ are incompatible with a being a relative extreme.

Suppose

$$f'(a) > 0, \qquad \text{i.e.,} \quad \lim_{x \to a} \frac{f(x) - f(a)}{x - a} > 0.$$

The sign of the difference quotient persists throughout some neighborhood of a, making $f(x) - f(a)$ and $x - a$ have the same sign throughout that neighborhood. Since $x - a$ changes sign in every neighborhood of a, so also must $f(x) - f(a)$; both $f(x) < f(a)$ and $f(x) > f(a)$ must occur in every neighborhood of a. Thus a is not a relative extreme.

The case $f'(a) < 0$ is similar.

EXERCISE 1. Supply the details.

This line of argument—from the sign of $f'(a)$, via persistence of sign of the difference quotient, to comparisons of values of f near a, has interesting additional consequences.

Proposition. *If f is differentiable on (a, b), then f' has the intermediate-value property on (a, b). That is, for any u, v in (a, b) and any value c between $f'(u)$ and $f'(v)$ there is a point γ between u and v such that $f'(\gamma) = c$.*

Proof. By considering $f(x) - cx$ we may change the intermediate value to zero, so we may suppose that $f'(u)$ and $f'(v)$ have opposite signs, where $a < u < v < b$, and show that $f'(\gamma) = 0$ must hold at some γ interior to $[u, v]$. But the two extreme values of f over $[u, v]$ exist and occur either at an endpoint or at a point where f' vanishes, under our assumptions on f. Thus the result follows if we show that it is impossible for both of these extremes to occur at endpoints. The indicated kind of argument achieves this, as the reader is invited to show in

EXERCISE 2. Take the case $f'(u) < 0 < f'(v)$ and show that:

(a) the min cannot occur at u; and
(b) if the max occurs at u then the min cannot occur at v.

The Relative Extreme Theorem can be rephrased informally as: In order to be a candidate for a relative extreme point, a number c must have one of the properties:

c is not interior to dom f,

f' does not exist at c, and

$f'(c) = 0$.

One recognizes the need for a test that tells whether *a* given candidate c is *in fact* a relative extreme, and, if so, whether it is a relative maximum or minimum.

By sketching some continuous graphs that include various points of the above three kinds, one observes that in each interval between an adjacent

pair of such points, the function is strictly monotonic. Moreover, if this is true in general then the desired test is an immediate consequence; e.g., increasing to the left of c and decreasing to the right implies c is a relative maximum.

Suppose a and b denote two **adjacent** candidates for relative extremes of f (i.e., $[a, b] \subset \operatorname{dom} f$ and no point of (a, b) is a candidate). This gives the conditions:

$$f \text{ is continuous on } [a, b],$$

$$f' \text{ exists on } (a, b), \text{ and}$$

$$f' \text{ never zero in } (a, b).$$

We wish to investigate the monotonicity of f under these hypotheses[1] by choosing any u and v in $[a, b]$ with $u < v$ and examining the sign of $f(v) - f(u)$. Since the knowledge at hand about f concerns f', we see the need for a theorem relating $f(v) - f(u)$ to f'.

We have the important

Mean-Value Theorem. *If f is continuous on $[u, v]$ and differentiable on (u, v) there is a $\xi \in (u, v)$ such that*

$$f(v) - f(u) = f'(\xi)(v - u).$$

Remarks. 1. The strict monotonicity of f between adjacent candidate points, and the desired test, the First Derivative Test, follow at once from this theorem.

EXERCISE 3. Write out the details: a statement and proof of sufficient conditions for an isolated candidate point to be a relative minimum, a relative maximum or neither.

2. In view of the above discussion, the weaker hypothesis, that f' exists on (u, v) rather than on $[u, v]$, is required for the applications of the theorem; it is not just a splitting of hairs, as one might suspect upon first encountering the theorem.

3. The theorem has a geometric interpretation (see figure), which makes it quite plausible. The quantity $[f(v) - f(u)]/(v - u)$ is the slope of the segment joining the points at the two ends of the graph of f over $[u, v]$. The conclusion is that the tangent to the graph must be parallel to that segment at some point (at least one) ξ interior to $[u, v]$. Note that this can only be avoided (experimentally!) by putting a corner in the graph.

4. The name Rolle's Theorem refers to the special case in which

[1] The third can be sharpened, in view of the intermediate value property of f', to: either $f'(x) < 0$ for all $x \in (a, b)$ or $f'(x) > 0$ for all $x \in (a, b)$).

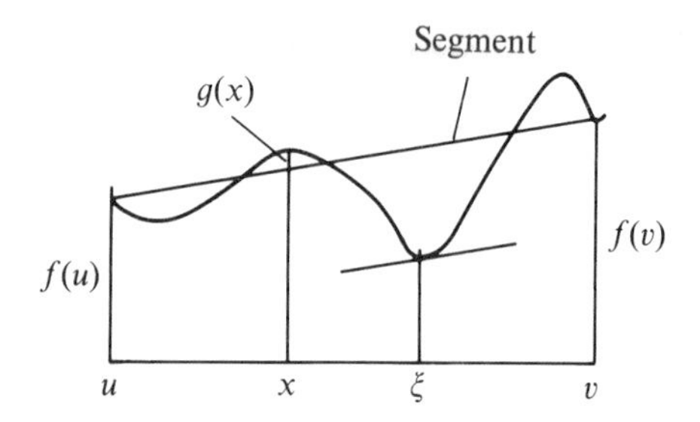

$f(u) = f(v)$ is assumed (sometimes both assumed $= 0$) and $f'(\xi) = 0$ for some $\xi \in (u, v)$ is inferred. The proof below begins by reducing the general case to this special case.

Proof. Consider the difference between the height of the graph of f at x and the height of the segment, namely,

$$g(x) = f(x) - \left[f(u) + \frac{f(v) - f(u)}{v - u}(x - u) \right].$$

Since

$$g'(x) = f'(x) - \frac{f(v) - f(u)}{v - u},$$

the conclusion we seek translates into

$$g'(\xi) = 0 \qquad \text{for some} \quad \xi \in (u, v).$$

Now $g(u) = g(v) = 0$ and g is continuous on $[u, v]$. The fact that g assumes its maximum and minimum values at points of $[u, v]$ tells us that either g is identically zero, or at least one of the extremes occurs at an interior point $\xi \in (u, v)$. In any case, there is a $\xi \in (u, v)$ at which $g'(\xi) = 0$.

Exercise 4. A candidate for a relative extreme of a function can be a limit point of the set of all candidates (there is no "adjacent" candidate). Show that this is the case for $c = 0$ and

$$f(x) = \begin{cases} x \sin(1/x), & x \neq 0, \\ 0, & x = 0. \end{cases}$$

If a candidate point c has another candidate d adjacent to it then we have shown that f is strictly increasing on $[c, d]$ (or $[d, c]$ as the case may be) if and only if $f'(\xi) > 0$ for some $\xi \in (c, d)$ (if and only if $f'(\xi) > 0$ for *every* $\xi \in (c, d)$) and similarly for "decreasing," and "< 0." Thus a single evaluation of just the sign of f' at convenient points on either side of c will reveal the nature of the graph of f near c. The possible appearances of the graph include the following:

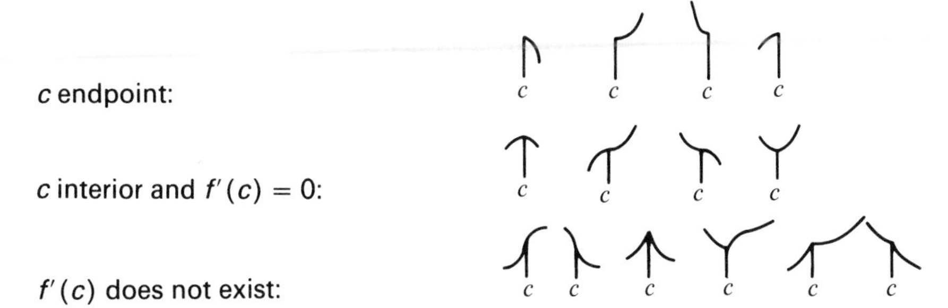

The First Derivative Test of a candidate for a relative extreme consists of inferring the appearance of the graph near $(c, f(c))$ from the nature of c and the nearby values of the sign of f' in this manner. By applying it and evaluating f at a few points, one can often sketch the major features of the graph of f using only a modest amount of computational effort. The term used for such graphical analyses via derivatives is "curve sketching."

We have commented on the role of Analysis in applications of mathematics as a source of mathematical models for various phenomena. The output of such models is often a formula for a function, representing the variations of an important quantity. An equally important application of Analysis is that of extracting from these formulas the relevant information about the variation of the quantities. The curve sketching techniques are powerful tools for extracting and representing such information; making drawings from formulas.

EXERCISE 5. An automobile's suspension system is modeled by a problem in differential equations, which yields the solution

$$d(t) = 8e^{-6t} \sin 3t, \qquad t > 0,$$

to describe the vertical displacement $d(t)$ of the chassis from its equilibrium position, t seconds after hitting a bump. Find the maximum displacement and when it occurs. Estimate the time beyond which the displacement remains less than 0.1.

The curve sketching techniques are augmented by exploiting the second derivative.

Definition. The graph of a differentiable function is **concave up (down) at** x_0 if the graph is *strictly* above (below) the tangent line at x_0 at all points of some neighborhood of x_0.

EXERCISE 6. Sketch some graphs (over an interval) that have tangents whose slopes increase with increasing x. Observe that the graph is always concave up, whether the graph itself increases or decreases. Thus justify the geometric terminology indicated by

$$f \text{ concave up} \leftrightarrow f' \text{ increasing.}$$

Similarly,

$$f \text{ concave down} \leftrightarrow f' \text{ decreasing.}$$

EXERCISE 7. Suppose f'' exists on (a, b). Show that f is concave up on (a, b) if and only if $f''(x) > 0$ for $x \in (a, b)$.

A point where the concavity changes is called a **point of inflection**. (Strictly speaking, "point" here refers to some $(c, f(c))$ on the graph, but no confusion results from referring to the "point" c.)

EXERCISE 8. Show that if f'' exists, it must vanish at an inflection point. Show that f'' can vanish at a noninflection point.

EXERCISE 9. Prove the **Second Derivative Test**: If $f'(a) = 0$ and a is interior to dom f' then if $f''(a)$ exists:

1. $f''(a) > 0$ implies a is a relative minimum;
2. $f''(a) < 0$ implies a is a relative maximum; and
3. $f''(a) = 0$ can occur for a a relative minimum, a relative maximum or neither.

EXERCISE 10. Show that two functions that have the same derivative on an interval (a, b) differ by a constant there. Hint: Apply the Mean Value Theorem to the difference.

EXERCISE 11. Suppose f is differentiable on an interval (a,b). Show that the continuity of f is uniform on (a, b) if the derivative f' is bounded there.

EXERCISE 12. Sketch the graph of

$$f(x) = \frac{\sqrt[3]{x - 2(x + 1)}}{x - 3}.$$

Newton's Method for solving an equation $f(x) = 0$ is the iterative procedure, or algorithm, based on:

1. Choosing a first guess x_1 (use curve sketching!).

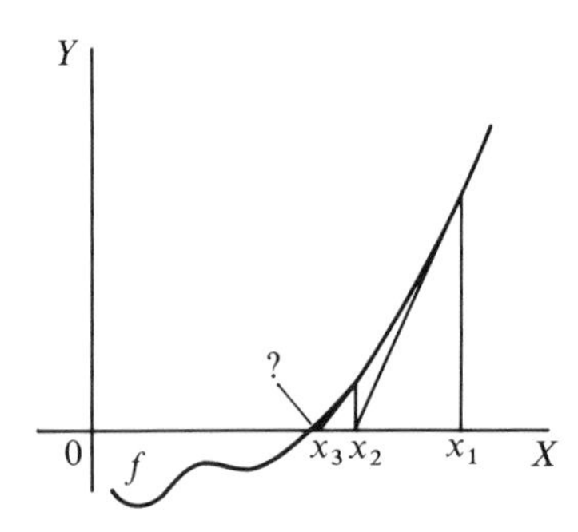

2. Following the tangent line at $(x_1, f(x_1))$ to the X-axis, to find a second approximation x_2 to the solution, possibly better than the first.
3. Iterating the procedure, to produce a sequence of successive approximations $\{x_n\}$ which may converge to the solution.

EXERCISE 13. Write the equation of the tangent line at $(x_1, f(x_1))$ and find its X intercept x_2. Iterate to obtain the formula

$$x_{n+1} = x_n - \frac{f(x_n)}{f'(x_n)}, \qquad n = 1, 2, 3, \ldots.$$

What is the geometric significance of the fact that the iteration breaks off when $f'(x_n) = 0$ for some n?

EXERCISE 14. Use $f(x) = x^2 - 2$ and $x_1 = 1$ to approximate $\sqrt{2}$ by Newton's Method. Compare with the sequence given by $x_{n+1} = \frac{1}{2}(x_n + 1/x_n)$.

EXERCISE 15. Comment on the outcome if f and x_1 are as shown.

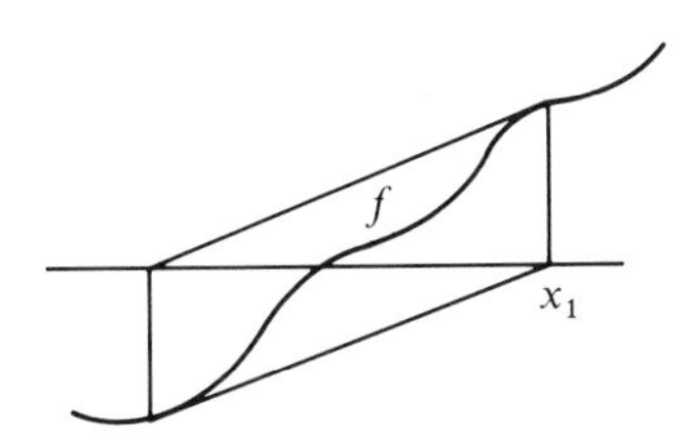

EXERCISE 16. Estimate the location of all solutions of

$$3x^4 - 20x^3 + 48x^2 - 48x + 3 = 0$$

by curve sketching.

EXERCISE 17. A curve sketching problem on a Calculus test is to have the answer shown. What is the question (a polynomial of degree 3)?

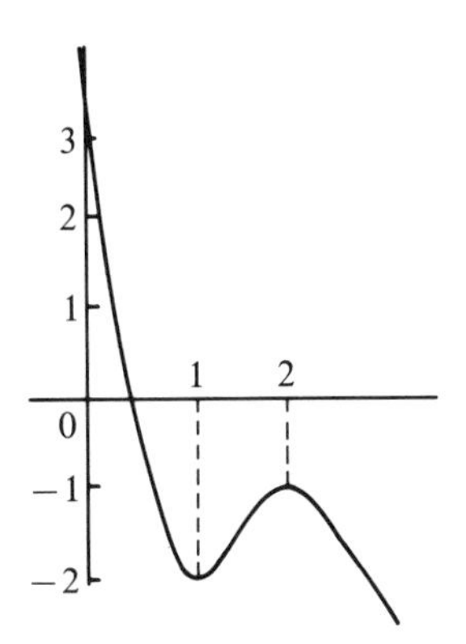

§4. Taylor's Theorem

Most applications of mathematics ultimately require the precise evaluation of functions by numerical means at some points of their domains. The curve sketching techniques can often provide a rough graph using only coarse numerical evaluations. That rough graph, in fact, often serves to identify the precise calculations that are worth doing—ranges of x on which the expenditure of computational resources finding $f(x)$ accurately would be a waste, are easily identified, at the very least.

The task of computing a value $f(x)$ to a prescribed accuracy calls for a combination of Arithmetic and Approximation; the approximation of functions by rational functions.

We begin our discussion of this subject by observing that the division operation can be bypassed in the arithmetic, and the question reduced to that of approximating functions by polynomials. This is because the reciprocal function can be approximated by polynomials, in view of the identity

$$\frac{1}{1-x} = 1 + x + x^2 + \cdots + x^n + \frac{x^{n+1}}{1-x}, \qquad n \in \mathbb{N}.$$

For any nonzero $a \in \mathbb{R}$,

$$\frac{1}{y} = \frac{1}{a-(a-y)} = \frac{1}{a}\sum_{k=0}^{n}\left(1-\frac{y}{a}\right)^k + \frac{(1-(y/a))^{n+1}}{y}, \qquad y \neq 0,$$

expresses the reciprocal function as a polynomial plus the error term $(1-(y/a))^{n+1}/y$, which approaches 0 as $n \to \infty$, for y between 0 and $2a$.

EXERCISE 1. Verify this assertion.

Remark. This is local approximation near a, which will be our focus in what follows. There is also the notion of seeking approximation uniformly on a prescribed interval. Indeed, if f is continuous on that interval, the Weierstrass Approximation Theorem asserts the possibility of such an approximation of f by polynomials. Moreover, the Bernstein polynomials $B_n(f, x)$ (see p. 124) achieve this uniform approximation (other polynomials that do can also be found).

The defect in this resolution of the problem is that, in general, polynomials that approximate f uniformly on an interval can only be written through the use of values of f at points throughout that interval; a certain circularity in the task of computation arises. This reasonable observation is illustrated by the case of the Bernstein polynomials $B_n(f; x)$. The values of f at the $n+1$ equally spaced points in the interval are required in order to write $B_n(f; x)$. Thus, for example, to approximate sin 1 via the Bernstein polynomials of sin x on $[0, \pi]$, one needs to know all the values $\sin k\pi/2n$, $k = 0, 1, 2, \ldots, n$, for a number of indices n.

We shall see that the local approximation of f near a uses values of f and its derivatives at a. For many functions f, such information is available at various points a, as the trigonometric functions illustrate.

EXERCISE 2. Find the Bernstein polynomial on $[0, 1]$ of degree 4 for the function $1/(x-2)$.

In order to consider the local approximation of f near a, we put $x = a + h$, observing that polynomials in x become polynomials in h and vice versa.

Approximating $f(a+h)$ near $h = 0$ by the polynomial $a_0 + a_1 h + a_2 h^2 + \cdots + a_n h^n$ gives the error

$$e_n(h) = f(a+h) - (a_0 + a_1 h + \cdots + a_n h^n).$$

Local approximation

$$\lim_{h \to 0} e_n(h) = 0$$

translates into

$$a_0 = \lim_{h \to 0} f(a+h),$$

which is the first of the following series of observations:

1. $\lim_{h\to 0} e_n(h) = 0$ if and only if f is continuous at a and $a_0 = f(a)$.
2. Then
$$\frac{e_n(h)}{h} = \frac{f(a+h) - f(a)}{h} - (a_1 + a_2 h + \cdots + a_n h^{n-1})$$
and so $e_n(h) = o(h)$ if and only if $f'(a)$ exists and
$$a_1 = f'(a).$$
3. If f' exists on a neighborhood of a then, differentiating the formula for e_n gives
$$e_n'(h) = f'(a+h) - f'(a) - (2a_2 h + 3a_3 h^2 + \cdots + na_n h^{n-1})$$
and so step 2 gives
$$\frac{e_n'(h)}{h} = \frac{f'(a+h) - f'(a)}{h} - (2a_2 + 3a_3 h + \cdots + na_n h^{n-2}).$$
Thus $e_n'(h) = o(h)$ if and only if $f''(a)$ exists and
$$a_2 = \tfrac{1}{2} f''(a).$$
4. By induction, $e_n^{(m-1)}(h) = o(h)$ if and only if $f^{(m)}(a)$ exists and
$$a_m = \frac{1}{m!} f^{(m)}(a).$$

EXERCISE 3. Supply details.

Definition. For any f such that $f^{(n)}(a)$ exists, the n**th Taylor polynomial of** f **at** a is

$$T_n(f; a)(x) = f(a) + f'(a)(x - a) + \frac{f''(a)}{2!}(x - a)^2 + \cdots + \frac{f^{(n)}(a)}{n!}(x - a)^n.$$

The coefficients are the **Taylor coefficients of** f **at** a.

The error when f is approximated near a by T_n satisfies

$$e_n^{(k)}(h) = o(h), \qquad k = 0, 1, 2, \ldots, n - 1;$$

the function and its first $n - 1$ derivatives are approximated to first order near a by T_n and its first $n - 1$ derivatives, respectively. Moreover, T_n is the unique polynomial of degree n having this property. This summarizes the series of observations above.

The order of approximation here is clarified by the

Lemma. *If* $g(h)$, *with* $g(0) = 0$, *is differentiable in a neighborhood of* 0 *and* $g'(h) = o(h^k)$ *then* $g(h) = o(h^{k+1})$.

Proof. For $|h|$ small enough, the Mean-Value Theorem applies on the closed interval joining 0 and h. It gives

$$0 \leq \left|\frac{g(h)}{h^{k+1}}\right| = \left|\frac{g'(\xi)h}{h^{k+1}}\right| \qquad \text{for some } \xi \text{ between 0 and } h.$$

Thus

$$0 \leq \left|\frac{g(h)}{h^{k+1}}\right| = \left|\frac{g'(\xi)}{\xi^k}\left(\frac{\xi}{h}\right)^k\right| < \left|\frac{g'(\xi)}{\xi^k}\right|.$$

Now $h \to 0$ forces $\xi \to 0$ and the result follows.

Thus the fact that $e_n^{(n-1)}(h) = o(h)$ implies that $e_n(h) = o(h^n)$: *the* nth *Taylor polynomial of* f *at* a *approximates* f *of order* n *locally at* a.

Remarks. 1. Geometrically, the tangent line

$$f(a) + f'(a)(x - a),$$

which is $T_1(f; a)$, is the best *linear* approximation to f, locally at a. It shares value and slope with f at a. But unless f is itself linear, it is clear that some parabola with vertical axis, sharing value, slope, and concavity (or curvature) with f at a, must approximate f near a better than any line. The parabola which is the graph of the Taylor quadratic

$$f(a) + f'(a)(x-a) + \tfrac{1}{2}f''(a)(x-a)^2$$

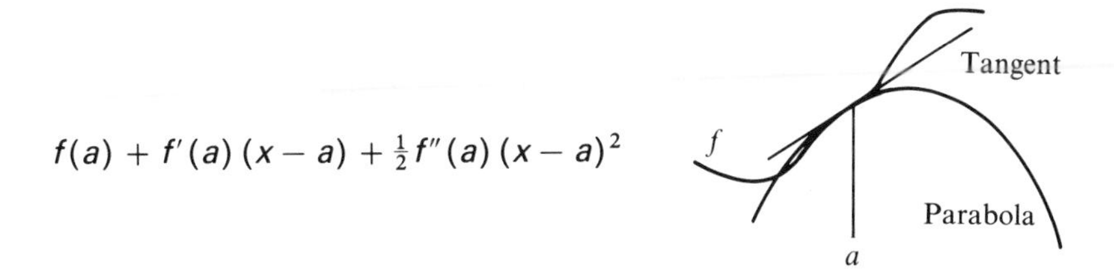

shares these features with f and gives a second-order approximation to f near a, in accord with this expectation.

2. If the values of f and its derivatives to order n are known at a then $T_n(f; a)$ expresses the arithmetic that will give approximate values for $f(x)$ when x is near a.

EXERCISE 4. Find sin 1 approximately, using a fourth-degree polynomial for the arithmetic and the values $\sin(\pi/3) = \sqrt{3}/2$, $\cos(\pi/3) = \frac{1}{2}$. Express your answer in terms of $\sqrt{3}$ and π.

EXERCISE 5. Contemplate the fact that the phrase "approximately equal," without an accompanying statement about the error or accuracy, lacks precise meaning. *Any* number is approximately equal to sin 1, most are just terribly inaccurate values for sin 1. The effort expended on the preceding exercise is only validated once a statement about the accuracy of the answer can be established.

EXERCISE 6. Find the general Taylor polynomial for $\log(1 + x)$ at $x = 0$. Derive six successive approximate values for log 2. Comment on their accuracies, given that $\log 2 = 0.693147\ldots$.

The mere fact of a local approximation, even of high order, does not mean that a "reasonable" approximation is achieved by a "reasonable" amount of arithmetic, as the preceding exercise shows. One faces a three-way tradeoff: between accuracy, the range of values x over which the desired accuracy holds and the amount of arithmetic (degree of polynomial) required to achieve it. The fact that f is locally approximated by T_n must be supplemented by usable estimates of the error, of a kind that will allow the tradeoff to be made.

Such estimates of the error result from analyses of the functions $e_n(h)$. We undertake one such analysis next.

Since $e_n(h) = o(h^n)$ and the simplest function that is $o(h^n)$ is h^{n+1}, it makes sense to seek an expression for $e_n(h)$ of the form $M(h)h^{n+1}$. This means that

$M(h)$ is defined by the equation

$$f(a+h) = f(a) + f'(a)h + \frac{f''(a)}{2!}h^2 + \cdots + \frac{f^{(n)}(a)}{n!}h^n + M(h)h^{n+1}.$$

Now, take $M(h)$ as the coefficient of t^{n+1} in forming the "augmented Taylor polynomial"

$$f(a) + f'(a)t + \frac{f''(a)}{2!}t^2 + \cdots + \frac{f^{(n)}(a)}{n!}t^n + M(h)t^{n+1}.$$

The error upon approximating $f(a+t)$ by this polynomial is

$$E(t) = f(a+t) - \left[f(a) + f'(a)t + \cdots + \frac{f^{(n)}(a)}{n!}t^n\right] - M(h)t^{n+1}.$$

This E clearly vanishes at $t = 0$, and the very meaning of $M(h)$ says that it vanishes at $t = h$; Rolle's Theorem applies to E on the closed interval joining 0 and h.

Remark. Observe also that the $(n+1)$st derivative of E is $f^{(n+1)} - (n+1)!\,M(h)$, so that an explicit formula for $M(h)$ will result if we show that this derivative vanishes at some value η of t;

$$M(h) = \frac{f^{(n+1)}(a+\eta)}{(n+1)!}.$$

Rolle's Theorem gives a ξ_1 between 0 and h for which $E'(\xi_1) = 0$. Since

$$E'(t) = f'(a+t) - \left[f'(a) + f''(a)t + \cdots + \frac{f^{(n)}(a)}{(n-1)!}t^{n-1}\right] - (n+1)M(h)t^n,$$

it is clear that $E'(0) = 0$, so that Rolle's Theorem applies to E' on the closed interval joining 0 and ξ_1. There is, therefore, a ξ_2 between 0 and ξ_1, hence also between 0 and h, such that $E''(\xi_2) = 0$.

Differentiating again, we see that $E''(0) = 0$, so that Rolle's Theorem applies to E'' on the closed interval joining 0 and ξ_2, to give $E'''(\xi_3) = 0$ for some ξ_3 between 0 and ξ_2, hence also between 0 and h.

Continuing, we obtain a ξ_n between 0 and h such that $E^{(n)}(\xi_n) = 0$. Calculation shows that $E^{(n)}(0) = 0$ so that Rolle's Theorem applies to $E^{(n)}$, on the assumption that $f^{(n+1)}$ exists between 0 and h, to finally yield the existence of an η between 0 and h such that $E^{(n+1)}(\eta) = 0$;

$$f^{(n+1)}(a+\eta) - (n+1)!\,M(h) = 0;$$

$$M(h) = \frac{f^{(n+1)}(a+\eta)}{(n+1)!}.$$

We have proved an important theorem, which we state using ξ for $a + \eta$.

Taylor's Theorem. *If f has n derivatives continuous on the closed interval joining a and $a + h$ and if $f^{(n+1)}$ exists on the interior, then there is a value ξ in the interior such that*

$$f(a + h) = f(a) + f'(a)h + \frac{f''(a)}{2!}h^2 + \cdots + \frac{f^{(n)}(a)}{n!}h^n + \frac{f^{(n+1)}(\xi)}{(n + 1)!}h^{n+1}.$$

The formula is called **Taylor's formula** with the Lagrange form of the remainder. In x notation it reads

$$f(x) = f(a) + f'(a)(x - a) + \cdots + \frac{f^{(n)}(a)}{n!}(x - a)^n + \frac{f^{(n+1)}(\xi)}{(n + 1)!}(x - a)^{n+1}.$$

The $n = 0$ case of Taylor's Theorem is just the Mean-Value Theorem; Taylor's Theorem gives generalizations of the Mean-Value Theorem.

An estimate of the error $e_n(h)$ is available from the Lagrange expression in case $f^{(n+1)}$ is bounded on an interval I containing a in its interior; namely,

$$|e_n(h)| \leq \frac{1}{(n + 1)!} \sup_{t \in I} |f^{(n+1)}(t)| \, |h|^{n+1}.$$

Other formulas for the remainder can be found and used to estimate $|e_n(h)|$ in other ways. An excellent reference is E. Hille [14].

Exercise 7. Find an upper bound for the error in the value of sin 1 computed in Exercise 4. Is that value correct to within an error of ± 0.0000005?

Exercise 8. Estimate the error in the value of log 2 obtained by using the nth Taylor polynomial of $\log(1 + x)$ about $x = 0$. Find a value of n that will insure an error no larger than ± 0.05.

Exercise 9. Suppose f'' exists in some neighborhood of a. Use Taylor's formula with $n = 1$ to show that $f''(a) > 0$ implies the graph is concave up at $(a, f(a))$.

The Mean-Value Theorem may be viewed as saying that the ratio of the change in $f(x)$ to that in x over an interval $[a, b]$, $[f(b) - f(a)]/(b - a)$, is realized by some value of f' taken in (a, b). A generalization of a kind different from that provided by Taylor's Theorem concerns replacing x by a $g(x)$ and considering the ratio $[f(b) - f(a)]/[g(b) - g(a)]$. With suitable hypotheses on f and g, the Mean-Value Theorem gives

$$\frac{f(b) - f(a)}{g(b) - g(a)} = \frac{f(b) - f(a)}{b - a} \cdot \frac{b - a}{g(b) - g(a)} = f'(\xi) \cdot \frac{1}{g'(\eta)}$$

for some ξ and η in (a, b). The usefulness of this observation is limited by the fact that ξ and η can be quite unrelated—we only know one thing about them; that they both lie in (a, b). The question arises, whether there is a single ξ in (a, b) giving, under suitable hypotheses, the result

$$\frac{f(b) - f(a)}{g(b) - g(a)} = \frac{f'(\xi)}{g'(\xi)}.$$

Rewriting this as

$$g'(\xi)\,[f(b) - f(a)] - f'(\xi)\,[g(b) - g(a)] = 0$$

presents the idea of applying Rolle's Theorem to the function

$$g(x)\,[f(b) - f(a)] - f(x)\,[g(b) - g(a)].$$

EXERCISE 10. The **Cauchy Mean-Value Theorem** has the conclusion

$$\frac{f(b) - f(a)}{g(b) - g(a)} = \frac{f'(\xi)}{g'(\xi)} \qquad \text{for some} \quad \xi \in (a, b).$$

Formulate suitable hypotheses and give a proof.

These generalizations of the Mean-Value Theorem provide techniques for evaluating indeterminate forms. The treatise of Hille (loc.cit.) discusses these ideas at some length. We shall only recall the statement of

L'Hôpital's Rule. *Let f and g be differentiable in a neighborhood of 0 and suppose $g(x) \neq 0$ there. Let $\lim_{x \to a} f(x) = \lim_{x \to a} g(x) = 0$ so that $\lim_{x \to a} [f(x)/g(x)]$ has the "indeterminate form 0/0". Then whenever $\lim_{x \to a} [f'(x)/g'(x)]$ exists, so does $\lim_{x \to a} [f(x)/g(x)]$ and they have the same value.*

EXERCISE 11. Provide the (brief) proof.

§5. Functions Defined Implicitly

The task of "solving equations" is a central one in Mathematics and its applications. If x and y are real variables, an equation to be solved for y in terms of x can be written $F(x, y) = 0$, for some function of two variables F. Analysis asks not only for the existence and uniqueness of a solution, but also *how* the solution y depends on the "data" x: continuously, differentiably, C^1, etc. We are interested in conditions on F which imply the existence of a solution function $y = f(x)$ defined and continuous (differentiable, C^1, etc.) on a "maximal" domain. (Recall that this approach to the definition and properties of the functions $\sqrt[n]{x}$ via $F(x, y) = y^n - x$ has been discussed in Chapter 3, §1.)

In order to state appropriate conditions on F, we must digress briefly to define continuity and differentiability for functions of two real variables, i.e., functions F: $X-Y$ plane $\to \mathbb{R}$, where the $X-Y$ plane refers to the set of all ordered pairs of real numbers; geometrically, all points of a plane in which a rectangular coordinate system has been chosen.

Definition. In the $X-Y$ plane, the **open disk** with center (x_0, y_0) and radius $\delta > 0$ is the set

$$\{(x, y): \sqrt{(x - x_0)^2 + (y - y_0)^2} < \delta\}.$$

Definition. A **neighborhood of** (x_0, y_0) is any set containing an open disk of positive radius with center (x_0, y_0).

Definition. An **open set** in the $X-Y$ plane is a set which is a neighborhood of each of its points.

EXERCISE 1. A set G is open if and only if for each $(x_0, y_0) \in G$ there is an open *rectangle* of center (x_0, y_0) contained in G, i.e., a set $\{(x, y): |x - x_0| < \delta_1$ and $|y - y_0| < \delta_2\}$ for some $\delta_1 > 0$ and $\delta_2 > 0$.

Definition. F: $X-Y$ *plane* $\to \mathbb{R}$ is **continuous at** $(x_0, y_0) \in \operatorname{dom} F$ if for every $\varepsilon > 0$ there is a δ such that

$$(x, y) \in \operatorname{dom} F \quad \text{and} \quad \sqrt{(x - x_0)^2 + (y - y_0)^2} < \delta$$

implies

$$|F(x, y) - F(x_0, y_0)| < \varepsilon.$$

F is **continuous** if F is continuous at every $(x_0, y_0) \in \operatorname{dom} F$.

EXERCISE 2. F is continuous if and only if the inverse image under F of every open set in $\mathbb{R}$ is open relative to dom F.

For any (x_0, y_0) such that each of its neighborhoods intersects dom F, it makes sense to consider the idea of the limit of F as (x, y) approaches (x_0, y_0). Assuming such an (x_0, y_0) we make the

Definition. L **is the limit of** F **as** (x, y) **approaches** (x_0, y_0), in symbols

$$L = \lim_{(x,y)\to(x_0,y_0)} F(x, y),$$

if for every $\varepsilon > 0$ there is a $\delta > 0$ such that $(x, y) \in \operatorname{dom} F \cap$ the δ disk at (x_0, y_0) implies $|F(x, y) - L| < \varepsilon$.

The continuity of F at (x_0, y_0) is then simply the property

$$F(x_0, y_0) = \lim_{(x,y)\to(x_0,y_0)} F(x, y).$$

The persistence of sign holds, as the

Proposition. *If* $\lim_{(x,y)\to(x_0,y_0)} F(x, y) > 0$ *then there is a* $\delta > 0$ *such that* $(x, y) \in$ dom F *and* $\sqrt{(x - x_0)^2 + (y - y_0)^2} < \delta$ *implies* $F(x, y) > 0$.

EXERCISE 3. Give a proof.

EXERCISE 4. Define continuity for a function of the form F: X–Y plane → U–V plane. Suppose $F(x, y) = (u, v)$, so that u and v are real-valued functions of (x, y), **the coordinate functions of** F, $u(x, y)$ and $v(x, y)$. Discuss the connection between the continuity of u and v and that of F.

It is immediate that a continuous F: X–Y plane → $\mathbb{R}$ is **separately continuous** in the sense that both:

1. for x fixed, the function $y \to F(x, y)$ is a continuous function of $\mathbb{R}$ into $\mathbb{R}$; and
2. for y fixed, the function $x \to F(x, y)$ is a continuous function of $\mathbb{R}$ into $\mathbb{R}$.

EXERCISE 5. Make these statements precise as regards the domains and prove the assertion.

Separately continuous does not imply continuous however, as the following example shows.

EXAMPLE. The function $\Phi(x, y)$ given by

$$\Phi(x, y) = \begin{cases} \dfrac{xy}{x^2 + y^2} & \text{for } (x, y) \neq (0, 0), \\ 0 & \text{for } (x, y) = (0, 0) \end{cases}$$

is separately continuous, but is not continuous at (0, 0).

EXERCISE 6. Prove this assertion. Hint: The value is $\frac{1}{2}$ at all points (x, x) except (0, 0).

Remark. This observation shows that the continuity of $F(x, y)$ is an "essentially two-dimensional" property, not just the conjunction of two one-dimensional properties. As we proceed to the differentiability, we shall find that a similar remark applies.

Differentiability is defined to embody the idea of approximating the increments in the value of F, due to increments h and k in x and y, by a *linear* expression in those increments: $ah + bk$, for constants a and b. The approximation is required to be of "order one" in the distance $\sqrt{h^2 + k^2}$, the "size" of the increment in (x, y). Thus, the

Definition. Let (x_0, y_0) be contained in some open disk contained in dom F. F **is differentiable at** (x_0, y_0) if there is a linear expression $ah + bk$ such that

$$\lim_{(h,k)\to(0,0)} \frac{[F(x_0 + h, y_0 + k) - F(x_0, y_0)] - (ah + bk)}{\sqrt{h^2 + k^2}} = 0.$$

This is two dimensional. By fixing one of the variables and considering the differentiation of the resulting function of the other variable, we obtain the two one-dimensional ideas, the **partial derivatives**, of F at (x_0, y_0):

$$F_x(x_0, y_0) = \lim_{h\to 0} \frac{F(x_0 + h, y_0) - F(x_0, y_0)}{h}$$

and

$$F_y(x_0, y_0) = \lim_{k\to 0} \frac{F(x_0, y_0 + k) - F(x_0, y_0)}{k},$$

when the limits exist. The differentiation formulas, Mean-Value Theorem, etc., are all at our disposal in dealing with the partial derivatives.

Suppose F is differentiable at (x_0, y_0) and take the limit $(h, k) \to (0, 0)$ in the special way in which k is always $= 0$. The differentiability condition reduces to the statement that $F_x(x_0, y_0)$ exists and that $a = F_x(x_0, y_0)$. Similarly, the special way $(0, k) \to (0, 0)$ shows that $F_y(x_0, y_0)$ exists and gives the value of b. Thus, the only possible linear expression $ah + bk$ for which the differentiability can hold is **the differential of** F **at** (x_0, y_0):

$$F_x(x_0, y_0)h + F_y(x_0, y_0)k.$$

We denote this linear expression by $dF_{(x_0, y_0)}(h, k)$. It can be written whenever the two partials exist and it fulfills the differentiability requirement whenever F is differentiable at (x_0, y_0).

It is natural to ask whether the existence of both partials at (x_0, y_0) insures the differentiability there. (Do the two one-dimensional properties imply the two-dimensional property?)

EXERCISE 7. Show that the function Φ of the example above has both partials at $(0, 0)$ equal to zero, hence that $d\Phi_{(0,0)}(h, k) \equiv 0$.

EXERCISE 8. Show that the differentiability of Φ at $(0, 0)$ reduces to the requirement that

$$\lim_{(h,k)\to(0,0)} \frac{hk}{(h^2 + k^2)^{3/2}} = 0.$$

That this is false is just Exercise 6 again!

Having seen by this example that the differentiability at (x_0, y_0) does not follow from the mere existence of the two partials there (two one-dimensional properties) we proceed to show that it does follow from mild two-dimensional assumptions.

Proposition. *If F_x and F_y both exist on a neighborhood of (x_0, y_0) and are both continuous at (x_0, y_0) then F is differentiable at (x_0, y_0).*

Proof. We choose a closed rectangle (with sides parallel to the axes and of positive length) on which both F_x and F_y exist. Taking $(x_0 + h, y_0 + k)$ in this rectangle, we estimate the difference

$$[F(x_0 + h, y_0 + k) - F(x_0, y_0)] - [F_x(x_0, y_0)h + F_y(x_0, y_0)k]$$

between the increment in F and the differential. By adding and subtracting the term $F(x_0, y_0 + k)$:

$$[F(x_0 + h, y_0 + k) - F(x_0, y_0 + k)] + [F(x_0, y_0 + k) - F(x_0, y_0)] - [F_x(x_0, y_0)h + F_y(x_0, y_0)k],$$

two terms appear which can be expressed, via the Mean-Value Theorem, in terms of F_x and F_y. There exist ξ and η satisfying $|\xi| < |h|$ and $|\eta| < |k|$ such that the expression can be written

$$[F_x(\xi, y_0 + k)h + F_y(x_0, \eta)k] - [F_x(x_0, y_0)h + F_y(x_0, y_0)k].$$

The differentiability requirement takes the form

$$\lim_{(h,k)\to(0,0)} \left\{[F_x(\xi, y_0 + k) - F_x(x_0, y_0)]\frac{h}{\sqrt{h^2 + k^2}} + [F_y(x_0, \eta) - F_y(x_0, y_0)]\frac{k}{\sqrt{h^2 + k^2}}\right\} = 0.$$

Since $(h, k) \to (0, 0)$ implies $(\xi, \eta) \to (0, 0)$, the assumed (two-dimensional!) continuity of F_x and F_y at (x_0, y_0) gives this result, since the factors $h/\sqrt{h^2 + k^2}$ and $k/\sqrt{h^2 + k^2}$ remain in $[-1, 1]$.

Definition. F is **continuously differentiable**, or C^1, on an open set G if F_x and F_y exist and are continuous on G.

EXERCISE 9. Show that f is differentiable everywhere and that f_x and f_y are not continuous at (0, 0), where

$$f(x, y) = \begin{cases} x^2 \sin\frac{1}{x} + y^2 \sin\frac{1}{y} & \text{if } xy \neq 0, \\ x^2 \sin\frac{1}{x} & \text{if } x \neq 0 \text{ and } y = 0, \\ y^2 \sin\frac{1}{y} & \text{if } x = 0 \text{ and } y \neq 0, \\ 0 & \text{if } (x, y) = (0, 0). \end{cases}$$

This differentiable f is not C^1 on any neighborhood of $(0, 0)$.

We resume our discussion of implicitly defined functions with a formal definition.

Definition. A function $f: \mathbb{R} \to \mathbb{R}$ **is defined implicitly by** F (or by the equation $F(x, y) = 0$) if

$$x \in \operatorname{dom} f \quad \text{implies} \quad F(x, f(x)) = 0.$$

This says that the graph of f is contained in the set

$$N_F = \{(x, y): F(x, y) = 0\}$$

called the **nullset of** F. The problem can be viewed as the search for graphs contained in nullsets: Find conditions on F that insure the existence of subsets of N_F which are the graphs of functions f, continuous, differentiable, C^1, etc., on a "maximal" domain.

Some idea of the variety of plane sets which can occur as N_F, even for simple functions F, is seen in the

EXERCISE 10. Sketch N_F for:

(i) $F(x, y) = x^2 + y^2$;
(ii) $F(x, y) = x^2 - 1$;
(iii) $F(x, y) = y^2 - 1$;
(iv) $F(x, y) = x - \sin y$; and
(v) $F(x, y) = \sin x + \sec y$.

A sketch of a possible nullset helps to formulate the kind of theorem one should seek. Suppose N_F is a curve in the plane which has several "branches," i.e., sections cut only once by vertical lines, such as those

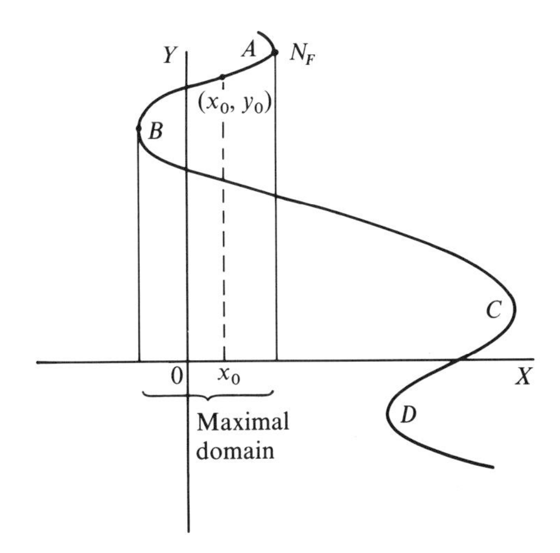

joining the labeled points in the figure. A graph could be chosen which "jumps between branches" as x varies, but not if we require continuity. By doing so, we focus on a single "branch" of N_F. The specification of the branch can be made by choosing a point (x_0, y_0) on it. This means the points like A, B, C, D would be avoided because they belong to two branches.

EXERCISE 11. Recall from Calculus that if F is differentiable then the gradient vector $(F_x(x_0, y_0), F_y(x_0, y_0))$ is orthogonal to N_F at (x_0, y_0), hence that the points to be avoided could be characterized by the condition (gradient horizontal there, or tangent vertical) $F_y(x_0, y_0) = 0$.

The fixing of (x_0, y_0) with $F_y(x_0, y_0) \neq 0$ selects one of possibly several implicit functions f and assigns its value at x_0: $f(x_0) = y_0$. It is natural to expect to be able to extend the definition *to a neighborhood of* x_0 by assuming a local property of F such as differentiability or C^1, i.e., it is reasonable to seek a *local* theorem, defining f near x_0. *Then* one would seek an extension to the maximal domain suggested in the figure, by applying the local theorem at points near x_0 where f is then defined, letting, say, some $(x_1, f(x_1))$ on the branch assume the role of (x_0, y_0) to extend to a neighborhood of x_1. The extension process can be repeated as long as the condition $F_y \neq 0$ continues to hold.

The indicated local result is the important:

Implicit Function Theorem. *If F has continuous partial derivatives on a neighborhood of a point (x_0, y_0) at which $F(x_0, y_0) = 0$ and $F_y(x_0, y_0) \neq 0$, then there are $s > 0$ and $t > 0$ such that the part of N_F in the rectangle $[x_0 - s, x_0 + s] \times [y_0 - t, y_0 + t]$ is the graph of a continuously differentiable function f. The derivative of f on $(x_0 - s, x_0 + s)$ is given by the implicit differentiation formula:*

$$f'(x) = -\frac{F_x(x, f(x))}{F_y(x, f(x))}.$$

Remark. Our proof of the existence of $\sqrt[n]{x}$ via the Intermediate-Value Theorem in Chapter 3, §1, suggests an approach to the proof of this result. Namely, obtain a rectangle $[x_0 - s, x_0 + s] \times [y_0 - t, y_0 + t]$ so that:

(a) $F(x, y)$ has opposite signs on the top and bottom; and
(b) $F(x, y)$, for each $x \in [x_0 - s, x_0 + s]$, is a strictly monotonic function of y.

The Intermediate-Value Theorem then applies at each x, to the function of y given by $F(x, y)$ on $[y_0 - t, y_0 + t]$. It gives the existence, by (a), and uniqueness, by (b), of the value $f(x)$ for which $F(x, f(x)) = 0$. The existence of the required rectangle follows by three applications of persistence of sign, as is seen in the

Proof. The hypotheses, F_y continuous and $F_y(x_0, y_0) \neq 0$, insure the persistence of the sign of $F_y(x_0, y_0)$ throughout a neighborhood of (x_0, y_0), hence throughout some square $[x_0 - t, x_0 + t] \times [y_0 - t, y_0 + t]$ with $t > 0$.

Since $F(x_0, y_0) = 0$, this means in particular that $F(x_0, y_0 - t)$ and $F(x_0, y_0 + t)$ are nonzero and have opposite signs ($F(x_0, y)$ is strictly monotonic in y).

The persistence of sign applies to both $F(x, y_0 - t)$ and $F(x, y_0 + t)$ as functions of x, giving them the signs they have at x_0 throughout open intervals containing x_0. The value of $s > 0$ can then be chosen so that $[x_0 - s, x_0 + s]$ is contained in both of these intervals. The rectangle $[x_0 - s, x_0 + s] \times [y_0 - t, y_0 + t]$ has the properties (a) and (b) above, so that $f(x)$ is defined for $x \in [x_0 - s, x_0 + s]$ by the Intermediate-Value Theorem.

Remark. Notice that the hypotheses hold again at the points $(x_0 \pm s, f(x_0 \pm s))$, so that the theorem provides extensions of f outside $[x_0 - s, x_0 + s]$ at each end. Repetition provides the extension to the maximal domain (in the figure, the open interval bounded by the x coordinates of A and B).

It remains to show that $f \in C^1$ and the formula holds.

For $\varepsilon > 0$, choose $t < \varepsilon$ in the above argument and call the corresponding value of s, from the argument, δ. This gives

$$|x - x_0| < \delta \quad \text{implies} \quad |f(x) - f(x_0)| < \varepsilon.$$

In view of this, f is continuous at x_0. But any x in the domain of f can play the role of x_0 here, so the same argument shows that in fact f is continuous. This continuity is used in our argument that f is C^1.

To examine the differentiability of f at x_0, consider an increment h in x and let $k = f(x_0 + h) - f(x_0)$, so that k/h is the difference quotient for f at x_0. We calculate it as follows. Since $(x_0, f(x_0))$ and $(x_0 + h, f(x_0 + h))$ belong to N_F,

$$0 = F(x_0 + h, y_0 + k) - F(x_0, y_0).$$

Add and subtract $F(x_0 + h, y_0)$;

$$0 = [F(x_0 + h, y_0 + k) - F(x_0 + h, y_0)] + [F(x_0 + h, y_0) - F(x_0, y_0)].$$

The Mean-Value Theorem provides a ξ between x_0 and $x_0 + h$ and an η between y_0 and $y_0 + k$ to rewrite the two brackets:

$$0 = F_y(x_0 + h, \eta)k + F_x(\xi, y_0)h.$$

Thus, using $F_y \neq 0$, the difference quotient is

$$\frac{k}{h} = -\frac{F_x(\xi, y_0)}{F_y(x_0 + h, \eta)}.$$

Now, when $h \to 0$ the continuity already shown gives $k \to 0$, which in turn insures

$$(\xi, y_0) \to (x_0, y_0) \quad \text{and} \quad (x_0 + h, \eta) \to (x_0, y_0).$$

The assumed continuity of F_x and F_y at (x_0, y_0) then yields the limit of the difference quotient:

$$f'(x_0) = -\frac{F_x(x_0, y_0)}{F_y(x_0, y_0)}.$$

This proof (that f' exists and is given by the formula at x_0) applies with x_0 replaced by any $x \in \operatorname{dom} f$ since all the hypotheses used at x_0 ($F(x_0, f(x_0)) = 0$, $F_y(x_0, f(x_0)) \neq 0$ and the continuity of F_x and F_y) also hold at x. It follows that f' exists throughout dom f. Finally, the continuity of f' follows because it is given by the formula, and F_x, $F_y(\neq 0)$, and f are continuous.

Remark. Calculus courses include so many exercises dealing with explicitly defined functions that a student gets the impression that implicitly defined functions occur only rarely. To correct this impression, one need only survey a text on Differential Equations, especially the applications. An excellent reference is Simmons [25].

EXERCISE 12. Use implicit differentiation to prove that the equation of the tangent line to the ellipse $x^2/a^2 + y^2/b^2 = 1$ at a point (x_0, y_0) on the ellipse is

$$\frac{x_0 x}{a^2} + \frac{y_0 y}{b^2} = 1.$$

Discuss hyperbolas.

EXERCISE 13. Derive the formulas for the derivatives of $\sin^{-1} x$ and $\log x$.

CHAPTER 5

Integration

The process of integration has ancient origins. Archimedes derived formulas for the areas of several kinds of figures by ingenious arguments based on the "method of exhaustion," whereby a figure is approximated by finite unions of simpler figures whose areas are known. Such methods ultimately evolved into a form of heuristic reasoning, which is still used in each of the sciences to establish mathematical expressions for various quantities; not only length, area, and volume in geometry, but also mass, moments, electrical charge, power, yield from an investment, population, and many others.

With the emergence of a formal theory of integration, this heuristic reasoning survives as a shorthand for detailed underlying definitions and proofs: the process of "setting up" the integral. Some mastery of the process is an essential part of "Integral Calculus" and it plays an important guiding role in the formulation of the theory; in laying the foundations.

The reader will recognize the process to which we refer, the heuristic reasoning, from an example. An important quantity in Physics is the *work* done in moving an object by overcoming a force F through a distance d (e.g., lifting an object from the floor to a table). It is defined to be the product (in suitable units); $W = Fd$.

A problem appears when the force, rather than being fixed, varies with the position; F is a function of x, say, for $a \le x \le a + d$. The work W required to move the object from a to $a + d = b$ against a force $F(x)$ is to be defined.

The heuristic reasoning imagines W as the sum of many small parts, each of which is the work done in moving the object across a small range,

x_{i-1} to x_i, where

$$a = x_0 < x_1 < x_2 < \cdots < x_{i-1} < x_i < \cdots < x_{n-1} < x_n = b,$$

and the force there is taken to be constant and equal to any value $F(u_i)$, for u_i in (x_{i-1}, x_i). The sum of these small amounts of work

$$\sum_{i=1}^{n} F(u_i)(x_i - x_{i-1})$$

is supposed to approximate the sought-for quantity: the work. In other words, the work is *defined* as the number, if such exists, that is approximated by the set of all such sums.

In any science, there are many fundamental quantities that are expressed as the product of two other quantities, $Q = xy$, and the full meaning of the quantity Q must embrace the case in which the one factor y is a function f of the other factor x. The above reasoning applies, provided only that the nature of Q is that it "adds" in the same sense that work does: work from a to b + work from b to c = work from a to c. Q is then a "limit" of sums, an integral

$$Q = \int_a^b f \quad \text{or} \quad \int_a^b f(x)\,dx.$$

Since xy has, for x and y positive, an interpretation as the area of a rectangle, every such quantity Q has an area interpretation: the geometrical problem of (defining and) computing the "area" under the graph of a positive function f over an interval $[a, b]$ is seen to be the prototype of a great variety of problems in all sciences.

Here, the heuristic reasoning leads to the approximation of the region by a finite union of rectangles. We shall draw upon this case for the intuition to guide our general definitions and arguments.

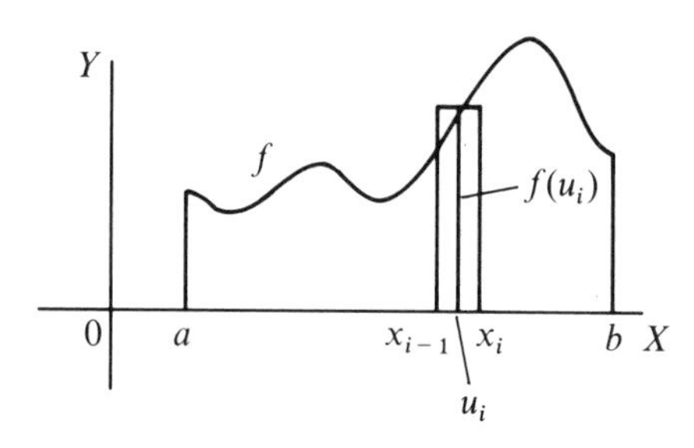

Once the appropriate definitions have been made, so that certain functions f on certain intervals $[a, b]$ give rise to a number $\int_a^b f$, the rules governing the manipulation of these numbers must be established.

Finally, the task of calculating the number $\int_a^b f$ must be considered, and, in view of its importance, reduced as much as possible to a routine. Thus, the "Integral Calculus" consists of three parts: setting-up, manipulating, and calculating integrals.

It was B. Riemann who, in 1854, gave the rigorous definition and development of the integral that embodies what we have called the heuristic reasoning process. Later, G. Darboux presented a more elegant definition of the integral and showed it to be equivalent to that of Riemann. His approach is used in most contemporary presentations.

All of this is done in the case of a bounded function on a compact interval. The extension to certain more general cases constitutes the study of so-called improper integrals.

The integration process provides a new way to define functions, a fact that has many important applications. Of these, we shall only present an example: the important gamma function, which generalizes the factorial idea.

§1. Definitions. Darboux Theorem

Guided by the intuitive approach to the area problem, we proceed to define integrability for a bounded function f on a compact interval $[a, b]$, using the Darboux approach. The definition of the integral $\int_a^b f$ of an integrable function on $[a, b]$ accompanies that of integrability.

The selection of approximating rectangles begins with the partitioning of $[a, b]$ into a finite number of subintervals—the bases of the rectangles.

Definition. A **partition** of $[a, b]$ is a finite set of points in $[a, b]$ that includes a and b.

The subintervals are, of course, the intervals between successive points of the finite set. If p is a partition, we usually denote its points in the order of their magnitude:

$$p = \{t_0, t_1, \ldots, t_n\} \qquad \text{where} \quad a = t_0 < t_1 < \cdots < t_{n-1} < t_n = b.$$

In this notation, the (closed) ith subinterval is $[t_{i-1}, t_i]$ and $(t_i - t_{i-1})$ is its length, $i = 1, 2, \ldots, n$.

Using the boundedness of f, we introduce, as heights of rectangles,

$$m_i = \inf\{f(t): t_{i-1} \leq t \leq t_i\} \quad \text{and} \quad M_i = \sup\{f(t): t_{i-1} \leq t \leq t_i\},$$

and form:

the **lower** (Darboux) **sum** for f and p,

$$L(f, p) = \sum_{i=1}^{n} m_i(t_i - t_{i-1}),$$

and
the **upper** (Darboux) **sum** for f and p,

$$U(f, p) = \sum_{i=1}^{n} M_i(t_i - t_{i-1}).$$

Clearly, $L(f, p) \le U(f, p)$ for any f and p.

Any partition of $[a, b]$ that contains p is called a **refinement** of p. Two partitions p and q have common refinements; $p \cup q$ is one.

The main fact about Darboux sums is that *every* lower sum is $\le$ *every* upper sum:

Lemma. *For partitions p and q of $[a, b]$,*

$$L(f, p) \le U(f, q).$$

Proof. We shall show that under refinement, lower sums increase (wide sense) and upper sums decrease. This gives the conclusion since then

$$L(f, p) \le L(f, p \cup q) \le U(f, p \cup q) \le U(f, q).$$

Suppose a point u is adjoined to (t_{k-1}, t_k). The term $m_k(t_k - t_{k-1})$ is replaced, in forming the lower sum for the refinement $\{t_0, \ldots, t_{k-1}, u, t_k, \ldots, t_n\}$, by

$$\inf\{f(t)\colon t_{k-1} \le t \le u\}(u - t_{k-1}) + \inf\{f(t)\colon u \le t \le t_k\}(t_k - u).$$

But m_k is $\le$ each of these infima, since it is the infimum over the larger range $[t_{k-1}, t_k]$ of t. Thus, this expression exceeds

$$m_k(u - t_{k-1}) + m_k(t_k - u) = m_k(t_k - t_{k-1}),$$

the term it replaces when u is adjoined. Since any refinement may be achieved one point at a time, this shows that lower sums increase under refinement.

Similarly, upper sums decrease under refinement.

EXERCISE 1. Supply the details for upper sums.

If we denote by $\mathscr{L}$ the set of all lower sums for f on $[a, b]$ and by $\mathscr{U}$ the set of upper sums, the lemma can be abbreviated $\mathscr{L} \le \mathscr{U}$.

In the intuitive, area, case, lower sums are underestimates and upper sums are overestimates. The area is defined precisely when there is no gap between the two sets of estimates; that is, when there is exactly one number that is approximated by both sets, $\mathscr{L}$ and $\mathscr{U}$ (within ε, for any $\varepsilon > 0$). Guided by this intuition; no ε gap for any $\varepsilon > 0$, we make the

Definition. The bounded function f is **integrable on** $[a, b]$ if, for every $\varepsilon > 0$, there exist partitions p_ε and q_ε such that

$$U(f, q_\varepsilon) - L(f, p_\varepsilon) < \varepsilon.$$

Remark. By considering common refinements, it follows easily that the definition can be reformulated as:

For any $\varepsilon > 0$ there is a partition p_ε of $[a, b]$ such that

$$U(f, p_\varepsilon) - L(f, p_\varepsilon) < \varepsilon.$$

EXERCISE 2. Prove this.

Thus, integrability is investigated by examining the finite sums

$$U(f, p) - L(f, p) = \sum_{i=1}^{n} (M_i - m_i)(t_i - t_{i-1}).$$

For any bounded f on $[a, b]$, we define the **lower integral**, $\underline{\int_a^b} f$, and the **upper integral** $\overline{\int_a^b} f$ by

$$\underline{\int_a^b} f = \sup_p L(f, p) = \sup \mathscr{L}$$

and

$$\overline{\int_a^b} f = \inf_p U(f, p) = \inf \mathscr{U}.$$

Proposition. $\underline{\int_a^b} f \leq \overline{\int_a^b} f$ *and equality holds if and only if* f *is integrable on* $[a, b]$.

EXERCISE 3. Prove the proposition.

Definition. The **integral** of the integrable function f on $[a, b]$ is the common value of its upper and lower integrals. It is denoted $\int_a^b f$, or perhaps $\int_a^b f(x)\,dx$, using a "dummy variable" x.

We shall see that many functions are integrable. Of more interest at first is the fact that *some* functions are *not* integrable:

EXAMPLE. Let

$$f(x) = \begin{cases} 1 & \text{if } x \text{ is rational,} \\ 0 & \text{if } x \text{ is irrational.} \end{cases}$$

For any $a < b$, f is not integrable on $[a, b]$.

EXERCISE 4. Prove that every lower sum for f is zero, while every upper sum is $b - a$.

Two sufficient conditions for integrability: continuity and monotonicity, are easily shown.

Proposition. *If f is continuous on $[a, b]$ then f is integrable on $[a, b]$.*

Proof. Since f is continuous uniformly on $[a, b]$, there is, given $\varepsilon > 0$, a δ such that any partition p, all of whose subintervals are of length $< \delta$, has all $M_i - m_i < \varepsilon/(b - a)$. Hence,

$$U(f, p) - L(f, p) = \sum_{i=1}^{n} (M_i - m_i)(t_i - t_{i-1})$$

$$< \frac{\varepsilon}{b - a} \sum_{i=1}^{n} (t_i - t_{i-1}) = \varepsilon.$$

Proposition. *If f is monotonic on $[a, b]$ then f is integrable on $[a, b]$.*

Proof. For an increasing f and a partition p,

$$M_i - m_i = f(t_i) - f(t_{i-1}),$$

and so

$$U(f, p) - L(f, p) = \sum_{i=1}^{n} [f(t_i) - f(t_{i-1})](t_i - t_{i-1}).$$

But

$$\sum_{i=1}^{n} [f(t_i) - f(t_{i-1})] = f(b) - f(a)$$

(a "telescoping" sum), and it is clear that any partition p_ε into subintervals satisfying

$$(t_i - t_{i-1}) < \frac{\varepsilon}{f(b) - f(a)}, \qquad i = 1, 2, \ldots, n,$$

(we assume the nontrivial case, $f(b) > f(a)$) yields

$$U(f, p_\varepsilon) - L(f, p_\varepsilon) < \frac{\varepsilon}{f(b) - f(a)} \sum_{i=1}^{n} [f(t_i) - f(t_{i=1})] = \varepsilon,$$

proving the result in the case of an increasing f.

EXERCISE 5. Supply the details in the case of a decreasing f.

Remark. The trick of estimating the one kind of factors so as to get a telescoping sum of the other kind of factors, which gives these two results, fails to shed light on the nature of integrability. In particular, it is not at all clear what common property of continuous and monotone functions insures their integrability. Section 3 below contains a discussion of the nature of integrability. These two propositions cover all functions encountered in ordinary applications of Calculus.

The Darboux choices, m_i and M_i, for the heights of the approximating rectangles are the extremes of the values that it makes sense to consider. Their use leads to the $\mathscr{L} \leq \mathscr{U}$ observation and the simplicity of the definitions of integrability and integral which it yields. The other possible values for the heights are the values $f(u_i)$ that f actually assumes at points u_i in $[t_{i-1}, t_i]$. It is such choices which occur in the heuristic reasoning upon which Riemann based his discussion.

Definition. A **Riemann sum** for the bounded function f and the partition p of $[a, b]$ is any number of the form

$$R(f, p) = \sum_{i=1}^{n} f(u_i)(t_i - t_{i-1}),$$

where for each $i = 1, 2, \ldots, n$, u_i denotes an arbitrary number in $[t_{i-1}, t_i]$.

EXERCISE 6. When are the Darboux sums Riemann sums?

Because of the arbitrariness in the choice of the points u_i, it is not easy or natural to study Reimann sums using inequalities (comparing them, taking infima of them, etc.). Rather, it is indicated that a form of "limit" of Riemann sums be introduced, in agreement with the idea underlying the heuristic reasoning: a number that is approximated within a given $\varepsilon > 0$ by *any* Riemann sum that is based on a sufficiently "fine" partition.

The "fineness" of a partition is measured by the length of its longest subinterval:

Definition. For a partition p of $[a, b]$, the **mesh of** p is the number

$$\|p\| = \max_{1 \leq i \leq n} (t_i - t_{i-1}),$$

where $p = \{t_0, t_1, \ldots, t_n\}$.

Definition. L is the **limit** of the Riemann sums of f on $[a, b]$ provided, for every $\varepsilon > 0$, there is a δ such that $|R(f, p) - L| < \varepsilon$, for any p with $\|p\| < \delta$ and *any* Riemann sum $R(f, p)$.

EXERCISE 7. Show that at most one such L can exist.

The use of the Darboux approach to the definitions of integrability and integral, with their simplicity, is justified by the

Darboux Theorem. *The limit of the Riemann sums of the bounded function f*

on $[a, b]$ *exists if and only if* f *is integrable on* $[a, b]$. *When the limit exists, it is* $\int_a^b f$.

Remarks. 1. This statement is often conveyed by the equation

$$\int_a^b f(t)\, dt = \lim_{\|p\| \to 0} \sum_{i=1}^{n} f(u_i)\,(t_i - t_{i-1})$$

in the sense that one side exists if and only if the other does, and then they are equal. This equation also serves as a shorthand expression of the heuristic reasoning used in "setting up" integrals.

2. The Darboux approach confines all of the nastiness inherent in dealing with Riemann sums in proofs, to the single proof of this theorem.

Proof. Suppose a limit L exists. Fix $\varepsilon > 0$ and choose δ so that

$$\|p\| < \delta \quad \text{implies} \quad |R(f, p) - L| < \frac{\varepsilon}{4} \quad \text{for all } R(f, p).$$

Fix $p_\varepsilon = \{t_0, t_1, \ldots, t_n\}$ of mesh less than δ. By the definitions of m_i and M_i, there are points u_i and v_i in (t_{i-1}, t_i) giving f values near these extremes; for $i = 1, 2, \ldots, n$,

$$f(u_i) < m_i + \frac{\varepsilon}{4(b-a)} \quad \text{and} \quad f(v_i) > M_i - \frac{\varepsilon}{4(b-a)}.$$

Multiplying by $(t_i - t_{i-1})$ and adding, we get

$$\sum_{i=1}^{n} f(u_i)(t_i - t_{i-1}) < L(f, p_\varepsilon) + \frac{\varepsilon}{4} \quad \text{and} \quad \sum_{i=1}^{n} f(v_i)(t_i - t_{i-1}) > U(f, p_\varepsilon) - \frac{\varepsilon}{4}.$$

But these are Riemann sums, hence are within $\varepsilon/4$ of L, giving

$$\begin{aligned} L - \frac{\varepsilon}{2} &< \sum_{i=1}^{n} f(u_i)(t_i - t_{i-1}) - \frac{\varepsilon}{4} < L(f, p_\varepsilon) \le U(f, p_\varepsilon) \\ &< \sum_{i=1}^{n} f(v_i)(t_i - t_{i-1}) + \frac{\varepsilon}{4} < L + \frac{\varepsilon}{2}. \end{aligned}$$

It follows that

$$U(f, p_\varepsilon) - L(f, p_\varepsilon) < \varepsilon$$

and that f is integrable. Moreover, the integral, $\int_a^b f$, which then exists, fits in the middle of the above string of inequalities;

$$L(f, p_\varepsilon) \le \int_a^b f \le U(f, p_\varepsilon).$$

We conclude that

$$\left| \int_a^b f - L \right| < \varepsilon.$$

Since $\varepsilon > 0$ is arbitrary, $L = \int_a^b f$.

For the converse, suppose that f is integrable. Given $\varepsilon > 0$, fix p_ε so that

$$U(f, p_\varepsilon) - \frac{\varepsilon}{4} < \int_a^b f < L(f, p_\varepsilon) + \frac{\varepsilon}{4}.$$

We shall show that there is a δ such that

$$\circledast \qquad \begin{cases} \|p\| < \delta \text{ implies, for any } R(f, p), \\ R(f, p) - \dfrac{\varepsilon}{4} < U(f, p_\varepsilon) \quad \text{and} \quad L(f, p_\varepsilon) < R(f, p) + \dfrac{\varepsilon}{4}, \end{cases}$$

which allows the continuation of the preceding inequalities at each end, to get

$$R(f, p) - \frac{\varepsilon}{2} < U(f, p) - \frac{\varepsilon}{4} < \int_a^b f < L(f, p_\varepsilon) + \frac{\varepsilon}{4} < R(f, p) + \frac{\varepsilon}{2},$$

whenever $\|p\| < \delta$. That is,

$$\|p\| < \delta \quad \text{implies} \quad \left| R(f, p) - \int_a^b f \right| < \varepsilon \quad \text{for any } R(f, p),$$

which is the desired conclusion.

The proof of $\circledast$ begins with the notations

$$p = \{s_0, s_1, \ldots, s_m\}$$

and

$$u_j \in [s_{j-1}, s_j],$$

so that

$$R(f, p) = \sum_{j=1}^{n} f(u_j)(s_j - s_{j-1})$$

expresses the general Riemann sum.

For the first inequality in $\circledast$, we wish to estimate this Riemann sum in terms of $U(f, p) = \sum_{i=1}^{n} M_i(t_i - t_{i-1})$ when $\|p\|$ is small.

Now, if u_j falls in $[t_{i-1}, t_i]$ then the estimate $f(u_j) \le M_i$ is valid. This suggests grouping the terms of the sum for $R(f, p)$ into subsums according to which interval $[t_{i-1}, t_i]$ the points u_j belong to.[1]

The notation is simplified if we require, at first, that $\|p\| < \frac{1}{2}\|p_\varepsilon\|$, for then every subinterval $[t_{i-1}, t_i]$ of p_ε will contain at least one u_j.

The index set $\{0, 1, 2, \ldots, m\}$ then separates into nonempty, disjoint subsets $\{k_{i-1} + 1, \ldots, k_i\}$, for $i = 1, 2, \ldots, n$ (taking $k_0 = 0$ and $k_n = m$), such that

$$u_j \le M_i \quad \text{precisely when} \quad j \in \{k_{i-1} + 1, \ldots, k_i\}$$

[1] This is the nasty part.

for $i = 1, 2, \ldots, n$. Thus,

$$\begin{aligned} R(f, p) &= \sum_{i=1}^{n} \sum_{j=k_{i-1}+1}^{k_i} f(u_j)(s_j - s_{j-1}) \\ &\leq \sum_{i=1}^{n} M_i \sum_{j=k_{i-1}+1}^{k_i} (s_j - s_{j-1}) = \sum_{i=1}^{n} M_i(s_{k_i} - s_{k_{i-1}}) \end{aligned}$$

(the inner sum telescopes).

But it is evident that

$$s_{k_i} - s_{k_{i-1}} \leq (t_i - t_{i-1}) + 2\|p\|, \qquad i = 1, 2, \ldots, n,$$

since the points s_{k_i} and $s_{k_{i-1}}$ cannot be outside $[t_{i-1}, t_i]$ by more than a distance $\|p\|$.

Multiplying by M_i and adding gives, in view of the preceding estimate,

$$R(f, p) < U(f, p_\varepsilon) + 2\|p\| \sum_{i=1}^{n} M_i.$$

Now, $\sum_{i=1}^{n} M_i$ can be ≤ 0, in which case

$$R(f, p) < U(f, p_\varepsilon) + \frac{\varepsilon}{4}$$

holds without further restriction on $\|p\|$. Otherwise, the restriction

$$\|p\| < \frac{\varepsilon}{8 \sum_{i=1}^{n} M_i}$$

gives it.

By a similar estimate;

$$R(f, p) > \sum_{i=1}^{n} m_i(s_{k_i} - s_{k_{i-1}})$$

and

$$s_{k_i} - s_{k_{i-1}} > (t_i - t_{i-1}) - 2\|p\|,$$

we find

$$R(r, p) > L(f, p_\varepsilon) - 2\|p\| \sum_{i=1}^{n} m_i.$$

If $\sum_{i=1}^{n} m_i \leq 0$, the other half of ⊛,

$$R(f, p) > L(f, p_\varepsilon) - \frac{\varepsilon}{4},$$

is immediate, and otherwise the restriction

$$\|p\| < \frac{\varepsilon}{8 \sum_{i=1}^{n} m_i}$$

gives it. The existence of δ giving ⊛ is then clear, completing the proof of the Darboux theorem.

§2. Foundations of Integral Calculus. The Fundamental Theorem of Calculus

We use the term "Integral Calculus" to refer to three things:

1. *Setting up* integrals to represent various quantities. The heuristic reasoning we have mentioned, together with things like disk and shell methods, constitute this aspect of the subject. The Darboux Theorem (the definition in Riemann's approach) provides the foundation.

2. *Manipulation* of expressions involving integrals, e.g., integration of sums term-by-term, integrals of limits as limits of integrals, estimates of integral quantities. A number of propositions constitute the foundations of this aspect of the subject. They are established in the first part of this section.

3. *Computation* of integrals. Attempting the exact evaluation of $\int_a^b f$ leads us to one of the versions of the Fundamental Theorem of Calculus. The two theorems which go by this name constitute the foundation of this aspect of the subject: the techniques for finding antiderivatives, including substitution, integration by parts and by partial fractions. These foundations are proved in the second part of this section.

Let us denote by $R[a, b]$ the class of all Riemann integrable functions on $[a, b]$, a collection of bounded functions. The properties of $R[a, b]$ and the properties of $\int_a^b f$ as a numerical-valued function on $R[a, b]$ are summarized in the

Theorem. *If f and g are in $R[a, b]$ and α is real, then:*

1. $f + g \in R[a, b]$ *and* $\int_a^b (f + g) = \int_a^b f + \int_a^b g$.
2. $\alpha f \in R[a, b]$ *and* $\int_a^b \alpha f = \alpha \int_a^b f$.
3. $f(x) \leq g(x)$ *for all* $x \in [a, b]$ *implies* $\int_a^b f \leq \int_a^b g$.
4. $|f| \in R[a, b]$ *and* $|\int_a^b f| \leq \int_a^b |f|$.
5. $fg \in R[a, b]$ *and* $|\int_a^b fg| \leq (\int_a^b f^2)^{1/2}(\int_a^b g^2)^{1/2}$.

Proof. 1. For any partition p of $[a, b]$ and any choice of the points u_i in the subintervals, the resulting Riemann sums clearly satisfy

$$R(f + g, p) = R(f, p) + R(g, p).$$

By hypothesis, for $\varepsilon > 0$ there is a δ such that $\|p\| < \delta$ gives both

$$\left|R(f, p) - \int_a^b f\right| < \frac{\varepsilon}{2} \quad \text{and} \quad \left|R(g, p) - \int_a^b g\right| < \frac{\varepsilon}{2}.$$

Thus, $\|p\| < \delta$ gives

$$\left|R(f + g, p) - \left(\int_a^b f + \int_a^b g\right)\right| \leq \left|R(f, p) - \int_a^b f\right| + \left|R(g, p) - \int_a^b g\right| < \varepsilon.$$

The conclusion follows.

2. It is again obvious that, using the same points u_i to form both Riemann sums,

$$R(\alpha f, p) = \alpha R(f, p).$$

The case $\alpha = 0$ being obvious, we suppose $\alpha \neq 0$ and choose, for $\varepsilon > 0$, a δ such that $\|p\| < \delta$ gives

$$\left| R(f, p) - \int_a^b f \right| < \frac{\varepsilon}{|\alpha|}.$$

Then,

$$\left| R(\alpha f, p) - \alpha \int_a^b f \right| = |\alpha| \left| R(f, p) - \int_a^b f \right| < \varepsilon,$$

when $\|p\| < \delta$. The conclusion follows.

3. By 1 and 2, $g - f \in R[a, b]$. But, under the hypothesis, any Riemann sum for $g - f$ is ≥ 0. Hence $\int_a^b (g - f) \geq 0$. Applying 1 and 2 once more gives the conclusion.

4. If we show that $|f| \in R[a, b]$, the inequality follows from 3 in view of

$$-|f(x)| \leq f(x) \leq |f(x)|,$$

i.e., by 3,

$$-\int_a^b |f| \leq \int_a^b f \leq \int_a^b |f|,$$

which says,

$$\left| \int_a^b f \right| \leq \int_a^b |f|.$$

In the triangle inequality,

$$|f(x)| - |f(y)| \leq |f(x) - f(y)|,$$

take $\sup_{t_{i-1} \leq x \leq t_i}$ holding y fixed in $[t_{i-1}, t_i]$, to get

$$\begin{aligned} \sup_{t_{i-1} \leq x \leq t_i} |f(x)| - |f(y)| &\leq \sup_{t_{i-1} \leq x \leq t_i} |f(x) - f(y)| \\ &= M_i - f(y), \end{aligned}$$

then take $\inf_{t_{i-1} \leq y \leq t_i}$ to obtain a term-by-term comparison of the sums, to show that

$$U(|f|, p) - L(|f|, p) \leq U(f, p) - L(f, p)$$

for any partition p.

The integrability of $|f|$, assuming that of f, then follows directly from the definition.

5. Since

$$fg = \tfrac{1}{4}\{(f + g)^2 - (f - g)^2\},$$

if we prove

$$f \in R[a, b] \quad \text{implies} \quad f^2 \in R[a, b]$$

then, using 1 and 2, the conclusion $fg \in R[a, b]$ follows. But, if M is an upper bound for $|f|$ on $[a, b]$, then

$$f^2(x) - f^2(y) = [|f(x)| + |f(y)|][|f(x)| - |f(y)|] \le 2M|f(x) - f(y)|$$

and the same argument as in 4 gives

$$U(f^2, p) - L(f^2, p) \le 2M[U(f, p) - L(f, p)],$$

and the integrability of f^2 is seen to result from that of f.

6. For any real α, $\int_a^b (\alpha g - f)^2 \ge 0$ holds (by 1, 2, 3 and the previous result). By 1 and 2, the left side expands to a quadratic expression in α:

$$\left(\int_a^b g^2\right)\alpha^2 - \left(2\int_a^b fg\right)\alpha + \left(\int_a^b f^2\right) \ge 0, \qquad \text{all } \alpha.$$

A quadratic in α is always ≥ 0 precisely when its discriminant is ≤ 0 (roots are equal or complex). The inequality to be shown is just the statement that the discriminant of this quadratic is ≤ 0.

EXERCISE 1. Give an example to show that $\int_a^b fg$ and $(\int_a^b f)(\int_a^b g)$ are not necessarily equal.

EXERCISE 2. Prove that $A\alpha^2 + 2B\alpha + C \ge 0$ for all $\alpha \in \mathbb{R}$ if and only if $B^2 - AC \le 0$.

EXERCISE 3. Prove that $f, g \in R[a, b]$ and $1/g$ bounded on $[a, b]$ implies $f/g \in R[a, b]$.

EXERCISE 4. Let $\min(f, g)$ be defined at x to be $\min[f(x), g(x)]$. Show that

$$\min(f, g) = \tfrac{1}{2}[(f + g) - |f - g|]$$

and comment on the integrability of $\min(f, g)$. Discuss $\max(f, g)$.

Remarks. 1. The results 1 and 2 are often summarized in the statement: $R[a, b]$ is a "vector space" of functions on $[a, b]$ and $f \to \int_a^b f$ is a "linear" function on $R[a, b]$ with values in $\mathbb{R}$. The additional property in 5 earns $R[a, b]$ the name of an "algebra" of functions on $[a, b]$. The inequality in 5 is the Cauchy–Schwarz inequality, also associated with the name of Bunyiakowski (all three discovered it, in different contexts). It is very important.

2. The triangle inequality for n numbers,

$$\left|\sum_{i=1}^{n} x_i\right| \le \sum_{i=1}^{n} |x_i|,$$

has, when applied to Riemann sums, the "limiting" form $|\int_a^b f| \le \int_a^b |f|$ in 3.

3. It can happen that $|f|$ is integrable but f is not; an awkwardness, if not a defect, of the integration process discussed here. Lebesgue's 1902 refinement of the process removed this defect and others.

EXERCISE 5. Prove that

$$f(t) = \begin{cases} -1, & x \in [0, 1] \text{ and } x \text{ rational}, \\ 1, & x \in [0, 1] \text{ and } x \text{ irrational}, \end{cases}$$

is an example of this defect.

If the infimum and supremum of f on $[a, b]$ are denoted, respectively, by m and M, then, by integrating the inequality $m \leq f(x) \leq M$, we obtain (supposing $a < b$)

$$m \leq \frac{1}{b-a}\int_a^b f \leq M.$$

This quantity has a natural interpretation as the average or mean value of f on $[a, b]$. (The average of numbers, $y_1, y_2, \ldots, y_n$, namely,

$$\frac{1}{n}\sum_{i=1}^n y_i = \sum_{i=1}^n y_i \Big/ \sum_{i=1}^n 1,$$

has the "limiting" form $\int_a^b f/\int_a^b 1$.) Now, if f is continuous on $[a, b]$, then m and M are values of f in $[a, b]$ (by the Extreme-Value Theorem) and the Intermediate-Value Theorem gives the following result, called the Mean-Value Theorem of Integral Calculus.

Proposition. *If f is continuous on $[a, b]$, there exists a $\xi \in (a, b)$ such that*

$$\frac{1}{b-a}\int_a^b f = f(\xi).$$

Next, we consider the dependence of $\int_a^b f$ on the interval of integration $[a, b]$.

Theorem. *If $c \in (a, b)$ and f is integrable on both $[a, c]$ and $[c, b]$ then $f \in R[a, b]$ and*

$$\int_a^b f = \int_a^c f + \int_c^b f.$$

Proof. Fix $\varepsilon > 0$. Choose a partition p_ε of $[a, c]$ and a partition q_ε of $[c, b]$ such that

$$U(f, p_\varepsilon) - L(f, p_\varepsilon) < \frac{\varepsilon}{2} \quad \text{and} \quad U(f, q_\varepsilon) - L(f, q_\varepsilon) < \frac{\varepsilon}{2},$$

so that

$$[U(f, p_\varepsilon) + U(f, q_\varepsilon)] - [L(f, p_\varepsilon) + L(f, q_\varepsilon)] < \varepsilon.$$

But $p_\varepsilon \cup q_\varepsilon$ is a partition of $[a, b]$ whose upper and lower sums are the bracketed quantities here. The result $f \in R[a, b]$ follows.

The inequalities

$$L(f, p_\varepsilon) \le \int_a^c f \le U(f, p_\varepsilon)$$

and

$$L(f, q_\varepsilon) \le \int_c^b f \le U(f, q_\varepsilon)$$

can be added, giving

$$L(f, p_\varepsilon \cup q_\varepsilon) \le \int_a^c f + \int_c^b f \le U(f, p_\varepsilon \cup q_\varepsilon),$$

for every $\varepsilon > 0$. But the only number for which this is true is $\int_a^b f$. Consequently,

$$\int_a^b f = \int_a^c f + \int_c^b f.$$

If the lower limit of integration a is larger than the upper limit of integration b, the symbol $\int_a^b f$ has no meaning as yet. (We have favored one of the two possible orientations of an interval (i.e., left to right) in our discussions so far.) The definition

$$\int_a^b f = -\int_b^a f \qquad \text{if} \quad b < a$$

removes this restriction in a way that preserves the formula, the **additivity**

$$\int_a^b f = \int_a^c f + \int_c^b f$$

even when c is not between a and b.

An additional natural question concerns the interaction of integration with limits of sequences of functions.

Example. If $r_1, r_2, \ldots$ is an enumeration of the rationals in $[0, 1]$ and

$$f_n(s) = \begin{cases} 1 & \text{if } x \in \{r_1, r_2, \ldots, r_n\}, \\ 0 & \text{otherwise,} \end{cases}$$

then each f_n is integrable and $\int_0^1 f_n = 0$. However, the pointwise limit is not integrable; it is our example of a function that is not integrable.

It is also possible to have $\lim f_n$ integrable while having

$$\int_a^b \lim f_n \ne \lim \int_a^b f_n$$

as in the

Example. Define f_n on $[0, 1]$ to have as its graph a right triangle over $[0, 1/n]$

whose area is 1. Clearly, $\int_0^1 f_n = 1$ for $n = 1, 2, 3, \ldots$, so that

$$1 = \lim \int_0^1 f_n \neq \int_0^1 \lim f_n = \int_0^1 0 = 0.$$

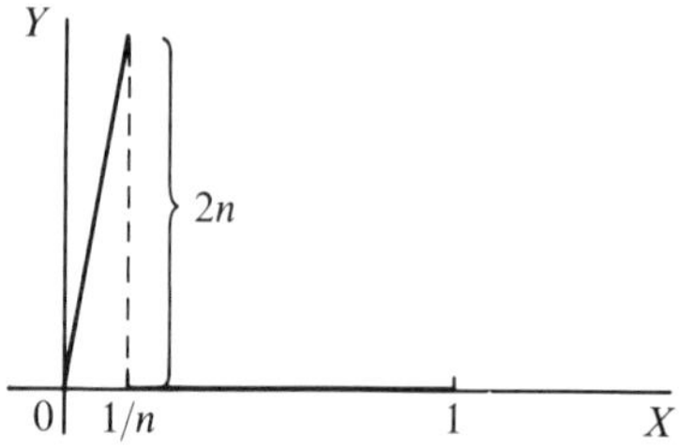

This failure of the integral to interchange with limits—especially the failure in general of the limit function to be integrable—must be considered a serious technical defect of the Riemann integral. It was among the motivations for the deeper analysis of the integration process which culminated in the Lebesgue integral, a refinement of the process which is needed in more advanced theories and applications. The Lebesgue integral exists and equals our $\int_a^b f$ whenever $f \in R[a, b]$.

A positive result for the Riemann integral is the

Theorem. *If* $f_n \in R[a, b]$, $n = 1, 2, 3, \ldots$, *and* $f_n \to f$ *on* $[a, b]$, *then the conclusion*

$$f \in R[a, b] \quad \text{and} \quad \int_a^b f = \lim \int_a^b f_n$$

holds provided the convergence is uniform on $[a, b]$.

Proof. (We suppose $a < b$.) First, we show that $\{\int_a^b f_n\}$ is a Cauchy sequence. For $\varepsilon > 0$, choose N from the assumed uniform convergence, so that

$$m, m \geq N \quad \text{and} \quad x \in [a, b] \quad \text{implies} \quad |f_m(x) - f_n(x)| < \frac{\varepsilon}{b - a}.$$

Then

$$m, n \geq N \quad \text{gives} \quad \left| \int_a^b f_m - \int_a^b f_n \right| = \left| \int_a^b (f_m - f_n) \right| < \int_a^b |f_m - f_n| < \varepsilon.$$

Thus, $\lim \int_a^b f_n$ exists.

Consider any Riemann sum $R(f, p)$. If the same points of evaluation u_i are used,

$$R(f, p) = R(f - f_m, p) + R(f_m, p).$$

This is used to obtain the following estimate:

$$\begin{aligned}
&\left| R(f, p) - \lim \int_a^b f_n \right| \\
&\quad = \left| R(f - f_m, p) + R(f_m, p) - \int_a^b f_m + \int_a^b f_m - \lim \int_a^b f_n \right| \\
&\quad \leq \sum_{i=1}^n |f(u_i) - f_m(u_i)|(t_i - t_{i-1}) + \left| R(f_m, p) - \int_a^b f_m \right| \\
&\quad\quad + \left| \int_a^b f_m - \lim \int_a^b f_n \right|.
\end{aligned}$$

Given $\varepsilon > 0$, m can be chosen so that

$$|f(x) - f_m(x)| < \frac{\varepsilon}{3(b - a)} \qquad \text{for all} \quad x \in [a, b],$$

by the uniform convergence, and also so that $|\int_a^b f_m - \lim \int_a^b f_n| < \varepsilon/3$. There is a δ such that

$$\|p\| < \delta \quad \text{implies} \quad \left| R(f_m, p) - \int_a^b f_m \right| < \frac{\varepsilon}{3}.$$

Thus,

$$\|p\| < \delta \quad \text{implies} \quad \left| R(f, p) - \lim \int_a^b f_n \right| < \varepsilon,$$

which shows that $f \in R[a, b]$ and $\int_a^b f = \lim \int_a^b f_n$.

We consider next the task of computing $\int_a^b f$.

The exact computation of $\int_a^b f = \lim_{\|p\| \to 0} \sum_{i=1}^n f(u_i)(t_i - t_{i-1})$ begins with the observation that $f(u_i)(t_i - t_{i-1})$ has the form that appears in the Mean-Value Theorem: If f is the derivative of some function F that satisfies the hypotheses of the Mean-Value Theorem, then we can rewrite the general term:

$$f(u_i)(t_i - t_{i-1}) = F'(u_i)(t_i - t_{i-1}).$$

The computation advances by changing u_i to some $v_i \in (t_{i-1}, t_i)$ that gives

$$F'(v_i)(t_i - t_{i-1}) = F(t_i) - F(t_{i-1}),$$

since then the sum

$$\sum_{i=1}^n F'(v_i)(t_i - t_{i-1}) = \sum_{i=1}^n [F(t_i) - F(t_{i-1})] = F(b) - F(a)$$

telescopes.

The Darboux Theorem insures that the limit of Riemann sums (existence and value) is independent of the points used in forming the sums; changing u_i

to v_i has no effect on the limit. We conclude that if $f \in R[a, b]$, then

$$\int_a^b f = F(b) - F(a),$$

provided $F' = f$ on (a, b) and F satisfies the hypotheses of the Mean-Value Theorem (F is continuous on $[a, b]$ and differentiable on (a, b)).

We have proved the

Theorem. *If F is continuous on $[a, b]$ and F' exists in (a, b) and coincides there with some $f \in R[a, b]$, then, writing $\int_a^b F'$ for $\int_a^b f$,*

$$\int_a^b F' = F(b) - F(a).$$

(Note: Since different functions $f \in R[a, b]$ that agree with F' on (a, b) can only differ at the two endpoints, the meaning of $\int_a^b F'$ here is unambiguous.)

The term Fundamental Theorem of Calculus is applied to theorems asserting that differentiation and integration are operations on functions which "reverse" each other. In the order "d/dx followed by $\int^x$" the result is only determined up to an added constant (p. 152, Exercise 10). If c and x are in (a, b), this theorem gives

$$F(c) + \int_c^x F' = F(x),$$

which says that $\int_c^x$ (plus "evaluation at c") reverses differentiation; it is a Fundamental Theorem of Calculus.

The theorem is, of course, the basis for the "techniques of integration," in which the problem of calculating $\int_a^b f$ is reduced to that of finding an antiderivative F of f. Since every derivative formula can be written as an antiderivative formula, the "techniques" consist of reducing given problems to ones that involve recognizable antiderivatives:

Partial Fractions. By reversing the algebraic process of adding fractions, *any* rational function is expressible as a finite sum of simpler rational functions, the partial fractions, each of which has a recognizable antiderivative (involving polynomials, logarithms, arctangents, etc.).

EXERCISE 6. Evaluate $\int_1^2 (3x - 2)/x(x^2 + 1)\, dx$.

Integration by Parts. The formula

$$\int_a^b uv' = u(b)v(b) - u(a)v(a) - \int_a^b u'v,$$

which results from applying the above theorem to the product rule, allows the problem $\int_a^b f$ to be reduced to a solvable problem for each interpretation of f as uv' in which v' and $u'v$ have recognizable antiderivatives.

EXERCISE 7. State and prove a theorem that embodies integration by parts.

EXERCISE 8. Evaluate $\int_0^1 x^3/(1 + x^2)^3\, dt$.

Integration by Substitution. The chain rule, viewed as an antiderivative formula, suggests the formula

$$\int_a^b f(u(x))u'(x)\,dx = \int_{u(a)}^{u(b)} f(t)\,dt.$$

(If F is an antiderivative of f then $(F \circ u)'(x) = f(u(x))u'(x)$, by the chain rule, so that

$$\int_a^b f(u(x))u'(x)\,dx = (F \circ u)(b) - (F \circ u)(a)$$

$$= F(u(b)) - F(u(a)) = \int_{u(a)}^{u(b)} f.)$$

The reader will require no reminding of how pervasively this idea is used in the techniques of integration. The formulation and proof of a theorem to justify the method involves only one subtle point; the existence of an antiderivative F of f.

Note that this same fundamental question; To find conditions on an integrable f which insure that it is a derivative, is raised by the very statement of the above Fundamental Theorem. We shall show that every *continuous* function on $[a, b]$ is a derivative there, as part of the second fundamental theorem, which follows.

EXERCISE 9. Is the substitution formula valid under the hypotheses $f \in C[a, b]$ and $u \in C^1[J]$, for some open interval J containing $[a, b]$?

Theorem. 1. *$f \in R[a, b]$ implies $\int_a^x f = F(x)$ is continuous on $[a, b]$.*

2. *At points x_0 where f is continuous, F is differentiable and $F'(x_0) = f(x_0)$.*

Remarks. 1. The theorem asserts in specific terms that $\int_a^x$ is a "smoothing" operation on a function f; it yields a continuous result when applied to an integrable (hence possibly discontinuous) f and a differentiable result when applied to a continuous f.

2. The construction of an antiderivative of a continuous function is thus accomplished by the integration process.

3. The integration process on $C[a, b]$ is "reversed" by differentiation; this is a Fundamental Theorem of Calculus.

Proof. 1. Since for $a \leq x < y \leq b$,

$$|F(y) - F(x)| = \left| \int_a^y f - \int_a^x f \right| = \left| \int_x^y f \right| \leq \int_x^y |f| \leq \sup_{a \leq x \leq b} |f(x)| \cdot (y - x),$$

the continuity of F (uniformly on $[a, b]$) is clear.

2. The difference quotient of F at x_0 is

$$\frac{F(x) - F(x_0)}{x - x_0} = \frac{1}{x - x_0} \int_{x_0}^x f.$$

An expression for the prospective limit, $f(x_0)$, in compatible form is

$$\frac{1}{x - x_0} \int_{x_0}^x f(x_0)$$

(the integrand is constant).

We have,

$$\left| \frac{F(x) - F(x_0)}{x - x_0} - f(x_0) \right| = \left| \frac{1}{x - x_0} \int_{x_0}^x [f - f(x_0)] \right|$$

$$\leq \begin{cases} \dfrac{1}{x - x_0} \displaystyle\int_{x_0}^x |f - f(x_0)|, & \text{if } x_0 < x \\ \dfrac{1}{x_0 - x} \displaystyle\int_x^{x_0} |f - f(x_0)|, & \text{if } x < x_0 \end{cases}$$

$$\leq \sup\{|f(t) - f(x_0)|: t \text{ between } x_0 \text{ and } x\}.$$

The continuity of f at x_0 insures that the last expression can be made arbitrarily small by choosing $|x - x_0|$ sufficiently small. It follows that $F'(x_0)$ exists (one-sided if x_0 is a or b) and equals $f(x_0)$.

The interchangeability of limits and derivatives was not considered in the previous chapters, where a straightforward result on the subject belongs. Consider the

EXAMPLE. The sequence $f_n(x) = \sin nt/\sqrt{n}$, $n = 1, 2, 3, \ldots$, converges to 0 uniformly on $\mathbb{R}$, but the sequence of derivatives $f_n'(x) = \sqrt{n} \cos nx$ is unbounded at every $x \in \mathbb{R}$.

EXERCISE 10. Prove these assertions. Hint: If $|\cos mx| < \frac{1}{2}$ then $|\cos 2mx| > \frac{1}{2}$.

With enough hypotheses, the fundamental theorems can be applied, to give a result which is sometimes useful.

Theorem. *If $f_n \in C^1(a, b)$, $n = 1, 2, 3, \ldots$, and if $f_n \to f$ and $f_n' \to g$ uniformly on every closed subinterval of (a, b), then f is differentiable on (a, b) and $f' = g$.*

Proof. For c and $x \in (a, b)$, the first fundamental theorem gives $\int_c^x f_n' = f_n(x) - f_n(c)$. Letting $n \to \infty$, since $f_n' \to g$ uniformly, we have

$$\int_c^x g = f(x) - f(c).$$

As the uniform limit of continuous functions, g is continuous. Hence the second fundamental theorem applies: $\int_c^x g$, hence also $f(x)$, is an antiderivative of g. In other words, f is differentiable and $f' = g$.

§3. The Nature of Integrability. Lebesgue's Theorem

The question of the integrability on $[a, b]$ of a function f is that of whether the partition p_ε can be chosen to make the sum

$$U(f, p_\varepsilon) - L(f, p_\varepsilon) = \sum_{i=1}^{n} (M_i - m_i)(t_i - t_{i-1})$$

less than the prescribed $\varepsilon > 0$.

In case f is continuous on $[a, b]$ we saw that the uniformity allows p_ε to be chosen so that *every* first, $(M_i - m_i)$, factor can be estimated by $\varepsilon/(b - a)$.

On the other hand, if f is monotonic then p_ε may be chosen so that *every* second factor, $(t_i - t_{i-1})$, is less than $\varepsilon/|f(b) - f(a)|$, to yield integrability (see p. 174).

The integrability can also result from *some* of the terms $(M_i - m_i)(t_i - t_{i-1})$ having the first factor small, while the remaining terms have a small sum of the second factor, as we now see.

For $\varepsilon > 0$ fixed and any partition p, denote by I the subset of the indices $\{1, 2, \ldots, n\}$ for which $M_i - m_i < \varepsilon/2(b - a)$, so that

$$\begin{aligned} U(f, p) - L(f, p) &= \sum_{i \in I} (M_i - m_i)(t_i - t_{i-1}) + \sum_{i \notin I} (M_i - m_i)(t_i - t_{i-1}) \\ &< \frac{\varepsilon}{2(b - a)} \sum_{i \in I} (t_i - t_{i-1}) + \sum_{i \notin I} (M_i - m_i)(t_i - t_{i-1}) \\ &< \frac{\varepsilon}{2} + \sum_{i \notin I} (M_i - m_i)(t_i - t_{i-1}). \end{aligned}$$

For the $i \notin I$ terms, the $M_i - m_i$ factors are not small; they can be estimated crudely, say by $M - m$ for any upper bound M and lower bound m of f on $[a, b]$. This gives

$$U(f, p) - L(f, p) < \frac{\varepsilon}{2} + (M - m) \sum_{i \notin I} (t_i - t_{i-1})$$

and reduces the question of the integrability of f to the question whether p_ε can be chosen so that

$$\sum_{i \notin I} (t_i - t_{i-1}) < \frac{\varepsilon}{2(M - m)}.$$

To formalize the separation of the index set $\{1, 2, \ldots, n\}$ into the subset I and its complement, the following definition is useful.

Definition. For a bounded function f and an interval J that intersects dom f,

$$\omega(f, J) = \sup\{f(x): x \in J\} - \inf\{f(x): x \in J\}$$

is called the **oscillation of f on J**.

The foregoing discussion leads to the

Integrability Criterion. For every $\varepsilon > 0$ there is a p_ε such that the subintervals of p_ε on which $\omega[f, [t_{i-1}, t_i]] \geq \varepsilon/2(b - a)$ have total length $< \varepsilon/2(M - m)$.

It is especially easy to evaluate the oscillations in case the function only has finitely many distinct values, and this case gives some insight.

Definition. For any set S of reals, the **characteristic function of S** is χ_S, where

$$\chi_S(x) = \begin{cases} 1 & \text{if } x \in S, \\ 0 & \text{if } x \notin S. \end{cases}$$

If f takes only the values $y_1, y_2, \ldots, y_k$ and S_i is the set of points where f takes the value y_i, $i = 1, 2, \ldots, k$, then

$$f = \sum_{i=1}^{k} y_i \chi_{S_i}.$$

Our earlier example of a nonintegrable function on $[a, b]$ is just the restriction to $[a, b]$ of $\chi_{\mathbb{Q}}$, where $\mathbb{Q}$ is the set of all rationals.

Since both $\mathbb{Q}$ and its complement are everywhere dense, the oscillation of this function on *any* interval is 1.

In general, the oscillation of any χ_S on an interval I is zero unless I intersects both S and its complement, in which case it is 1. Thus the integrability criterion focuses attention on those points with the property that every interval containing the point meets both S and its complement.

Definition. Let S be any set of reals. If every neighborhood of x_0 has points in common with both S and its complement, x_0 is called a **boundary point** of S. The set of all boundary points of S is called the **boundary** of S and is denoted ∂S.

EXERCISE 1. Give examples of sets S for which:

(a) $\partial S = \mathbb{R}$;
(b) $\partial S = \phi$;
(c) $\partial S = S$;
(d) $\partial S \subsetneqq S$; and
(e) $\partial S \cap S = \phi$.

EXERCISE 2. S is closed if and only if $\partial S \subset S$.

Now if $S \subset [a, b]$ we see that $\chi_S \in R[a, b]$ if and only if, for $\varepsilon > 0$, there are finitely many subintervals of $[a, b]$ such that their union contains ∂S and the sum of their lengths is $< \varepsilon/2$. For, such intervals, together with their complementary intervals in $[a, b]$ constitute a partition p_ε as required by the integrability criterion.

Definition. A set of reals **has content zero** if, given $\varepsilon > 0$, there is a covering of the set by finitely many intervals, the sum of whose lengths is $< \varepsilon$.

In this language, χ_S is integrable if and only if ∂S has content zero.

EXERCISE 3. Formulate a necessary and sufficient condition for the integrability of $f = \sum_{i=1}^{k} y_i \chi_{S_i}$, the general function taking only finitely many distinct values in $[a, b]$.

Remark. If we say that a set $S \subset [a, b]$ **has content** when ∂S has content zero, then $\int_a^b \chi_S$ provides a measure of the size of S (the "content" of S) which generalizes the idea of "length" from subintervals to arbitrary sets having content.

EXERCISE 4. Does $\mathbb{Q} \cap [a, b]$ have content?

EXERCISE 5. Show that the value set of a convergent sequence has content zero. Generalize.

EXERCISE 6. Let C denote the Cantor set; the set remaining in $[0, 1]$ after the removal, successively, of $(\frac{1}{3}, \frac{2}{3})$, then $(\frac{1}{9}, \frac{2}{9})$ and $(\frac{7}{9}, \frac{8}{9})$, etc. *ad* infinitum. Show that $\partial C = C$.

EXERCISE 7. Show that C has content zero. Hint: By calculating a geometric sum, evaluate

$$\frac{1}{3} + \frac{2}{9} + \frac{4}{27} + \cdots + \frac{2^{n-1}}{3^n}$$

and show that it has limit $1 =$ length of $[0, 1]$.

EXERCISE 8. Since $\partial C = C$ has content zero, χ_C is integrable. What is $\int_0^1 \chi_C$?

Remark. Since C is uncountable, we see that there are sets of content zero with all possible cardinalities; finite, countable, uncountable.

A modification of the Cantor construction yields sets that are their own boundaries but do not have content zero, hence some new examples of nonintegrable functions: their characteristic functions.

Choose any α with $0 < \alpha < 1$ and remove from $[0, 1]$, successively,

1. the central open interval of length $\alpha/3$;
2. the central open intervals of length $\alpha/9$ from each of the two remaining closed intervals; and
3. the central open interval of length $\alpha/3^3$ from each of the four remaining closed intervals, etc., ad infinitum.

Call the set that remains C_α.

The total length removed from $[0, 1]$ in forming C_α is

$$\sum_{n=1}^{\infty} \frac{2^{n-1}}{3^n}\alpha = \alpha.$$

Hence C_α cannot be covered by a finite number of subintervals of $[0, 1]$ whose total length is less than $1 - \alpha$; C_α does not have content zero.

Exercise 9. Show that $\partial C_\alpha = C_\alpha$ to complete the proof that χ_{C_α} is not integrable.

Exercise 10. Show that C_α is uncountable. Hint: Imitate the proof that $[0, 1]$ is uncountable which is based on the Nested Intervals Theorem (see p. 55).

Suppose a function assumes only the finitely many distinct values $y_1, y_2, \ldots, y_k$, and the set of points in $[a, b]$ at which the value y_m occurs is a finite union of subintervals of $[a, b]$, with $[a, b]$ the union of all such subintervals. Such a function is called a **step function** on $[a, b]$. Obviously, every step function is integrable.

Exercise 11. Let $R(f, p)$ be any Riemann sum for a bounded function on $[a, b]$. Define a step function whose integral equals $R(f, p)$.

Exercise 12. Show that if f is integrable there exists a sequence of step functions $\{f_n\}$ such that $\int_a^b f_n \to \int_a^b f$.

Remark. An alternative way to introduce the integral is to define it first for step functions and then extend it via limits to more general functions. See, for example, E. Hille [14].

Consider next some examples of functions that assume a countable infinity of distinct values.

EXERCISE 13. Apply the integrability criterion to

$$f(x) = \begin{cases} 1/n, & 1/(n+1) < x \leq 1/n, n = 1, 2, 3, \ldots, \\ 0, & x = 0. \end{cases}$$

EXAMPLE. Let[2]

$$f(x) = \begin{cases} 1/q & \text{if } x = p/q \text{ in lowest terms,} \\ 0 & \text{if } x \text{ is irrational or } x = 0. \end{cases}$$

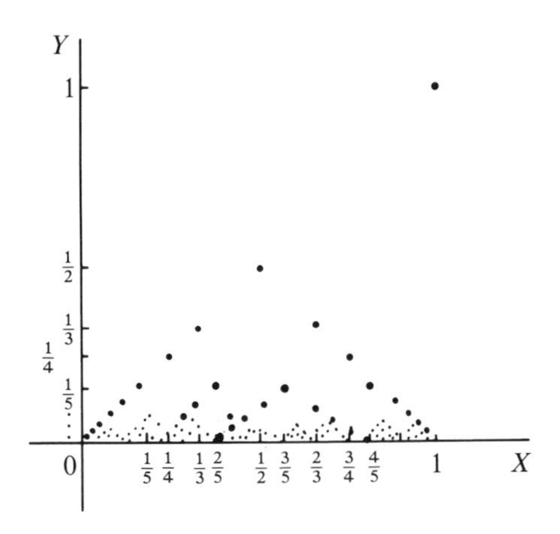

Since $f(x) = 0$ happens in every open interval I that intersects $[0, 1]$:

$f(x) \geq \varepsilon/2$ holds if and only if $\omega(f, I) \geq \varepsilon/2$, for every open interval I containing x; or

$f(x) \geq \varepsilon/2$ if and only if $\inf_I \{\omega(f, I): x \in I, I \text{ open}\} \geq \varepsilon/2$.

Now, the set of points $x \in [0, 1]$ at which $f(x) \geq \varepsilon/2$ is obviously finite. Cover it by subintervals of total length $< \varepsilon/2$ and adjoin all complementary subintervals to form a partition p_ε satisfying the integrability criterion.

We conclude that $f \in R[0, 1]$.

EXERCISE 14. What is $\int_0^1 f$?

In this example, the integrability results from the fact that the sets

$$\{x: f(x) \geq C\}, \qquad \text{for} \quad C > 0,$$

(being finite) have content zero, or that for every $C > 0$,

$$\{x: \inf\{\omega(f, I): x \in I, I \text{ open}\} \geq C\}$$

has content zero.

[2] This f is called the "popcorn function" by Agnew [1].

We formulate the

Definition. For a bounded function f and a point $x \in \operatorname{dom} f$, the **saltus of** f **at** x is $\omega(f, x) = \inf_I \{\omega(f, I): x \in I, I \text{ open}\}$.

With this definition, the example suggests that integrability is the result of the saltus exceeding C, for each $C > 0$, only on a set of content zero.

EXERCISE 15. Suppose f is a step function. Find the saltus at every x.

EXERCISE 16. Find the saltus $\omega(f, 0)$ if

$$f(x) = \begin{cases} \sin(1/x), & x \neq 0, \\ 0, & x = 0. \end{cases}$$

Theorem. *The bounded function f on $[a, b]$ is integrable if and only if*

$$\{x \in [a, b]: \omega(f, x) \geq C\}$$

has content zero for every $C > 0$.

Proof. Suppose $f \in R[a, b]$ and $C > 0$. For any $\varepsilon > 0$, we must show the above set can be covered by a finite union of intervals with total length $< \varepsilon$. These intervals come from the subintervals of a partition.

Choose p_ε so that $U(f, p_\varepsilon) - L(f, p_\varepsilon) < C\varepsilon$ and consider those subintervals containing a point x at which $\omega(f, x) \geq C$. Denote by J the indices of those subintervals. Then, fixing points x_j at which $\omega(f, x_j) \geq C$ in each $[t_{j-1}, t_j]$ for $j \in J$,

$$\begin{aligned} C \sum_{j \in J} (t_j - t_{j-1}) &\leq \sum_{j \in J} \omega(f, x_j)(t_j - t_{j-1}) \\ &\leq \sum_{j \in J} (M_j - m_j)(t_j - t_{j-1}) \leq U(f, p_\varepsilon) - L(f, p_\varepsilon) < C\varepsilon. \end{aligned}$$

The condition is thus seen to be necessary.

For the proof that it is sufficient, fix $\varepsilon > 0$ and apply it to cover the set

$$\left\{x \in [a, b]: \omega(f, x) \geq \frac{\varepsilon}{4(b - a)}\right\}$$

by subintervals $I_1, I_2, \ldots, I_k$ with total length $\sum_{j=1}^k |I_j| < \varepsilon/2(M - m)$. (Note that these intervals can be taken to be open since any that are not can be slightly enlarged to open intervals without exceeding the total length $\varepsilon/2(M - m)$. Thus the part of $[a, b]$ not covered by $I_1, I_2, \ldots, I_k$ can be supposed compact.)

The intervals $I_1, I_2, \ldots, I_k$ will be some of the subintervals of a partition p_ε we shall construct, in order to obtain $U(f, p_\varepsilon) - L(f, p_\varepsilon) < \varepsilon$. The contribution to $U(f, p_\varepsilon) - L(f, p_\varepsilon)$ of the terms corresponding to these subintervals is at most $\sum_{j=1}^k (M - m)|I_j| < \varepsilon/2$.

The rest of $[a, b]$; the set $A = [a, b] - \bigcup_{j=1}^{k} I_j$, consists of points x for which $\omega(f, x) < \varepsilon/4(b-a)$. But if $\omega(f, x) < \varepsilon/4(b-a)$ then some open interval I_x containing x gives $\omega(f, I_x) < \varepsilon/2(b-a)$.

Any such collection $\{I_x: x \in A\}$ is an open covering of the compact set A. There is a finite collection which covers A, by the Heine–Borel Theorem. Take all the endpoints of these subintervals which lie in A as the additional endpoints of the partition p_ε. Now, any subinterval $[t_{i-1}, t_i]$ of p_ε that is contained in A has

$$M_i - m_i \leq \omega(f, I_x) \quad \text{for any } I_x \text{ containing it.}$$

Hence

$$M_i - m_i < \frac{\varepsilon}{2(b-a)}$$

for all such subintervals of p_ε, and the corresponding terms of $U(f, p_\varepsilon) - L(f, p_\varepsilon)$ contribute at most $\varepsilon/2(b-a)$ times their total length, or $< \varepsilon/2$. This completes the proof.

This theorem relates integrability to the points where a nonzero saltus occurs. These points are seen, in turn, to be points of discontinuity, in the

Proposition. $\omega(f, x) = 0$ *if and only if* f *is continuous at* x.

EXERCISE 17. Give the proof.

Definition. An interior point x of dom f is a **jump discontinuity** of f if the one-sided limits

$$\lim_{\substack{h \to 0 \\ h > 0}} f(x+h) \quad \text{and} \quad \lim_{\substack{h \to 0 \\ h < 0}} f(x+h)$$

exist and are unequal. The **size of the jump** is the absolute value of their difference.

EXERCISE 18. If f has a jump of size J at x then $\omega(f, x) = J$.

Denote by $\Delta(f)$ the set of all points of discontinuity of f.

EXERCISE 19.

$$\Delta(f) = \bigcup_{n=1}^{\infty} \left\{x \in [a, b]: \omega(f, x) \geq \frac{1}{n}\right\}.$$

EXERCISE 20. For any $c > 0$, the set $\{x \in [a, b]: \omega(f, x) \geq c\}$ is closed. Hint: It is easy to see that $\{x \in [a, b]: \omega(f, x) < c\}$ is open relative to $[a, b]$.

Thus $\Delta(f)$ is a countable union of closed sets.

In case f is integrable, these observations, together with the preceding theorem, reveal a property of $\Delta(f)$:

For each $n = 1, 2, 3, \ldots$, cover the set $\{x \in [a, b]: \omega(f, x) \geq 1/n\}$ by a finite number of subintervals with total length $< \varepsilon/2^{n+1}$, where $\varepsilon > 0$ is fixed.

The collection of all these intervals, for all n, is a countable collection of intervals which covers $\Delta(f)$ and has total length (a convergent series with sum) $\leq \sum_{n=1}^{\infty} \varepsilon/2^{n+1} = \varepsilon/2 < \varepsilon$.

The set $\Delta(f)$ of points of discontinuity of an integrable function is thus seen to be a set of measure zero under the

Definition. A set of reals has **measure zero** if for every $\varepsilon > 0$ it can be covered by a countable collection of intervals whose total length (converges as a series and) is $< \varepsilon$.

The converse is also true, giving Lebesgue's characterization of Riemann integrable functions, in which we say f is **continuous almost everywhere** when $\Delta(f)$ has measure zero:

Lebesgue's Theorem. $f \in R[a, b]$ *if and only if* f *is continuous almost everywhere.*

Proof. The necessity has been shown. For the sufficiency, suppose $\Delta(f)$ has measure zero. Fix $\varepsilon > 0$ and cover $\Delta(f)$ by intervals $J_1, J_2, \ldots$ with $\sum_{n=1}^{\infty} |J_n| < \varepsilon/2$. Exending each J_n to an open interval, by adding length $< \varepsilon/2^{n+1}$ as needed, we obtain open intervals $\{I_n\}$ with $\sum_{n=1}^{\infty} |I_n| < \varepsilon$ covering $\Delta(f)$.

Since for any $c > 0$,

$$\{x \in [a, b]: \omega(f, x) \geq c\}$$

is compact and contained in $\Delta(f)$, there is a finite subfamily of $\{I_n\}$ covering this set, by the Heine–Borel Theorem. Thus each such set has content zero; f is integrable.

EXERCISE 21. A countable union of sets of measure zero also has measure zero.

EXERCISE 22. The Cantor set C, an uncountable set, has measure zero.

EXERCISE 23. The modified Cantor set C_α, $0 < \alpha < 1$, does not have measure zero.

The fact that monotonic functions are integrable means that their sets of points of discontinuity have measure zero. In fact, they are countable sets, as we see next.

EXERCISE 24. If f is increasing on $[a, b]$ and $x \in (a, b)$ then

$$\omega(f, x) = \inf_{x<t} f(t) - \sup_{t<x} f(t)$$

and

$$\inf_{x<t} f(t) = \lim_{\substack{h\to 0\\ h>0}} f(x+h) \quad \text{and} \quad \sup_{t<x} f(t) = \lim_{\substack{h\to 0\\ h<0}} f(x+h).$$

Similarly for decreasing f.

This exercise shows that every discontinuity of a monotonic function is a jump.

Since the sum of the sizes of all jumps cannot exceed $|f(b) - f(a)|$ there can be only a finite number of jumps of size $\geq c$ for any $c > 0$. Letting c be $1, \frac{1}{2}, \frac{1}{3}, \ldots$, we see that $\Delta(f)$ is a countable union of finite sets, which proves the

Proposition. *A monotonic function f on $[a, b]$ has only jump discontinuities and $\Delta(f)$ is countable.*

Hence, "$\Delta(f)$ has measure zero" is the common property of continuous and monotonic functions that insures their integrability.

EXERCISE 25. Show that a set S has measure zero if and only if it can be covered by a countable family of intervals of *finite* total length, in such a way that every $x \in S$ is in infinitely many intervals of the family.

§4. Improper Integral

The definition of integral by the Riemann–Darboux approach makes essential use of the artificial restriction to bounded functions and compact intervals of integration. The program is not complete until a reasonable and useful extension of the meaning of $\int_a^b f$ is made, to include cases where f is unbounded, where $a = -\infty$ or $b = \infty$ or both, or a combination of these occurs. The most direct approach is to use limits of integrals of bounded functions on compact intervals to define such "improper integrals." It turns out that this approach yields a useful extension, as we shall indicate by brief discussions of the Gamma Function and the Laplace Transform.

We shall only touch upon the substance of these topics enough to expose the main ideas. Fully developed, the theory, techniques, and applications of improper integrals is an extensive technical field. An excellent introduction is given in Widder [27].

Any extension of the meaning of $\int_a^b f$ must preserve the property of additivity with respect to the range of integration, i.e.,

$$\int_a^b f = \int_a^c f + \int_c^b f.$$

Thus, if there are points $c_1 < c_2 < \cdots < c_k$ such that f is unbounded in the neighborhoods of each, the integral $\int_{-\infty}^{\infty} f$ (as an example) should be defined by defining each term of the sum

$$\int_{-\infty}^{c_1} f + \int_{c_1}^{c_2} f + \cdots + \int_{c_{k-1}}^{c_k} f + \int_{c_k}^{\infty} f$$

and taking the sum as the value of $\int_{-\infty}^{\infty} f$.

Each integral here involves two "bad" points, assuming f is bounded on compact intervals not including any c_i. Further simplification results from choosing arbitrary intermediate points $d_0, d_1, \ldots, d_k$,

$$-\infty < d_0 < c_1 < d_1 < c_2 < \cdots < c_{k-1} < d_{k-1} < c_k < d_k < \infty,$$

and taking the sum of terms

$$\int_{-\infty}^{d_0} f + \int_{d_0}^{c_1} f + \int_{c_1}^{d_1} f + \int_{d_1}^{c_2} f + \cdots + \int_{c_k}^{d_k} f + \int_{d_k}^{\infty} f$$

as the integral $\int_{-\infty}^{\infty} f$; terms having a single "bad" point.

Thus, it is enough to discuss two kinds of improper integral:

First Kind: Range of integration a half-line and integrand bounded on compact subintervals.

Second Kind: Range of integration a compact interval and integrand unbounded in every neighborhood of exactly one of the endpoints; otherwise, bounded on compact subintervals.

It is clearly sufficient to focus on the two cases $\int_a^\infty f$ and $\int_a^b f$ where $f(x) \to \infty$ as $x \to a$, $x > a$. Obvious modifications cover the other two cases.

Definition. $\int_a^\infty f$ is defined whenever $f \in R[a, T]$ for every $T > a$ and $\lim_{T\to\infty} \int_a^T f$ exists. When this is the case, the value assigned to $\int_a^\infty f$ is this limit.

Definition. If $f(x) \to \infty$ as $x \to a$, $x > a$, then $\int_a^b f$ is defined whenever $f \in R[t, b]$ for all $t \in (a, b]$ and $\lim_{t\to a; (t>a)} \int_t^b f$ exists. This limit is then the value assigned to $\int_a^b f$.

Remarks. 1. It is useful to write the symbols for integrals and *then* consider whether they have meaning according to the definition, as is done also with infinte series, limits, etc. Hence, we adopt terminology as follows:

(i) The symbol $\int_a^b f$ is called an **improper integral** in case $a = -\infty$, $b = \infty$ or both are finite and f is unbounded in $[a, b]$.

(ii) The improper integral is **convergent** if it is a finite sum

$$\int_{-\infty}^{d_0} f + \int_{d_0}^{t_1} f + \cdots + \int_{d_k}^{\infty} f,$$

as above, in which every term is defined.[3] Otherwise, it is **divergent**.

It is, of course, the question of convergence which is central to the study of improper integrals.

2. The analogy between improper integrals of the first kind and infinite series, via $x \leftrightarrow n$, $f(x) \leftrightarrow a_n$, "partial integral" $\int_a^T f(x)\,dx \leftrightarrow$ "partial sum" $\sum_{n=a}^{T} a_n$, and $\int_a^\infty f(x)\,dx \leftrightarrow \sum_{n=a}^{\infty} a_n$, is a fruitful one for guidance in developing the theory of such integrals and their uses.

EXAMPLES. 1. Power functions on $[1, \infty)$.

$$\int_1^\infty x^\alpha\,dx = \lim_{T\to\infty} \begin{cases} x^{\alpha+1}/(\alpha+1)]_1^T, & \alpha \neq -1, \\ \log T, & \alpha = -1. \end{cases}$$

Hence, we observe that $\alpha \geq -1$ gives divergence and $\alpha < -1$ gives convergence to $-1/(\alpha+1)$. This result, and the analogy with the p-series $\sum_{n=1}^{\infty} 1/n^p$, suggest changing notation, putting $-\alpha = p$, to summarize:

$$\int_1^\infty \frac{1}{x^p}\,dx = \frac{1}{p-1} \qquad \text{if} \quad p > 1 \text{ and diverges if } p \leq 1.$$

2. Exponential functions on $[0, \infty)$.

$$\int_0^\infty e^{-\alpha x}\,dx = \lim_{T\to\infty} \begin{cases} -(1/\alpha)[e^{-\alpha T} - 1], & \alpha \neq 0, \\ T, & \alpha = 0. \end{cases}$$

Hence,

$$\int_0^\infty e^{-\alpha x}\,dx = \frac{1}{\alpha} \qquad \text{for} \quad \alpha > 0 \text{ and diverges for } \alpha \leq 0.$$

Remark. Since for positive integrands f, the area under the graph of f over $[a, b]$ is $\int_a^b f$, there is an obvious extension of the use of the word "area," to certain unbounded sets, which gives meaning to statements like

[3] Allowing *infinitely many* "bad" points c_i, and addressing the issue of convergence of the resulting infinite sum of integrals, is an obvious further extension. Since any open subset of $\mathbb{R}$ is the union of countably many open intervals, this leads to a definition of *integration over open sets* in general. See Goffman [9] and Hille [14] for discussions which incorporate this step.

"The area of the region indicated in the figure is 1."

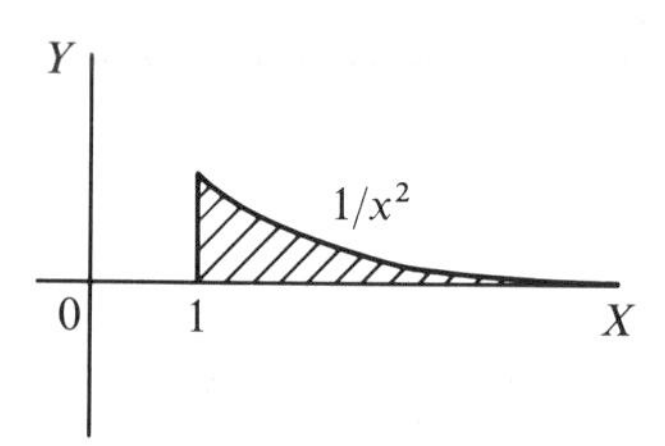

and

"The area of the region indicated in the figure is 1."

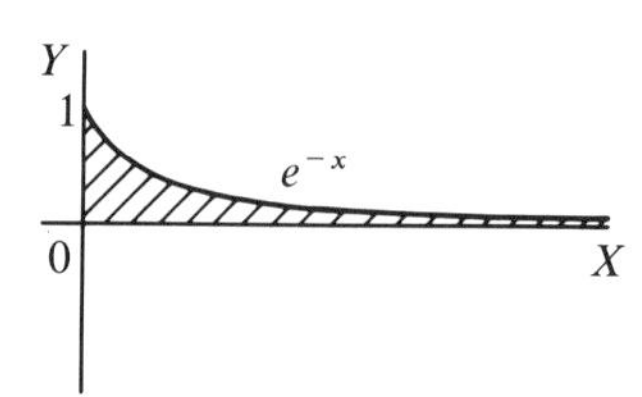

EXERCISE 1. Show that

$$\int_1^\infty \frac{e^{-\sqrt{x}}}{\sqrt{x}}\,dx = \frac{2}{e}.$$

EXERCISE 2. Discuss

$$\int_0^\infty \frac{x\,dx}{(1+x^2)^\alpha}.$$

EXERCISE 3. Criticize the calculation

$$\int_{-\infty}^\infty \frac{1}{x^2}\,dx = \lim_{T\to\infty}\left(\frac{-1}{x}\Big]_{-T}^{T}\right) = 0$$

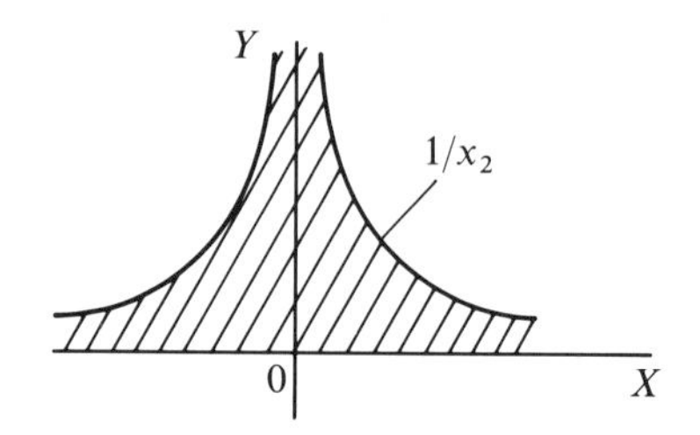

which assigns zero area to the region shaded in the figure.

EXERCISE 4. Show that

$$\int_{-\infty}^0 \frac{1}{1+x^2}\,dx = \frac{\pi}{2}.$$

EXERCISE 5. Discuss $\int_0^\infty \sin x\,dx$.

In each of the above integrals the "partial integrals" could be evaluated by the use of an antiderivative, and $\lim_{T\to\infty}$ examined directly. When an antiderivative is not at hand, one resorts to tests for convergence, separating the question of the existence from that of the value of the limit.

The simplest test is applicable to nonnegative integrands; it results at once from the consequent monotonicity in T of $\int_a^T f$.

Comparison Test. *If* $0 \leq f(x) \leq g(x)$ *for* $x > a$ *and* $\int_a^\infty g$ *converges, then* $\int_a^\infty f$ *converges and* $0 \leq \int_a^\infty f \leq \int_a^\infty g$.

(We assume, of course, that f and $g \in R[a, T]$ for all $T > a$.)

EXERCISE 6. Give the proof and an example showing that equality can hold in the conclusion even though $0 \leq f(x) < g(x)$ for all $x > a$.

EXERCISE 7. Does $\int_1^\infty (e^{1/x}/(1 + x^2))\, dx$ converge?

EXERCISE 8. The series analogy suggests that a necessary condition for the convergence of $\int_a^\infty f(x)\, dx$ is that $f(x) \to 0$ as $x \to \infty$, since a series $\sum_{n=0}^\infty a_n$ converges only if $a_n \to 0$ as $n \to \infty$. Show that this is false by constructing $f(x)$ to have value 1 for $x = 0, 1, 2, \ldots$ and so that $\int_{i-1}^i f = 1/2^i$ for $i = 1, 2, 3, \ldots$, hence that $\int_0^\infty f = 1$.

EXERCISE 9. State, with proof, a comparison test to infer the *divergence* of $\int_a^\infty f(x)\, dx$.

In case f takes values of both signs for arbitrarily large values of x, so that the comparison test is of no use, the simplest resort is to consider $|f|$ in place of f.

Definition. $\int_a^\infty f$ is **absolutely convergent** if $\int_a^\infty |f|$ is convergent.

This works because we have the

Proposition. *Absolute convergence implies convergence.*

Proof. Since

$$0 \leq |f(x)| - f(x) \leq 2|f(x)|,$$

the hypothesis and the comparison test insure the convergence of $\int_a^\infty (|f| - f)$. Since

$$\int_a^T f = \int_a^T \{|f| - (|f| - f)\} = \int_a^T |f| - \int_a^T (|f| - f),$$

for any $T > a$, the result follows by the result on the limit of a difference.

Integrals that converge but are not absolutely convergent are called **conditionally convergent** (as with series). A simple test for convergence, in the varying sign case where absolute convergence is not verified, exists in case the sign variations in $f(x)$ occur in a simple way—analogous to alternating signs for the terms of a series. Recall that an alternating series $\sum_{n=0}^\infty (-1)^n a_n$, where $a_n \geq 0$ for all n, converges if the sequence $\{a_n\}$ is decreasing and $\to 0$.

The analogy to alternating signs which is most immediate is sinusoidal oscillation: $f(x) \sin x$ with $f(x) \geq 0$. We have the

Proposition. *If f is decreasing and $\lim_{x\to\infty} f(x) = 0$ then $\int_a^\infty f(x) \sin x\, dx$ converges.*

Proof. Note that, under our assumptions, $f(x) \geq 0$ for $x > a$ and $f \in R[a, T]$ for $T > a$. We may take $a = 0$ for simplicity.

Fix T and choose $n \in \mathbb{N}$ so that $n\pi < T \leq (n+1)\pi$. Then

$$\begin{aligned}\int_0^T f(x) \sin x\, dx &= \sum_{k=0}^{n-1} \int_{k\pi}^{(k+1)\pi} f(x) \sin x\, dx + \int_{n\pi}^T f(x) \sin x\, dx \\ &= \sum_{k=0}^{n-1} (-1)^k \int_{k\pi}^{(k+1)\pi} |f(x) \sin x|\, dx + \int_{n\pi}^T f(x) \sin x\, dx.\end{aligned}$$

We consider the limit as $T \to \infty$, directly for the last term and by the alternating series test for the first term. First,

$$\begin{aligned}\left|\int_{n\pi}^T f(x) \sin x\, dx\right| &\leq \int_{n\pi}^{(n+1)\pi} |f(x)||\sin x|\, dx \leq f(n\pi) \int_{n\pi}^{(n+1)\pi} |\sin x|\, dx \\ &= 2f(n\pi)\end{aligned}$$

and the $f(x) \to 0$ hypothesis shows that the last term above has limit 0 as $n \to \infty$, i.e., as $T \to \infty$.

Next, the limit of the first term exists if and only if $\sum_{k=0}^\infty (-1)^k a_k$ converges, where $a_k = \int_{k\pi}^{(k+1)\pi} f(x)|\sin x|\, dx$. But

$$a_k = \int_{k\pi}^{(k+1)\pi} f(x)|\sin x|\, dx \geq \int_{(k+1)\pi}^{(k+2)\pi} f(x)|\sin x|\, dx = a_{k+1}$$

because f is decreasing, and, for the same reason,

$$a_k \leq f(k\pi) \int_{k\pi}^{(k+1)\pi} |\sin x|\, dx = 2f(k\pi) \to 0$$

as $k \to \infty$. The alternating series thus converges.

EXAMPLES. 1. $\int_1^\infty ((\sin x)/x^2)\, dx$ converges absolutely by the comparison

$$\left|\frac{\sin x}{x^2}\right| \leq \frac{1}{x^2}.$$

2. $\int_0^\infty ((\sin x)/x)\, dx$ converges conditionally. (Note: 0 is not a "bad" point since $\sin x/x \to 1$ as $x \to 0$.)

The proposition applies and shows the convergence. It remains to show that absolute convergence fails, which we do by reducing to a series,

$$\int_0^\infty \left|\frac{\sin x}{x}\right| dx = \sum_{k=0}^\infty \int_{k\pi}^{(k+1)\pi} \frac{|\sin x|}{x}\, dx$$

and showing a comparison with the divergent harmonic series:

$$\int_{k\pi}^{(k+1)\pi} \frac{|\sin x|}{x}\, dx \geq \frac{1}{(k+1)\pi} \int_{k\pi}^{(k+1)\pi} |\sin x|\, dx = \frac{2}{(k+1)\pi}.$$

EXERCISE 10. Show that $\int_0^\infty \sin x^2\, dx$ is conditionally convergent. (Hint: Substitute $t = x^2$.)

The comparison test has trivial generalizations. Namely, that $0 \leq f(x) \leq Kg(x)$ for all $x > M$, for some positive constants K and M, gives the conclusion:

$$\int_g^\infty g \text{ converges} \quad \text{implies} \quad \int_a^\infty f \text{ converges.}$$

Such conditions hold in case

$$\overline{\lim} \frac{f(x)}{g(x)} \geq 0.$$

EXERCISE 11. State and prove a "Limit Comparison Test" for convergence. For divergence.

Remark. The integrals $\int_2^\infty x\, dx/\sqrt{x^5+1}$ and $\int_2^\infty x\, dx/\sqrt{x^5-1}$ look alike, and both look like $\int_2^\infty (1/x^{3/2})\, dx$ (the ± 1 is "irrelevant"). A comparison is easy to get for the first one:

$$\frac{x}{\sqrt{x^5+1}} < \frac{x}{\sqrt{x^5}} = \frac{1}{x^{3/2}} \qquad \text{for} \quad x > 2.$$

A comparison for the second one is less apparent. The limit form gets at the result more easily:

$$\lim_{x\to\infty} \frac{x^{5/2}}{\sqrt{x^5 \pm 1}} = \lim_{x\to\infty} \frac{1}{\sqrt{1 \pm x^{-5/2}}} = 1.$$

EXERCISE 12. Test for convergence.

(a) $\displaystyle\int_0^\infty \frac{dx}{\sqrt{1+2x^2}}$. (b) $\displaystyle\int_2^\infty \frac{dx}{x(\log x)^\alpha}$. (c) $\displaystyle\int_0^\infty \frac{\cos x}{\sqrt{1+x^3}}$.

EXERCISE 13. Discuss $\int_1^\infty x^\alpha e^{-x}\, dx$.

We shall not present details of the corresponding ideas in the case of improper integrals of the second kind. We do, however, call attention to the results for power functions, which are complementary to those in the "first kind" case—the analogy of the p-series. (In both cases, the results provide comparison integrals for applying comparison tests.)

EXAMPLE

$$\int_a^b \frac{1}{(x-a)^p}\,dx = \frac{(b-a)^{1-p}}{1-p} \qquad \text{for} \quad -\infty < p < 1$$

and diverges for $p \geq 1$.

EXERCISE 14. The integral $\int_0^1 x^\alpha e^{-x}\,dt$ is proper if $\alpha \geq 0$. Show that when $-1 < \alpha < 0$, it is a convergent improper integral.

The last two exercises show that

$$\Gamma(x) = \int_0^\infty t^{x-1}e^{-t}\,dt$$

is defined for all $x > 0$. This is the **Gamma Function**. The reason it is of special interest is revealed by an integration-by-parts (integrating the t^{x-1} factor):

$$\Gamma(x) = \lim_{\substack{T\to\infty\\ s\to 0}} \left(\int_s^T t^{x-1}e^{-t}\,dt \right) = \lim_{\substack{T\to\infty\\ s\to 0}} \left(\frac{-t^x e^{-t}}{x}\Bigg|_s^T + \frac{1}{x}\int_s^T t^x e^{-t}\,dt \right)$$

$$= \frac{1}{x}\Gamma(x+1);$$

Γ satisfies the functional equation $\Gamma(x+1) = x\Gamma(x)$.

In particular, since $\Gamma(1) = 1$, an induction shows that

$$\Gamma(n+1) = n! \qquad \text{for} \quad n = 0, 1, 2, \ldots;$$

Γ provides a generalization of "factorial" to a continuous variable. (It can be shown to be the *only* such generalization satisfying certain mild regularity conditions. See Artin [2].)

The creation of this new function makes possible the generalization of various formulas which involve factorials.

The convergence of the integral for $\Gamma(x)$ can be shown to be uniform on any compact interval of the positive axis (Proof?). This, in turn, leads to proofs that Γ is continuous and, in fact, infinitely differentiable, via general results about improper integrals depending on a parameter (Widder, loc. cit.).

We wish also to mention the Laplace Transform, again without delving beneath the surface, but only to suggest ways in which our definitions have proved to be useful.

The representation of a function $g(x)$ by a power series, the formula $g(x) = \sum_{n=0}^\infty a_n x^n$, has the analogue

$$g(x) = \int_0^\infty a(x)x^t\,dt.$$

Taking $x > 0$, the expression x^t has a positive value, $e^{t \log x}$. The standard notation uses this and writes $-s$ for $\log x$ and $f(s)$ for $g(e^{-s})$:

$$f(s) = \int_0^\infty a(t)e^{-st}\,dt.$$

This $f(s)$ is called the **Laplace Transform of** $a(t)$.

Clearly, questions of convergence determine the domain of f and its continuity–differentiability properties. Assuming these issues can be settled so that the formulas have meaning, we observe the consequences of an integration-by-parts, differentiating $a(t)$:

$$f(s) = \lim_{\substack{T\to\infty\\ \tau\to 0}} \left(-\frac{a(t)e^{-st}}{s}\bigg|_\tau^T + \frac{1}{s}\int_0^\infty a'(t)e^{-st}\,dt \right).$$

Writing $\mathscr{L}(a)$ for f, this gives (under mild assumptions) the formula

$$s\mathscr{L}(a) - a(0) = \mathscr{L}(a');$$

"Multiplication by the variable" is the operation on Laplace Transforms which corresponds (up to subtraction of $a(0)$) to "differentiation" for the functions.

This observation can be exploited to turn problems involving differentiation into *algebraic* problems, together with the passing between functions and their Laplace Transforms. Above all, the solving of differential equations can be reduced to routine computations in many cases—so-called *operational methods.*

Here is a simple example to illustrate the ideas. Suppose $y(x)$ is sought, so that

$$y'' + y = \sin 3x, \qquad y(0) = 0 \quad \text{and} \quad y'(0) = 1.$$

If y solves the problem, then taking Laplace Transforms and applying the above formula,

$$\begin{aligned} \mathscr{L}(y'') + \mathscr{L}(y) &= \mathscr{L}(\sin 3x), \\ [s\mathscr{L}(y') - y'(0)] + \mathscr{L}(y) &= \mathscr{L}(\sin 3x), \\ s^2\mathscr{L}(y) - 1 + \mathscr{L}(y) &= \mathscr{L}(\sin 3x), \\ \mathscr{L}(y) &= \frac{\mathscr{L}(\sin 3x) + 1}{s^2 + 1}. \end{aligned}$$

EXERCISE 15. Show that $\mathscr{L}(\sin px) = p/(x^2 + p^2)$.

Now, this gives

$$\mathscr{L}(y) = \frac{3}{(s^2+9)(s^2+1)} + \frac{1}{s^2+1},$$

and it remains only to identify the y having this Laplace Transform. By partial fractions,

$$\mathscr{L}(y) = -\frac{1}{8}\frac{3}{s^2+9} + \frac{8}{11}\frac{1}{s^2+1},$$

which gives, in view of the exercise,

$$y(x) = -\tfrac{1}{8}\sin 3x + \tfrac{11}{8}\sin x.$$

EXERCISE 16. Verify that this y does solve the initial-value problem posed.

The method is discussed in many differential equations texts, see Simmons [25], for example. For further operational methods, see Kaplan [17].

§5. Arclength. Bounded Variation

If we sketch an arc in the plane and ask, "How long is it?", an answer would call for definitions of arc and arclength, including a characterization of those arcs that *have length* (finite), and a usable procedure for the computation of the arclength.

A geometric intuition suggests itself at once: Since we know the lengths of line segments, hence also of **polygonal arcs**—finitely many segments joined in order end-to-end—one can imagine all possible polygonal arcs "inscribed" in the arc and take the least upper bound of their lengths as the proper measure of the length of the arc.

By fixing a rectangular coordinate system in the plane we can proceed to define arc and arclength analytically. (The unit of measurement of arclength is fixed by choosing the same unit on both the X and Y axes.)

Definition. An **arc** in the $X-Y$ plane is the image there of a compact interval in $\mathbb{R}$ under a continuous map of $\mathbb{R}$ into the plane.

An arc is **simple** if there is a *one–one* map (continuous) between it and a compact interval.

EXAMPLES. 1. The segment joining (x_0, y_0) and (x_1, y_1) is a simple arc since it is the range of the one–one function

$$t \to ([1-t]x_0 + tx_1, [1-t]y_0 + ty_1) \quad \text{on} \quad 0 \le t \le 1.$$

2. The unit circle is an arc, since, for example, it is the range of the function

$$t \to (\cos t, \sin t) \quad \text{on} \quad 0 \le t \le 2\pi.$$

It is not simple; any mapping of a *closed* interval onto the circle must map at least one pair of distinct points onto the same point.

EXERCISE 1. Prove this. Hint: It is enough to show that a continuous g: $[0, 2\pi] \to$ unit circle which is one–one on $(0, 2\pi)$ and maps 0 into $(1, 0)$ must also map 2π into $(1, 0)$.

3. *The graph* of any continuous function on $[a, b]$ with real values is a simple arc in the plane since

$$x \to (x, f(x)) \quad \text{on} \quad a \le x \le b$$

is a one–one continuous map having the graph as image.

4. The space-filling curves (see the appendix to Chapter 3) are arcs.

There are a variety of functions having a given arc as range: each is called a **parametric representation** of the arc. If one representation is $(x(t), y(t))$ on $a \le t \le b$, then any continuous, one–one function φ on an interval $[c, d]$ with range $[a, b]$ provides another parametric representation of the same arc: $(x(\varphi(s)), y(\varphi(s)))$ on $c \le s \le d$. Such a φ is called a **change of parameter**. We know that a change of parameter is strictly monotonic and has a continuous inverse. In the case of a simple arc, it is true, conversely, that any two representations are connected by a change of parameter φ, as we show next. Suppose f: $[a, b] \to X-Y$ plane and g: $[c, d] \to X-Y$ plane are two parametric representations of a simple arc Γ. Clearly, $\varphi = f^{-1} \circ g$ is a one–one function of $[c, d]$ onto $[a, b]$. For the continuity of φ, we need only examine that of f^{-1}, which can be expressed in the form: S open in $[a, b]$ implies $f(S)$ open relative to Γ. In view of the one–one property, which permits the interchange of the application of f with the taking of complements, it is the same to consider whether $f(S)$ is closed whenever S is closed. But a closed $S \subset [a, b]$ is compact, hence its continuous image is compact, hence closed (same as closed relative to Γ).

EXERCISE 2. Suppose a function f has domain A and range B. Show that f is one–one if and only if $f(A - S) = B - f(S)$ for every $S \subset A$.

The points $f(a)$ and $f(b)$ are called the **endpoints** of the arc given by f: $[a, b] \to X$–Y plane. If they coincide, the arc is called a **closed arc**. The endpoints and the closedness/nonclosedness are *geometric* properties of arcs, in the sense that they are independent of the choice of parametrization; are *invariant* under changes of parameter. Of course, the arclength must be given an invariant definition.

EXERCISE 3. The segment of Example 1 above has the representation on $c \leq s \leq d$

$$x(s) = \frac{s-d}{c-d}x_0 + \frac{s-c}{d-c}x_1, \qquad y(x) = \frac{s-d}{c-d}y_0 + \frac{s-c}{d-c}y_1.$$

(a) What is the change of parameter?

(b) Give a parametric representation of the polygonal arc joining, in order, $(x_0, y_0), (x_1, y_1), \ldots, (x_n, y_n)$.

EXERCISE 4. By eliminating the parameter, express each arc as a subset of the nullset of some $F(x, y)$:

(a) $x(t) = 4t^2, \quad y(t) = t^4, \quad 0 \leq t \leq 1.$

(b) $x(t) = 3 \cos t, \quad y(t) = 4 \sin t, \quad 0 \leq t \leq 2\pi.$

(c) $x(t) = \sqrt{1 - t^2}, \quad y(t) = t, \quad -1 \leq t \leq 1.$

A particular parametric representation of an arc, f: $[a, b] \to X$–Y plane, can be viewed as the *motion* of a point in the plane by thinking of the parameter as time. Such a motion could reverse directions and retrace parts of the arc in many complex ways as t increases from a to b. There is, for a given parametric representation, the question of the "length of the trip along the arc" as well as the length of the arc, the two being the same in case of a simple arc. For the sake of simplicity, we shall put aside the possibilities of "doubling back" and give the definition in the main case, that of a simple arc.

Definition. Let $t \to (x(t), y(t))$, $t \in [a, b]$, be a one–one representation of a simple arc Γ. If $p = \{t_0, t_1, \ldots, t_n\}$ is any partition of $[a, b]$, the polygonal arc joining, in order, the points $(x(t_0), y(t_0)), (x(t_1), y(t_1)), \ldots, (x(t_n), y(t_n))$ is said to be **inscribed in** Γ.

Of course, the length of this inscribed polygonal arc is

$$l(\Gamma, p) = \sum_{i=1}^{n} [(x(t_i) - x(t_{i-1}))^2 + (y(t_i) - y(t_{i-1}))^2]^{1/2}.$$

If the arc Γ is reparametrized, say by a strictly increasing $\varphi: [c, d] \to [a, b]$, then a partition of $[c, d]$ giving the same inscribed polygonal arc as the partition p of $[a, b]$ is given by $\{\varphi^{-1}(t_0), \varphi^{-1}(t_1), \ldots, \varphi^{-1}(t_n)\}$. If φ is decreasing, reversing the order yields such a partition.

We conclude that the set of all lengths of polygonal arcs inscribed in Γ is a set of reals $\pi(\Gamma)$ which is invariant under changes of parameter. Thus, we may make the

Definition. The simple arc Γ is **rectifiable** if $\pi(\Gamma)$ is bounded (above). In this case, the **length of** Γ is

$$l(\Gamma) = \sup \pi(\Gamma).$$

This definition gives precise expression to the intuition indicated at the beginning of the discussion.

Next, we characterize the simple arcs that are rectifiable (i.e., have (finite) length).

EXERCISE 5. Prove that for $u, v \in \mathbb{R}$,

$$|u| \le \sqrt{u^2 + v^2} \le |u| + |v|.$$

Applying this inequality to the terms of the sum $l(p, \Gamma)$ for an arbitrary partition p gives

$$\sum_{i=1}^{n} |x(t_i) - x(t_{i-1})| \le l(\Gamma, p) \quad \text{and} \quad \sum_{i=1}^{n} |y(t_i) - y(t_{i-1})| \le l(\Gamma, p)$$

and also

$$l(\Gamma, p) \le \sum_{i=1}^{n} |x(t_i) - x(t_{i-1})| + \sum_{i=1}^{n} |y(t_i) - y(t_{i-1})|.$$

It is then clear that the boundedness of the set $\pi(\Gamma)$ of all the numbers $l(\Gamma, p)$ is equivalent to the simultaneous requirement that the following sets be bounded:

$$\left\{\sum_{i=1}^{n} |x(t_i) - x(t_{i-1})| : p = \{t_0, t_1, \ldots, t_n\} \text{ a partition of } [a, b]\right\}$$

and

$$\left\{\sum_{i=1}^{n} |y(t_i) - y(t_{i-1})| : p = \{t_0, t_1, \ldots, t_n\} \text{ a partition of } [a, b]\right\}.$$

This observation is the characterization of rectifiability in terms of the parameter functions. We introduce terminology and state this as a theorem.

Definition. A real-valued function φ on $[a, b]$ is of **bounded variation**, in symbols, $\varphi \in \mathrm{BV}[a, b]$, if the set

$$\left\{\sum_{i=1}^{n} |\varphi(t_i) - \varphi(t_{i-1})| : p = \{t_0, t_1, \ldots, t_n\} \text{ a partition of } [a, b]\right\}$$

is bounded. For such a φ, we denote by $V(\varphi: [a, b])$, called the **total variation** of φ on $[a, b]$, the supremum of this set.

Theorem. *A simple arc Γ is rectifiable if and only if the coordinate functions of its parametric representations are of bounded variation.*

Before proceeding to the question of calculating arclength, it is in order to digress on the subject of functions of bounded variation, noting that no continuity is assumed in the definition as it is in the application to arclength.

EXERCISE 6. Every function of bounded variation is bounded.

The most obvious examples of functions in $\mathrm{BV}[a, b]$ are the monotonic functions on $[a, b]$; the sum $\sum_{i=1}^{n} |\varphi(t_i) - \varphi(t_{i-1})|$ telescopes, when φ is monotonic, to $|\varphi(b) - \varphi(a)|$, which is then $V(\varphi; [a, b])$.

Since a monotonic function can be discontinuous to the extent of having countably many jumps, the same extent of discontinuity is possible for functions in $\mathrm{BV}[a, b]$.

Are there continuous functions that are not of bounded variation? A function like $\sin \frac{1}{x}$ comes to mind as a way to get unbounded variation since its values swing from -1 to 1 infinitely often in $(0, a]$. But no continuous extension to $[0, a]$ exists! A factor x corrects this difficulty, and in fact produces an example, as we see in the

EXAMPLE. The continuous function on $[0, 1]$

$$f(x) = \begin{cases} x \sin \dfrac{\pi}{2x}, & 0 < x \leq 1, \\ 0, & x = 0, \end{cases}$$

is not of bounded variation. The fact that

$$f\left(\frac{1}{k}\right) = \begin{cases} 0, & k \text{ even } \varepsilon \mathbb{N}, \\ \pm\dfrac{1}{k}, & k \text{ odd } \varepsilon \mathbb{N}, \end{cases}$$

shows that for the partitions $p_n = \left\{0, \frac{1}{2n-1}, \frac{1}{2n-2}, \ldots, \frac{1}{3}, \frac{1}{2}, 1\right\}$

$$\sum_{i=1}^{2n-1} |f(x_i) - f(x_{i-1})| = \frac{2}{2n-1} + \frac{2}{2n-3} + \cdots + \frac{2}{5} + \frac{2}{3} + 1.$$

The claim follows because $1 + \frac{2}{3} + \frac{2}{5} + \cdots + \frac{2}{2n-1} > 1 + \frac{1}{2} + \frac{1}{3} + \cdots + \frac{1}{n}$, which shows the sums to be unbounded.

When φ is continuous on $[a, b]$ and differentiable on (a, b) the Mean-Value Theorem gives points $\xi_i \in (t_{i-1}, t_i)$ in every subinterval of a partition so that

$$\sum_{i=1}^{n} |\varphi(t_i) - \varphi(t_{i-1})| = \sum_{i=1}^{n} |\varphi'(\xi_i)|(t_i - t_{i-1}).$$

It is then clear that if φ' is bounded, say $|\varphi'(t)| \leq M$ on $[a, b]$, then $\varphi \in \mathrm{BV}[a, b]$ and $V(\varphi\colon [a, b]) \leq M(b - a)$.

In particular, $C^1[a, b] \subset \mathrm{BV}[a, b]$ and for $\varphi \in C^1$, $V(\varphi; [a, b]) = \sup_{a \leq t \leq b} |\varphi'(t)|$.

Exercise 7. Is $g \in \mathrm{BV}[0, 1]$ for

$$g(x) = \begin{cases} x^2 \sin \dfrac{\pi}{2x}, & 0 < x \leq 1, \\ 0, & x = 0? \end{cases}$$

Exercise 8. $\mathrm{BV}[a, b]$ is a vector space of functions on $[a, b]$. Prove, and show that $V(f + g\colon [a, b]) \leq V(f\colon [a, b]) + V(g\colon [a, b])$ and $V(\alpha f\colon [a, b]) = |\alpha| V(f\colon [a, b])$ for α real.

Exercise 9. If f and $g \in \mathrm{BV}[a, b]$ then $fg \in \mathrm{BV}[a, b]$ (BV is an algebra of functions). Find an inequality which estimates $V(fg\colon [a, b])$ in terms of $V(f\colon [a, b])$ and $V(g\colon [a, b])$.

Exercise 10. Discuss quotients of functions in $\mathrm{BV}[a, b]$.

It is obvious that $f \in \mathrm{BV}[a, b]$ implies that f is also of bounded variation on any closed subinterval of $[a, b]$. We need the fact that the total variation of f is an additive function of the subinterval.

Proposition. *For any $c \in [a, b]$ (and taking $V(f\colon [x, x])$ to be zero)*

$$V(f\colon [a, b]) = V(f\colon [a, c]) + V(f\colon [c, b])$$

Proof. We may suppose $c \in (a, b)$. For any partition p, let $[t_{k-1}, t_k)$ contain c and use

$$|f(t_k) - f(t_{k-1})| \leq |f(c) - f(t_{k-1})| + |f(t_k) - f(c)|$$

to see that

$$\begin{aligned} \sum_{i=1}^{n} |f(t_i) - f(t_{i-1})| &\leq \left\{ \sum_{i=1}^{k-1} |f(t_i) - f(t_{i-1})| + |f(c) - f(t_{k-1})| \right\} \\ &\quad + \left\{ \sum_{i=k+1}^{n} |f(t_i) - f(t_{i-1})| + |f(t_k) - f(c)| \right\} \\ &\leq V(f\colon [a, c]) + V(f\colon [c, b]), \end{aligned}$$

hence that, taking suprema,

$$V(f: [a, b]) \leq V(f: [a, c]) + V(f: [c, b]).$$

For the reverse inequality, fix $\varepsilon > 0$ and choose $t_0, t_1, \ldots, t_n$ so that

$$a = t_0 < t_1 < \cdots < t_m = c \quad \text{and} \quad \sum_{i=1}^{m} |f(t_i) - f(t_{i-1})| > V(f: [a, c]) - \frac{\varepsilon}{2},$$

and

$$c = t_m < t_{m+1} < \cdots < t_n = b \quad \text{and} \quad \sum_{i=m+1}^{n} |f(t_i) - f(t_{i-1})| > V(f: [c, b]) - \frac{\varepsilon}{2}.$$

Then

$$V(f: [a, b]) \geq \sum_{i=1}^{n} |f(t_i) - f(t_{i-1})| > V(f: [a, c]) + V(f: [c, b]) - \varepsilon,$$

and, $\varepsilon > 0$ being arbitrary, the reverse inequality and the assertion follow.

In view of this proposition, the function

$$V_f(x) = V(f: [a, x]), \qquad x \in [a, b],$$

is an increasing function.

But $V_f - f$ is also an increasing function on $[a, b]$ because $x < x'$ gives

$$f(x') - f(x) \leq |f(x') - f(x)| \leq V(f: [x, x']) = V_f(x') - V_f(x),$$

hence $V_f(x) - f(x) \leq V_f(x') - f(x')$.

Since $f = V_f - (V_f - f)$, we have proved the

Theorem. *Every function of bounded variation can be expressed as the difference of two increasing functions.*

Remark. The expression is not unique. *Any* increasing function g can be added to both parts to get another such expression:

$$f = (V_f + g) - (V_f + g - f).$$

EXERCISE 11. Every $f \in \text{BV}[a, b]$ can be expressed as:

(a) the difference of two decreasing functions; and
(b) the sum of an increasing function and a decreasing function.

The terms increasing and decreasing here could be replaced by strictly increasing and strictly decreasing. Why?

The possibilities for discontinuity for functions of bounded variation are seen to be limited to jump discontinuities, at most countable in number, in view of the theorem.

Corollary. $\text{BV}[a, b] \subset R[a, b]$.

EXERCISE 12. Show that the inclusion is proper.

Proposition. *If f is a continuous function of bounded variation then V_f is continuous. Hence, a continuous function of bounded variation is expressible as the difference of two continuous increasing functions.*

Proof. It is enough to show that V_f inherits one-sided continuity from f. We show this for continuity from the left at x_0 for any $x_0 \in (a, b]$, the other case being similar.

Fix $\varepsilon > 0$ and choose δ so that

$$x \in (x_0 - \delta, x_0) \quad \text{gives} \quad |f(x) - f(x_0)| < \frac{\varepsilon}{2}.$$

Choose a partition of $[a, x_0]$, $a = t_0 < t_1 < \cdots < t_{n-1} < t_n = x_0$, so that $t_{n-1} \in (x_0 - \delta, x_0)$ and

$$\sum_{i=1}^{n} |f(t_i) - f(t_{i-1})| > V_f(x_0) - \frac{\varepsilon}{2}.$$

Subtract the last term of the sum to get

$$\sum_{i=1}^{n-1} |f(t_i) - f(t_{i-1})| > V_f(x_0) - \frac{\varepsilon}{2} - |f(t_0) - f(t_{n-1})| > V_f(x_0) - \varepsilon.$$

Now for any $x \in (t_{n-1}, x_0)$,

$$\begin{aligned} V_f(x) &\geq \sum_{i=1}^{n-1} |f(t_i) - f(t_{i-1})| + |f(x) - f(t_{n-1})| \\ &\geq \sum_{i=1}^{n-1} |f(t_i) - f(t_{i-1})| > V_f(x_0) - \varepsilon. \end{aligned}$$

The continuity from the left of V_f at x_0 follows.

The deepest fact about functions of bounded variation, discovered by Lebesgue as a consequence of his integration theory, is that they are *differentiable almost everywhere*. A proof which does not draw upon any such background knowledge was subsequently developed. It is the *first* topic presented in the modern classic, *Functional Analysis*, by F. Riesz and B. Sz. Nagy [22].

The remaining consideration about arclength, its computation, focuses on the formula

$$l(\Gamma) = \int_a^b \sqrt{[x'(t)]^2 + [y'(t)]^2}\, dt$$

for the length of the rectifiable simple arc Γ having the parametric representation $t \to (x(t), y(t))$ on $a \leq t \leq b$. Lebesgue's Theorem indicates that this expression is quite generally meaningful in view of the fact that sets of measure zero are negligible for Lebesgue integration. We shall consider the formula in the case where x and $y \in C^1[a, b]$.

Choosing partitions of $[a, b]$, with $\|p\| \to 0$, so that $l(\Gamma, p) \to l(\Gamma)$;

$$l(\Gamma) = \lim_{\|p\| \to 0} \sum_{i=1}^{n} [(x(t_i) - x(t_{i-1}))^2 + (y(t_i) - y(t_{i-1}))^2]^{1/2},$$

the C^1 assumption provides, by the Mean-Value Theorem, points ξ_i and $\eta_i \in (t_{i-1}, t_i)$, $i = 1, 2, \ldots, n$, such that

$$\begin{aligned} l(\Gamma) &= \lim_{\|p\| \to 0} \sum_{i=1}^{n} [[x'(\xi_i)(t_i - t_{i-1})]^2 + [y'(\eta_i)(t_i - t_{i-1})]^2]^{1/2} \\ &= \lim_{\|p\| \to 0} \sum_{i=1}^{n} \sqrt{[x'(\xi_i)]^2 + [y'(\eta_i)]^2}(t_i - t_{i-1}). \end{aligned}$$

This is almost the above formula. It would be if $\xi_i = \eta_i$, $i = 1, 2, \ldots, n$, were true, because of the Darboux Theorem. The needed variation of the Darboux Theorem is called Du Hamel's Principle, i.e., the

Lemma. *Let $x, y \in C[a, b]$ and $\varphi(x, y)$ be continuous on the closed rectangle [range x] $\times$ [range y], so that $I = \int_a^b \varphi(x(t), y(t))\, dt$ exists. Then for any partitions $p = \{t_0, t_1, \ldots, t_n\}$ of $[a, b]$ and any points ξ_i and $\eta_i \in (t_{i-1}, t_i)$, $i = 1, 2, \ldots, n$,*

$$I = \lim_{\|p\| \to 0} \sum_{i=1}^{n} \varphi(x(\xi_i), y(\eta_i))(t_i - t_{i-1}).$$

Proof. In view of the Darboux Theorem, it suffices to show that given $\varepsilon > 0$ there is a δ such that whenever $\|p\| < \delta$,

$$\left| \sum_{i=1}^{n} \varphi(x(\xi_i), y(\eta_i))(t_i - t_{i-1}) - \sum_{i=1}^{n} \varphi(x(\xi_i), y(\xi_i))(t_i - t_{i-1}) \right| < \varepsilon.$$

This expression is at most

$$\sum_{i=1}^{n} |\varphi(x(\xi_i), y(\eta_i)) - \varphi(x(\xi_i), y(\xi_i))|(t_i - t_{i-1})$$

and the uniform continuity of φ on the rectangle provides a δ such that

$$|\xi_i - \eta_i| < \delta \quad \text{implies} \quad |\varphi(x(\xi_i), y(\eta_i)) - \varphi(x(\xi_i), y(\xi_i))| < \frac{\varepsilon}{b - a}$$

for all $i = 1, 2, \ldots, n$. The result follows.

Remarks. 1. This validates the speculation we made (see p. 145) using the "infinitesimal triangle," to get $ds^2 = dx^2 + dy^2$.

2. In the case of the *graph* of a continuous function f on $[a, b]$, parametrized by $x \to (x, f(x))$, we have the length of the graph given, in the C^1 case, by $\int_a^b \sqrt{1 + [f'(x)]^2}\, dx$.

3. The usefulness of the formula for the computation of $l(\Gamma)$ is subject to the need to know an antiderivative of the integrand. Textbook problems are usually contrived to yield a perfect square for $[x']^2 + [y']^2$. The integral formula is extremely valuable for theory and approximate computation, regardless of whether an antiderivative is available.

EXERCISE 13. Find the length of the graph of:

(a) $\log \cos x$ on $0 \le x \le \pi/4$; and
(b) $x^{2/3}$ on $0 \le x \le 8$.

4. Our discussion is readily generalized to arcs in spaces of higher dimension. The method, however, *does not* generalize to the problem of finding the area of surfaces in space. Finite unions of triangles joined along their edges can play the role of polygonal arcs as simple surfaces with known area. The meaing of "inscribed" causes trouble, as H. Schwarz showed in 1890 (see Widder [27], and Goffman [9] for a discussion of this "Schwarz paradox"). An extensive discussion of the measurement of geometric quantities in connection with the school curriculum is to be found in Lebesgue [20].

A Word About the Stieltjes Integral and Measure Theory

The entire development of the integral can be carried out from a more general beginning than that of area, via length. The transitional observation is that the "elementary areas" $f(x)(x_i - x_{i-1})$ can be expressed, using the distance $d(x)$ = distance to the right of a fixed origin, of x, as $f(x)[d(x_i) - d(x_{i-1})]$ and then substituting for d a more general function φ. Thus, a notion of "integral of f with respect to φ," in the form

$$\int_a^b f(t)\, d\varphi(t) = \lim_{\|p\|\to 0} \sum_{i=0}^{n} f(\xi_i)[\varphi(t_i) - \varphi(t_{i-1})],$$

appears as a possibility.

Of course, there was solid motivation, in terms of applications and the unification of knowledge, underlying the development of this kind of integral—now known as the Stieltjes Integral. We only take note of the following:

(a) A mechanical application: Think of a wire occupying a portion of space which is described by parametric equations with parameter t, and suppose the material the wire is made from *varies* in density. The φ embodies the precise description of how the density varies: $\varphi(b) -$

$\varphi(a) =$ the mass of the portion occupied by the piece of arc corresponding to $a \leq t \leq b$.

With the constant integrand $f(t) = 1$, the Stieltjes integral represents the mass of the piece of wire. With other integrands it represents such quantities as the moments, about various points, axes, or planes, of the object; moments from which the centroid can be found, for example. The motion under various forces can then be described.

(b) Random variables: To describe a variable quantity X that takes its values in a random manner, i.e., governed by some probabilistic, as opposed to deterministic, underlying cause, a function φ of a parameter is invoked, to make specific the "random mechanism" by requiring that $\varphi(b) - \varphi(a) =$ the specified probability of a value in $[a, b]$, and X is made a function of the same real parameter t.

Finitely many values occurring ("randomly") according to a certain φ would be described by two step functions: $X(t) = \sum y_i \chi_{[t_{i-1}, t_i]}$, and a φ giving $\varphi(t_i) - \varphi(t_{i-1}) =$ probability of the value y_i. Then the "expected value $\bar{X}$ of X" is the sum of the possible values, each weighted by its probability;

$$\bar{X} = \sum X(\xi_i)\,[\varphi(t_i) - \varphi(t_{i-1})], \qquad \xi_i \in (t_{i-1}, t_i).$$

The symbol $\int_{-\infty}^{\infty} X(t)\, d\varphi(t)$ would indicate a Stieltjes integral way of writing this and suggest the use of Stieltjes integrals to express the "expected value" (and other crucial statistical quantities) associated with general "random variables."

As these comments suggest, the theory of random variables employs a developed theory of Stieltjes integration.

(c) Generality. The above examples use functions φ which are increasing functions of t. Other applications (e.g., electrical charge) call, more generally, for φ to be of bounded variation, to admit decreases as well as increases, in quite general combinations.

The concept of integrable function must be modified to one of "function f integrable with respect to φ" and corresponding theorems considered. In particular, a desired result would have the form:

If f is integrable with respect to φ and if φ' exists and gives $f\varphi'$ integrable (ordinary sense), then

$$\int_a^b f(t)\, d\varphi(t) = \int_a^b f(t)\varphi'(t)\, dt.$$

A straightforward imitation of the development of the Riemann integral *almost* suffices to develop a satisfactory theory of the Stieltjes integral. However, technical difficulties arise, e.g., in case f and φ have discontinuities at common points. Traditionally, these difficulties are overcome via ad hoc hypotheses. See, for example, Widder [27] or Ross [23]. (Note: Ross has recognized how the theorems can be made to lose awkward ad hoc hypotheses and gain elegance and usefulness by incorporating jump informaton about φ explicitly in defining the upper and lower sums. He appends these observations to his presentations of the traditional discussion.)

The full generality needed for the concept of the integral of one function with respect to another, calls for the incorporation of the Lebesgue kind of extension and the extension from the real line to higher-dimensional Euclidean spaces—and even more general settings. The abstract subject of "Measure Theory" extracts the crucial properties of the functions of intervals $[a, b] \to \varphi(b) - \varphi(a)$ which admit the development of a theory of "measure and integral" sufficiently abstract to embrace all of the above particular interpretations and applications—in a form free of the distractions that go along with the special contexts. The theory is developed in Halmos [16].

CHAPTER 6

Infinite Series

The Calculus, from its very beginnings, has featured the idea of "adding" infinitely many numbers. Formal expressions indicating such "additions," i.e., infinite series, were found to offer symbolic solutions to various problems. One such problem: to describe the motions of a stretched elastic string, such as a violin string, was the focus of a major dispute, which proved to be a watershed event in the history of Mathematics. The dispute was about a series solution to the problem and its effect was, ultimately, the clarification of some very basic concepts of Mathematics and its applications. [See González-Velasco [13] or Kline [18], for the history.)

We shall begin with a quick discussion of the vibrating string, to illustrate the importance of the technical discussion of infinite series which constitutes the bulk of the chapter.

An alternative motivation for that discussion is the search for classes of well-understood functions, to include some nonelementary functions. For, an obvious device for defining such classes is the "addition" of infinitely many functions of a simple kind. Thus, choosing the simple functions a_0, a_1x, $a_2x^2, \ldots$, i.e., forming "polynomials of infinite degree," leads to the subject of Power Series. Choosing the simple functions to be $\frac{1}{2}a_0$,[1] $a_1 \cos x + b_1 \sin x$, $a_2 \cos 2x + b_2 \sin 2x$, ... (in the complex setting, the exponentials $c_n e^{inz}$, $n \in \mathbb{Z}$) broaches the subject of Fourier Series. There are many other examples. We shall present brief treatments of these two examples, after the technical issues of convergence of, and computation with, general infinite series have been considered.

[1] Introducing the $\frac{1}{2}$ artificially in the constant function simplifies some subsequent formulas.

§1. The Vibrating String

A flexible string is stretched between two points and set into motion by being displaced from its equilibrium, into some shape, and released. The tension produces forces which tend to restore the string to its equilibrium, but it overshoots equilibrium and, as a result, begins a vibratory motion. If x denotes the distance of a point on the string from one end at equilibrium, and t denotes elapsed time after release, then knowing the displacement $y(x, t)$ from equilibrium for each x and t constitutes full knowledge of the motion.

The choice of unit for x that makes the length of the string π turns out to be nice. The initial displacement can then be presented as a function $f(x)$ on $[0, \pi]$. For example, "plucking the string in the middle" might be realized by choosing, for some $a > 0$,

$$f(x) = \begin{cases} \dfrac{2a}{\pi}x, & 0 \le x \le \dfrac{\pi}{2}, \\ a - \dfrac{2a}{\pi}\left(x - \dfrac{\pi}{2}\right), & \dfrac{\pi}{2} \le x \le \pi. \end{cases}$$

The problem is to find the $y(x, t)$ resulting from the $f(x)$ for a quite general collection of initial displacements $f(x)$[2].

The solution $y(x, t)$ must satisfy the *boundary conditions* (string attached at the ends)

$$y(0, t) = y(\pi, t) = 0 \qquad \text{for all} \quad t \ge 0$$

and the *initial conditions*

$$y(x, 0) = f(x) \qquad \text{for} \quad 0 \le x \le \pi \quad \text{(displacement)}$$

and

$$\frac{\partial y}{\partial t}(x, 0) = 0 \qquad \text{for} \quad 0 \le x \le \pi \quad \text{(release).}$$

The motion is governed by Newton's Law, $F = ma$, which, under certain simplifying assumptions, takes the form of a partial differential equation that $y(x, t)$ must satisfy: the **Wave Equation**

$$\alpha^2 \frac{\partial^2 y}{\partial x^2} = \frac{\partial^2 y}{\partial t^2}, \qquad \alpha \text{ constant.}$$

First, the string is assumed to be uniform, in the sense that its mass per unit length is a constant δ throughout its length. Then, the tension T is assumed

[2] The more general problem, in which each point is given an initial velocity $g(x)$, rather than simply being released ($g(x) = 0$ for all x), is treated by modifications of the arguments that follow.

large enough that other forces (gravity, air resistance, friction) are negligible by comparison.

The net force acting on a piece of the string located at x, of length Δx, is then the change in $T \sin \theta$ between its ends, while

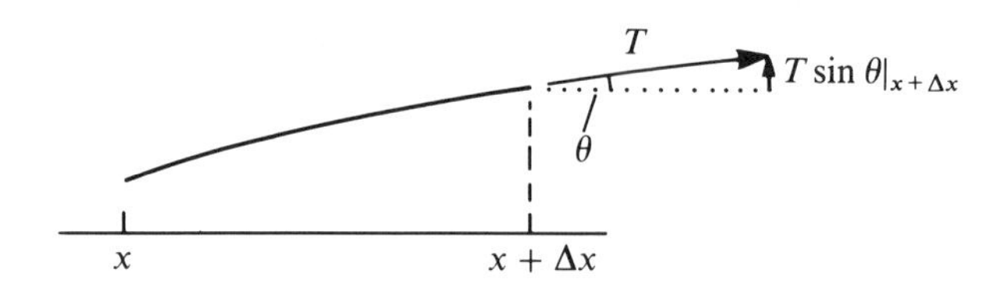

$$ma = \delta \Delta x \frac{\partial^2 y}{\partial t^2}(x, t)$$

for that piece. Thus,

$$F = T \sin \theta|_{x+\Delta x} - T \sin \theta|_x = \delta \Delta x \frac{\partial^2 y}{\partial t^2}(x, t) = ma.$$

Since θ is the inclination angle of the tangent to the graph of $y(x, t)$, t fixed, we have

$$\tan \theta = \frac{\partial y}{\partial x}(x, t).$$

For small angles, the difference between the sine and the tangent is small. Our final simplifying assumption is that the inclination angle θ never gets so large that this difference fails to be negligible, in which case $\partial y/\partial x$ can replace $\sin \theta$ to give

$$F = T\frac{\partial y}{\partial x}(x + \Delta x, t) - T\frac{\partial y}{\partial x}(x, t) = \delta \Delta x \frac{\partial^2 y}{\partial t^2}(x, t) = ma.$$

Dividing by $\delta \Delta x$, letting $\Delta x \to 0$, and writing α^2 for T/δ, we obtain the wave equation.

In summary, we accept the mathematical model of the vibrating string given by the initial-boundary-value problem for the wave equation:

$$\begin{cases} \alpha^2 \dfrac{\partial^2 y}{\partial x^2} = \dfrac{\partial^2 y}{\partial t^2}, & 0 \leq x \leq \pi, t \geq 0, \\ y(0, t) = y(\pi, t) = 0, & t \geq 0, \\ y(x, 0) = f(x), & \dfrac{\partial y}{\partial t}(x, 0) = 0, 0 \leq x \leq \pi. \end{cases}$$

Remark. Of course, the model would be absurd if it failed to have a unique solution since *there is* a unique motion in reality. Thus an essential part of *applied* mathematics is the proving of existence and uniqueness theorems, as partial (*but crucial*) validation of the models one studies. (Uniqueness in the present case is shown in Widder [27].)

The central observation about this problem is that, putting aside the condition $y(x, 0) = f(x)$, the remaining conditions it imposes have the property: *linear combinations of solutions are again solutions.*

EXERCISE 1. Verify this: that if y_1 and y_2 are solutions then $ay_1 + by_2$ is a solution, for any constants a and b.

This means that known solutions $y_1, y_2, y_3, \dots$ of the reduced problem yield new solutions $\sum c_n y_n$, which might then be made to satisfy the remaining condition, i.e., the coefficients c_n might be selected so that

$$f(x) = \sum c_n y_n(x, 0), \qquad 0 \leq x \leq \pi.$$

Moreover, this possibility can be pursued by starting with solutions $y_1, y_2, y_3, \dots$ of a *very special* form and seeking the generality through the variety of possible choices of $\{c_n\}$.

The special form of solution in which $y(x, t)$ is a product of a function of x alone and a function of t alone,

$$y(x, t) = X(x)T(t), \qquad 0 \leq x \leq \pi, t \geq 0,$$

has the virtue that the wave equation

$$\alpha^2 X''T = XT''$$

separates;

$$\frac{X''}{X} = \frac{1}{\alpha^2}\frac{T''}{T}, \qquad 0 \leq x \leq \pi, \quad t \geq 0.$$

This requires a function of x alone and a function of t alone to be identical, which can only happen if both functions *are* (the same) *constant* c:

$$X'' = cX, \quad 0 \leq x \leq \pi \quad \text{and} \quad T'' = c\alpha^2 T, \quad t \geq 0.$$

The boundary conditions on $y(x, t) = X(x)T(t)$, $t \geq 0$, translate into $X(0) = X(\pi) = 0$. Thus X satisfies the ordinary differential equation boundary-value problem:

$$X''(x) = cX(x), \quad 0 \leq x \leq \pi, \quad \text{and} \quad X(0) = X(\pi) = 0.$$

Seeking the special solutions in this form, and so reducing the partial differential equation to ordinary differential equations, constitutes the **Method of Separation of Variables.**

There is nothing lost if we require $c < 0$, say $c = -\lambda^2$, as long as a sequence of solutions is obtained, to preserve the expected generality in choices of $f(x)$ via

$$f(x) = \sum c_n y_n(x, 0).$$

EXERCISE 2. In fact, one can argue geometrically that $X''(x) = cX$ and $X(0) = X(\pi) = 0$ can *only* happen (with X not identically zero) in case $c < 0$. Describe such an argument. (Hint: $c > 0$ means the displacement and the concavity have the same sign at every x.)

We know that $\sin \lambda x$ and $\cos \lambda x$ satisfy $X'' = -\lambda^2 X$. The condition $X(0) = 0$ eliminates $\cos \lambda x$, and the condition $X(\pi) = 0$ imposes the requirement $\sin \lambda\pi = 0$ on the sine solutions. This holds in case $\lambda \in \mathbb{N}$; the functions $\sin x, \sin 2x, \sin 3x, \ldots$ are suitable X factors.

The time equation when $\lambda = n$ is $T'' = -(n\alpha)^2 T$, giving solutions $\sin n\alpha t$ and $\cos n\alpha t$. Of these, only the cosines can satisfy $\dfrac{\partial y}{\partial t}(x, 0) = 0$.

There remains the sequence of product solutions

$$y_n(x, t) = \sin nx \cos n\alpha t, \qquad n = 1, 2, 3, \ldots,$$

of the partial differential equation, the boundary conditions and the $\dfrac{\partial y}{\partial t}(x, 0) = 0$ condition. The symbolic solution of the vibrating string problem is then

$$y(x, t) = \sum_{n=1}^{\infty} c_n \sin nx \cos n\alpha t. \qquad (*)$$

In summary: The problem is solved by $(*)$ whenever the initial displacement $f(x)$ has a series representation ($t = 0$)

$$f(x) = \sum_{n=1}^{\infty} c_n \sin nx, \qquad 0 \leq x \leq \pi,$$

in which the coefficients $c_1, c_2, c_3, \ldots$ are such that $(*)$ has meaning (converges!) and can be differentiated term-by-term as needed to verify the wave equation.

The historical debate was about whether such a sum of sines could give an "arbitrary" function $f(x)$. It was fueled by the fact that a solution for "arbitrary" f had already been proposed, and the fact that terms like function, differentiable, and convergence were quite vague at the time; their ultimate clarification was the means of resolution of the debate.

EXERCISE 3. Extend f to all of $\mathbb{R}$ by:

1. Putting $f(x) = -f(-x)$ for $x \in [-\pi, 0]$.
2. Extending the resulting (odd) function on $[-\pi, \pi]$ to be periodic of period 2π; $f(x \pm 2\pi) = f(x)$, all $x \in \mathbb{R}$.

For any $t \geq 0$, it then makes sense to consider $f(x - \alpha t)$ and $f(x + \alpha t)$ for all $x \in \mathbb{R}$ (α being the constant in the wave equation).

Show that the average,

$$y(x, t) = \tfrac{1}{2}\{f(x - \alpha t) + f(x + \alpha t)\},$$

solves the vibrating string problem with initial displacement $f(x)$, $0 \leq x \leq \pi$.

That this approach works for "arbitrary" f would seem obvious to one who has no hesitancy about taking derivatives of "arbitrary" functions.

EXERCISE 4. Carry out the separation of variables in case the string is given an initial velocity $g(x)$ at each $x \in [0, \pi]$, where $g(0) = g(\pi) = 0$. Show that the formal solution is

$$y(x, t) = \sum_{n=1}^{\infty} (c_n \cos n\alpha t + d_n \sin n\alpha t) \sin nx,$$

where the requirements

$$f(x) = \sum_{n=1}^{\infty} c_n \sin nx \quad \text{and} \quad g(t) = \sum_{n=1}^{\infty} (\alpha n d_n) \sin nx$$

define the coefficients $\{c_n\}$ and $\{d_n\}$.

Remarks. 1. The question of representation on $[0, \pi]$ by sine series occurs in this example. More generally, the question of representation on $[-\pi, \pi]$ (or on $[0, 2\pi]$) by series of the form

$$\frac{a_0}{2} + \sum_{n=1}^{\infty} \{a_n \cos nx + b_n \sin nx\}$$

arises; general Trigonometric Series.

2. Since we had already defined the trigonometric functions and found their properties, we were able to draw upon them to solve the ordinary differential equation $y'' = -\alpha^2 y$. An alternative approach is the following.

Prove the existence and uniqueness of solutions of initial-value problems for ordinary differential equations in sufficient generality to include the problems

$$\begin{cases} y'' + y = 0, \\ y(0) = 0, \\ y'(0) = 1, \end{cases} \quad \text{and} \quad \begin{array}{l} y'' + y = 0, \\ y(0) = 1, \\ y'(0) = 0, \end{array}$$

and *define* their solutions to be $\sin x$ and $\cos x$, respectively.

In this approach to the definitions of elementary functions, their familiar properties are derived from their defining initial-value problems. (See Simmons [25], p. 156 for more details.)

EXERCISE 5. If sin x and cos x are defined in this manner, prove that for all x

$$\frac{d}{dx}\sin x = \cos x,$$

$$\frac{d}{dx}\cos x = -\sin x,$$

and

$$\sin^2 x + \cos^2 x = 1.$$

3. Another approach to the definitions and properties of the elementary functions is via power series. That is, define each function by a power series (the Taylor series, e.g., $\sum_{n=0}^{\infty} x^n/n!$ defines e^x) and derive their properties from general properties of power series (see §5 below).

4. The theory of power series can be used to prove the existence and uniqueness result called for in Remark 2, thus combining 2 and 3 (See Simmons [25], p. 208).

5. The question: Which functions have power series representations (are "polynomials of infinite degree")?, is best considered (as are ordinary polynomials) in the complex number setting. The answer unfolds in the beautiful Theory of Functions of a Complex Variable. (See, e.g., Boas [5].)

§2. Convergence: General Considerations

The definition of convergence and related terminology were already given in Chapter 2, pp. 78–79. They are recalled here concisely as follows: The **series** $\sum_{k=1}^{\infty} a_k$, whose **terms** are $a_1, a_2, a_3, \ldots$, and whose **partial sums** are

$$s_n = a_1 + a_2 + \cdots + a_n = \sum_{k=1}^{n} a_k, \qquad n = 1, 2, 3, \ldots,$$

is **convergent** if lim s_n exists. Otherwise it is **divergent**. The **sum** of a convergent series is the limit $S = \lim s_n$, and the symbols for the series are then also used to denote the sum:

$$S = \sum_{k=1}^{\infty} a_k = a_1 + a_2 + a_3 + \cdots.$$

Given a series, the first issue is its convergence. Thus, we seek conditions which insure that a given series *is* convergent (*is* divergent). Since a series is usually specified by giving its terms, e.g., a formula expressing a_n as a function of n, the conditions are usually imposed on the sequence of terms $\{a_k\}$. The definition of convergence is expressed in this way simply by writing the

Cauchy criterion for $\{s_n\}$:

$$\sum_{k=1}^{\infty} a_k \text{ converges if and only if } S_{n+m} - S_n = \sum_{k=n+1}^{n+m} a_k \to 0$$

as $n \to \infty$, uniformly in $m \in \mathbb{N}$.

In particular, we have the familiar

Proposition. *If* $\sum_{k=1}^{\infty} a_k$ *converges then* $a_n \to 0$.

Indeed, $a_n \to 0$ is equivalent to the $m = 1$ case of the Cauchy criterion.

This also reveals the extent by which $a_n \to 0$ alone falls short of insuring the convergence of $\sum_{k=1}^{\infty} a_k$; it is, on its face, a much more severe demand on $\{a_k\}$ that

$$a_{n+1} + a_{n+2} + \cdots + a_{n+m} \to 0 \qquad \text{as} \quad n \to \infty,$$

uniformly in m.

EXERCISE 1. Show that the stronger conclusion $na_n \to 0$ holds for a convergent series $\sum_{k=1}^{\infty} a_k$ in which the terms are positive and form a decreasing sequence.

The Cauchy criterion suggests two intuitions:

(i) The additions in $a_{n+1} + \cdots + a_{n+m}$ could be overcome, and convergence hold, by having $a_n \to 0$ "*fast enough*."
(ii) If terms of opposite sign appear in a "timely" way the mutual *cancellations* of terms in $a_{n+1} + \cdots + a_{n+m}$ could bring about the convergence.

That such guiding thoughts might be helpful is seen when we try to apply the definition of convergence directly; it gives insight only in case the partial sums can be expressed in a way that *reveals* the effect of $n \to \infty$. There are two cases in which this is true.

Proposition. *The* **Geometric Series**:

$$\sum_{k=0}^{\infty} x^k = \frac{1}{1-x} \textit{ for } |x| < 1 \quad \textit{and} \quad \sum_{k=0}^{\infty} x^k \textit{ diverges for } |x| \geq 1.$$

This is evident from the formula

$$s_n = \frac{1 - x^{n+1}}{1 - x},$$

which exists because of the special form of the terms.

Telescoping Series: Series whose terms are expressible as the differences of successive terms of another sequence have simple partial sums. For example, $\sum_{k=1}^{\infty} 1/k(k+1) = 1$ is seen to hold because of the partial fraction representation

$$\frac{1}{k(k+1)} = \frac{1}{k} - \frac{1}{k+1},$$

which gives

$$\begin{aligned} s_n &= \left(1 - \frac{1}{2}\right) + \left(\frac{1}{2} - \frac{1}{3}\right) + \left(\frac{1}{3} - \frac{1}{4}\right) + \cdots + \left(\frac{1}{n} - \frac{1}{n+1}\right) \\ &= 1 - \frac{1}{n+1}. \end{aligned}$$

Remark. If a finite number of the terms of a series are changed, there is no effect on its convergence/divergence. For this reason, any convergence/divergence condition on the sequence of terms should only be required to hold "from some index N on." But the conditions are easier to understand when this verbiage is omitted ("for all n" is written) and such will be our practice. The reader will easily recognize the places where the more general condition could be used.

Similarly, if there are zeros among the terms of a series, the convergence/divergence is not affected by ignoring them. Also, changing the signs of all terms (from some N on!) does not alter the convergence/divergence.

EXERCISE 2. Prove the assertion about zeros.

Accordingly, we need only consider convergence/divergence for two cases:

(a) Series of strictly positive terms; $a_k > 0$, all k; and
(b) Series having infinitely many strictly positive terms and infinitely many strictly negative terms.

(Note: The "cancellation" intuition is relevant only in case (b), while both cases are subject to the intuition about "how fast $a_n \to 0$.")

In case (a), the partial sums form a strictly increasing sequence, so we can say that *a series of positive terms converges precisely when* $\{s_n\}$ *is bounded.*

Since we have shown that $\{1 + \frac{1}{2} + \frac{1}{3} + \cdots + 1/n\}$ is an unbounded sequence (p. 24, Ex. 33) we have the

Proposition. The **Harmonic Series**, $\sum_{k=1}^{\infty} 1/k$, diverges.

Remark. Here, the terms approach zero, but too slowly to yield convergence. The "how fast" intuition is sharpened by this example, as indeed it is by every specific example we settle (e.g., $1/2^k \to 0$ "fast enough" since $\sum_{k=0}^{\infty} 1/2^k$ is a convergent geometric series.)

The study of convergence in general is facilitated by the introduction of notations for the positive (π) and negative (ν) terms: put

$$\pi_k = \begin{cases} a_k & \text{if } a_k > 0, \\ 0 & \text{if } a_k < 0, \end{cases} \quad \text{and} \quad \nu_k = \begin{cases} 0 & \text{if } a_k > 0, \\ a_k & \text{if } a_k < 0, \end{cases} \quad k \in \mathbb{N}.$$

We have

$$s_n = \sum_{k=1}^{n} \pi_k + \sum_{k=1}^{n} \nu_k, \qquad n = 1, 2, 3, \ldots,$$

and there are three possible cases:

(I) $\sum_{k=1}^{\infty} \pi_k$ and $\sum_{k=1}^{\infty} \nu_k$ both converge;
(II) one converges and the other diverges; and
(III) both diverge.

In case (I), $\sum_{k=1}^{\infty} a_k$ converges, but more can be said. The equations

$$\sum_{k=1}^{n} |a_k| = \sum_{k=1}^{n} \pi_k - \sum_{k=1}^{n} \nu_k, \qquad n = 1, 2, 3, \ldots,$$

also hold, which shows that $\sum_{k=1}^{\infty} |a_k|$ also converges in this case. (The convergence of $\sum_{k=1}^{\infty} a_k$ in this case is *not* due to cancellations.)

Definition. $\sum_{k=1}^{\infty} a_k$ is **absolutely convergent** if $\sum_{k=1}^{\infty} |a_k|$ is convergent.

Proposition. *Absolute convergence implies convergence.*

EXERCISE 3. Give two proofs:

(i) Using the above formula and "bounded, monotonic sequences converge."
(ii) Using the Cauchy criterion.

Case (I) is thus the absolute convergence case. It has the property that the "order" of the terms is irrelevant, a form of "commutative law" for infinite sums, which is made precise by the following definition and proposition.

Definition. Let φ be any one–one map of $\mathbb{N}$ onto itself. The series $\sum_{k=1}^{\infty} a_{\varphi(k)}$ is a **rearrangement** of $\sum_{k=1}^{\infty} a_k$.

Proposition. *A series is absolutely convergent with sum S if and only if each of its rearrangements is absolutely convergent with sum S.*

EXERCISE 4. Proof?

In case (II) above, the series cannot converge; the partial sums approach ∞ or $-\infty$. We say $\sum_{k=1}^{\infty} a_k$ is **properly divergent**. The order of the terms is once more irrelevant.

Case (III) is interesting. Here, $\lim_n \sum_{k=1}^n \pi_k = \infty$ and $\lim_n \sum_{k=1}^n \nu_k = -\infty$, and the values of the partial sums can be manipulated by altering the order of the original terms, i.e., by controlling the "timing" of cancellations. In fact, we shall now describe how to choose a rearrangement φ, assuming $a_n \to 0$, so that $\{\sum_{k=1}^n a_{\varphi(k)}\}$ has *any* prescribed limit inferior λ and *any* prescribed limit superior $\Lambda \geq \lambda$.

EXERCISE 5. Show that $a_n \to 0$ if and only if $a_{\varphi(n)} \to 0$, where φ is any rearrangement function.

To fix the ideas, suppose $0 < \lambda \leq \Lambda < \infty$. Choose positive terms $a_{\varphi(k)} = \pi_k$ in their natural order until their sum first exceeds $\Lambda - 1$, then include negative terms, in order, until the partial sum first is less than $\lambda + \frac{1}{2}$, then adjoin positive terms until $\Lambda - \frac{1}{3}$ is first exceeded by the partial sum, negative terms until a partial sum $< \lambda + \frac{1}{4}$ first occurs, etc. Each step is possible because $\{\sum_{k=1}^n \pi_k\}$ and $\{\sum_{k=1}^n \nu_k\}$ are unbounded under the assumption (case (III)). The subsequence of partial sums of $\sum_{k=1}^\infty a_{\varphi(k)}$ at which "first exceeds" occurs converges to Λ since $a_{\varphi(n)} \to 0$, and no subsequence has a larger limit for the same reason. Thus

$$\overline{\lim_n} \sum_{k=1}^n a_{\varphi(k)} = \Lambda.$$

Similarly,

$$\underline{\lim_n} \sum_{k=1}^n a_{\varphi(k)} = \lambda.$$

EXERCISE 6. Discuss how to construct the rearrangement in order to achieve the results in case $-\infty < \lambda < \Lambda = \infty$ and in case $\lambda = -\infty$ and $\Lambda = \infty$.

EXERCISE 7. Can φ be chosen so that for every rational r in $[\lambda, \Lambda]$ some subsequence of $\{\sum_{k=1}^n a_{\varphi(k)}\}$ converges to r? Explain.

Those series in case (III) that are convergent (in their natural order; the order in which they are presented) are called **conditionally convergent**. (This is the same as "convergent, but not absolutely convergent," since all case (II) series diverge.) A special case of the foregoing observations ($\lambda = \Lambda$ case) is the

Proposition. *If $\sum_{k=1}^\infty a_k$ is conditionally convergent, its terms can be rearranged so as to yield* any *sum, including proper divergence to ∞ or $-\infty$.*

Remarks. 1. Thus the question of "commutativity" for infinite sums has the answer, that one of the two *extremes* must hold; the order of terms is irrelevant, or care must be taken in altering that order to avoid changing the sum or losing the convergence in various ways.

2. The analogue of "associativity" concerns the grouping of finitely many successive terms, or insertion of parentheses, in $a_1 + a_2 + a_3 + \cdots$; replacing

$\sum_{k=1}^{\infty} a_k$ by the series

$$\sum_{i=1}^{\infty} (a_{k_i} + a_{k_i+1} + \cdots + a_{k_{i+1}})$$

for some $1 \leq k_1 \leq k_2 \leq \cdots$ with $k_i \to \infty$.

EXERCISE 8. Prove that convergence is preserved with the same sum (a form of associativity holds).

For series with infinitely many terms of each sign (type (b) above) the first interest is in absolute convergence, but when that fails, the possibility of convergence (conditional!) remains ... convergence which results from cancellations. The study of these would reasonably begin with simple patterns of positive and negative terms, the simplest being alternation; $+, -, +, -, +, \ldots$ or $-, +, -, +, \ldots$.

Definition. An **alternating series** is any series of the form

$$a_1 - a_2 + a_3 - a_4 + \cdots + (-1)^{n+1} a_n + \cdots$$

in which all of the a_k's have the same sign.

We may suppose $a_k > 0$ for all k, as we write the Cauchy criterion to investigate convergence:

$$\begin{aligned} s_{n+m} - s_n &= (-1)^{n+2} a_{n+1} + (-1)^{n+3} a_{n+2} + \cdots + (-1)^{n+m+1} a_{n+m} \\ &= (-1)^n [a_{n+1} - a_{n+2} + a_{n+3} - \cdots + (-1)^{m+1} a_{n+m}]. \end{aligned}$$

Only the case $a_n \to 0$ need be considered, and if this happens monotonically; $a_k - a_{k+1} \geq 0$, for all k, this expression can be organized by pairs to obtain some insights:

1. The bracketed expression can be written as

$$(a_{n+1} - a_{n+2}) + (a_{n+3} - a_{n+4}) + \cdots$$
$$+ \begin{cases} (a_{n+m-1} - a_{n+m}) & \text{for } m \text{ even,} \\ (a_{n+m-2} - a_{n+m-1}) + a_{n+m} & \text{for } m \text{ odd,} \end{cases}$$

a sum of *positive* quantities.

2. The absolute value can therefore be written by shifting the pairings as

$$|s_{n+m} - s_n| = a_{n+1} - (a_{n+2} - a_{n+3}) - (a_{n+4} - a_{n+5}) - \cdots$$
$$- \begin{cases} (a_{n+m-2} - a_{n+m-1}) - a_{n+m}, & m \text{ even,} \\ (a_{n+m-1} - a_{n+m}), & m \text{ odd,} \end{cases}$$

which shows that

$$|s_{n+m} - s_n| \leq a_{n+1} \qquad \text{for all } n, m.$$

The convergence follows from $a_n \to 0$, proving the

Alternating Series Test. *For the convergence of an alternating series* $\sum_{n=1}^{\infty} (-1)^{n+1} a_n$ *it is sufficient that* $\{a_n\}$ *approach zero monotonically.*

In particular, we have the important example:

Proposition. *The* **Alternating Harmonic Series**

$$\sum_{n=1}^{\infty} \frac{(-1)^{n+1}}{n} = 1 - \frac{1}{2} + \frac{1}{3} - \frac{1}{4} + \cdots$$

is conditionally convergent.

Remark. The cancellations yield convergence, even though the terms approach zero "too slowly."

EXERCISE 9. Make up a convergent alternating series for which $\{a_n\}$ is not monotonic. Make up a conditionally convergent one.

EXERCISE 10. If $\{a_n\}$ is monotonic and $a_n \to 0$, the sum of the series $\sum_{k=1}^{\infty} (-1)^{k+1} a_k$ and its nth partial sum differ by at most $|a_{n+1}|$; the "size" of the first omitted term is a bound on the "error" incurred by taking the partial sum as an estimate of the sum. Proof?

We turn our attention to the most important series; those with positive terms.

§3. Convergence: Series of Positive Terms

Whatever "$a_n \to 0$ fast enough" might mean, it would include "faster than fast enough is fast enough," a comparative statement which can be made precise in a transparent way as the

Comparison Test. *If* $\sum_{k=1}^{\infty} a_k$ *and* $\sum_{j=1}^{\infty} b_j$ *are two series of positive terms for which a comparison holds in the sense that*

$$a_n \leq b_n \qquad \text{for all } n,$$

then:

(i) $\sum_{j=1}^{\infty} b_j$ *convergent implies* $\sum_{k=1}^{\infty} a_k$ *convergent; and*
(ii) $\sum_{k=1}^{\infty} a_k$ *divergent implies* $\sum_{j=1}^{\infty} b_j$ *divergent.*

EXERCISE 1. Provide a brief proof based on the fact that, in the present case, convergence is the same as the boundedness of the sequence of partial sums.

Remarks. 1. The term-by-term comparison is the immediate (transparent) interpretation of "faster." It need only hold from some index on, of course.

2. Statement (ii) says "slower than too slow is too slow."

EXAMPLE.

$$\sum_{n=1}^{\infty} \frac{1}{2^n + n} \quad \text{converges because} \quad \frac{1}{2^n + n} < \frac{1}{2^n} \text{ and } \sum_{n=1}^{\infty} \frac{1}{2^n} \text{ converges.}$$

The series that is known to be convergent (divergent) is referred to as the **comparison series**. The various geometric series are useful comparison series.

Sometimes a choice of comparison series seems evident, but the comparison is not as obvious as it was in the example. The above series is "like" $\sum 1/2^n$ in that the $+n$ is negligible compared to the 2^n, and the comparison validates this thought. The same thought occurs in the case of $\sum_{n=1}^{\infty} 1/(2^n - n)$ but no comparison is evident. Is there an alternative interpretation of "$a_n \to 0$ faster," more accessible in such examples?

It makes sense to say $a_n \to 0$ "faster" than $b_n \to 0$ when $a_n/b_n \to 0$; "slower" if $a_n/b_n \to \infty$; and "at the same speed" if $\{a_n/b_n\}$ is bounded and bounded away from zero. Choosing the wide sense of this "faster" means requiring $\{a_n/b_n\}$ to be bounded, or $\overline{\lim}\,(a_n/b_n) < \infty$. This suggests the

Limit Comparison Test. *If $\sum_{k=1}^{\infty} a_k$ and $\sum_{j=1}^{\infty} b_j$ are two series of positive terms for which*

$$\overline{\lim}\,\frac{a_n}{b_n} < \infty$$

then:

(i) *$\sum_{j=1}^{\infty} b_j$ convergent implies $\sum_{k=1}^{\infty} a_k$ convergent; and*
(ii) *$\sum_{k=1}^{\infty} a_k$ divergent implies $\sum_{j=1}^{\infty} b_j$ divergent.*

Proof. There exist $B > 0$ and $N \in \mathbb{N}$ such that $a_n/b_n < B$ for all $n \geq N$, which gives a term-by-term comparison between the series $\sum_{k=N}^{\infty} a_n$ and $\sum_{j=N}^{\infty} Bb_n$. The conclusions follow at once.

EXAMPLE. The calculation

$$\lim \frac{1/(2^n - n)}{1/2^n} = \lim \frac{1}{1 - n/2^n} = 1$$

establishes the (suspected) convergence of $\sum_{n=1}^{\infty} 1/(2^n - n)$.

Remarks. 1. Rather than *showing* an explicit comparison, the limit form shows that one *exists*, which, of course, is all that is required.

2. It is tempting to imagine that some "borderline speed" of $a_n \to 0$ exists which separates convergence from divergence; that some "universal" comparison series exists. However, it is not hard to modify *any* given convergent series $\sum_{k=1}^{\infty} a_k$, to produce a new convergent series whose terms approach zero strictly more slowly.

The modified series has the form

$$a_1 + a_2 + \cdots + a_{n_2-1} + 2a_{n_2} + 2a_{n_2+1} + \cdots + 2a_{n_3-1} + 3a_{n_3} + \cdots$$
$$+ 3a_{n_4-1} + 4a_{n_4} + \cdots$$
$$= \sum_{l=1}^{\infty} l[a_{n_l} + a_{n_l+1} + \cdots + a_{n_{l+1}-1}] \quad \text{(taking } n_1 = 1).$$

The terms here have ratios to the corresponding terms of $\sum a_k$ with values (1, 1, ..., 1, 2, ..., 2, 3, ..., 3, ...) approaching ∞; these terms $\to 0$ strictly more slowly than do the a_n terms. It remains to show that $1 = n_1 < n_2 < n_3 < \cdots$ can be chosen so that the new series converges. But the Cauchy criterion gives indices $n_2, n_3, \ldots$ such that, for each l, $n \geq n_l$ gives $a_n + a_{n+1} + \cdots + a_{n+m} < 1/l2^l$ for all m. We see that these indices produce a modified series which converges by comparison with $\sum_{l=1}^{\infty} 1/2^l$.

A second important family of comparison series is the so-called p-series. We prove the

Proposition. *The p-series $\sum_{k=1}^{\infty} 1/k^p$ converges if $p > 1$ and diverges if $p \leq 1$.*

Proof. When $p \leq 1$, the comparison

$$\frac{1}{k^p} \geq \frac{1}{k}, \qquad k = 1, 2, 3, \ldots$$

with the Harmonic Series holds, showing the divergence.

In case $p > 1$, we have, for $k = 1, 2, 3, \ldots,$

$$\frac{1}{k^p} \leq \frac{1}{x^p} \qquad \text{for any} \quad x \in [k-1, k].$$

Integrating over $[k-1, k]$ on both sides gives

$$\frac{1}{k^p} \leq \int_{k-1}^{k} \frac{1}{x^p} dx, \qquad k = 1, 2, \ldots,$$

and consequently for any $n \geq 3$,

$$\sum_{k=2}^{n} \frac{1}{k^p} \leq \int_1^n x^{-p}\, dx = \frac{x^{-p+1}}{-p+1}\Bigg]_1^n = \frac{1 - n^{1-p}}{p-1} < \frac{1}{p-1}.$$

The partial sums are bounded; the series converges.

Remark. The awkwardness of computing with a general power $p > 1$ is avoided here by going to the continuous variable and integrating, thereby making the Fundamental Theorem of Calculus available to achieve the required estimate.

The idea of this proof generalizes to give

The Integral Test. *If there is a positive and decreasing function f on $[1, \infty)$ whose values at the integers are the terms of $\sum_{k=1}^{\infty} a_k$, i.e., $f(k) = a_k$, $k = 1, 2, 3, \ldots$, then the improper integral $\int_1^{\infty} f(x)\,dx$ and the series $\sum_{k=1}^{\infty} a_k$ either both converge or both diverge.*

EXERCISE 2. Proof?

Remark. The obvious candidate for f results from putting x in place of k in the formula for a_k. Calculus is then used to check whether f is decreasing.

EXAMPLE. $\sum_{k=2}^{\infty} 1/k \log k$ diverges because

$$\frac{1}{x \log x} \text{ is } > 0 \text{ for } x \geq 2 \quad \text{and} \quad \frac{d}{dx}\frac{1}{x \log x} = -\frac{1 + \log x}{(x \log x)^2} < 0$$

and

$$\int_2^{\infty} \frac{1}{x \log x}\,dx = \log \log x \Big]_2^{\infty} = \infty \quad \text{(i.e., diverges).}$$

This example has terms which approach zero strictly faster than $1/n$, but not fast enough to produce convergence. However, we have the example of

EXERCISE 3. Show that

$$\sum_{k=2}^{\infty} \frac{1}{k(\log k)^p} \qquad \text{for} \quad p > 1$$

converges.

EXERCISE 4. Discuss the series

$$\sum_{k=3}^{\infty} \frac{1}{k \log k (\log \log k)^p}, \qquad p \geq 1,$$

and generalize.

EXERCISE 5. Use the tests derived so far to examine the following for convergence, absolute convergence:

(a) $\displaystyle\sum_{k=1}^{\infty} \frac{k^2}{k^3 + 1}$;

(b) $\displaystyle\sum_{k=1}^{\infty} \frac{1}{(k-1)2^k}$;

(c) $\displaystyle\sum_{k=1}^{\infty} ke^{-k}$;

(d) $\displaystyle\sum_{k=2}^{\infty} \frac{(-1)^k \log k}{\sqrt{k}}$.

The possibility of settling the convergence/divergence of a series of positive terms by a comparison with a geometric series turns out to be directly expressible as a condition on the terms. For, supposing

$$0 < a_n \le r^n \qquad \text{for} \quad n \ge \text{some } N, \quad \text{with } 0 < r < 1,$$

we have $\sqrt[n]{a_n} \le r$ for $n \ge N$ and so

$$\overline{\lim}\ \sqrt[n]{a_n} < 1.$$

Conversely, if this condition holds, then choosing any r for which,

$$\overline{\lim}\ \sqrt[n]{a_n} < r < 1$$

gives a comparison of $\sum_{k=1}^{\infty} a_k$ with the convergent geometric series $\sum_{k=1}^{\infty} r^k$. On the other hand, if $\overline{\lim}\ \sqrt[n]{a_n} > 1$ a comparison with a divergent geometric series follows (the divergence of $\sum_{k=1}^{\infty} a_k$ in this case is more directly evident because $a_n \to 0$ clearly fails).

The Root Test. *The series of positive terms* $\sum_{k=1}^{\infty} a_k$ *converges if* $\overline{\lim}\ \sqrt[n]{a_n} < 1$ *and diverges if* $\lim \sqrt[n]{a_n} > 1$. *Nothing can be inferred from* $\overline{\lim}\ \sqrt[n]{a_n} = 1$.

EXERCISE 6. Show that for $a_n = 1/n^p$ with any $p \ge 1$, $\lim \sqrt[n]{a_n} = 1$, thus proving the last statement in the Root Test. (Recall that $n^{1/n} \to 1$, see p. 84).

EXERCISE 7. Test for convergence, absolute convergence:

(a) $\displaystyle\sum_{k=2}^{\infty} \frac{1}{(\log k)^k}$; (b) $\displaystyle\sum_{k=2}^{\infty} \frac{(1-k)^k}{k^{k+2}}$.

Of course, the computation of $\sqrt[n]{a_n}$ is often too difficult for easy application of the root test. We consider an alternative computation next: the ratio a_{n+1}/a_n of a term to the previous term.

This ratio has an obvious relevance to whether, and how, $a_n \to 0$, since a strict decrease from a_n to a_{n+1} means $a_{n+1}/a_n < 1$. Convergence tests based on the long-term tendency of these ratios are called **Ratio Tests**.

The simplest, and crudest, long-term requirement to place on the ratios, seeking to force the convergence of $\sum_{n=1}^{\infty} a_n$, is that they remain *bounded away from* 1 *from below*:

$$\text{for some} \quad r < 1, \qquad \frac{a_{n+1}}{a_n} < r \quad \text{for all } n.$$

This does, in fact, imply the convergence by comparison with the geometric series $a_1 \sum_{n=0}^{\infty} r^n$:

$$a_2 < ra_1,$$

$$a_3 < ra_2 < r^2 a_1,$$

$$a_4 < ra_3 < r^3 a_1,$$

etc.

This first ratio test can be formulated more concisely. The condition need only hold for $n \geq$ some N, and in that form it can be restated $\overline{\lim}\, a_{n+1}/a_n < 1$. Thus

D'Alembert's Ratio Test. *The series of* (strictly) *positive terms* $\sum_{k=1}^{\infty} a_k$ *converges if* $\overline{\lim}\, a_{n+1}/a_n < 1$. *It diverges if* $\underline{\lim}\, a_{n+1}/a_n > 1$.

(The second statement is obvious since it forces a strict increase from a_n to a_{n+1} infinitely often.)

Those series which are not settled by this test, those for which

$$\overline{\lim}\, \frac{a_{n+1}}{a_n} = 1,$$

will require a more subtle ratio test. D'Alembert's test is said to "fail" when the test quantity $\overline{\lim}\, a_{n+1}/a_n$ comes out to be 1. This is true, in particular, for the p-series for every $p \geq 1$, which shows that both convergent and divergent series can yield the value 1.

EXERCISE 8. Test for convergence;

$$\text{(a)} \sum_{k=1}^{\infty} \frac{n!}{2^n}; \qquad \text{(b)} \sum_{n=1}^{\infty} \frac{n^3}{n!}; \qquad \text{(c)} \sum_{n=1}^{\infty} \frac{n^n}{n!}.$$

Remark (Connection with Root Test.) We show here that $\overline{\lim}\, \sqrt[n]{a_n} \leq \overline{\lim}\, a_{n+1}/a_n$ for any sequence of strictly positive terms. It follows that any series testing convergent by D'Alembert's test would also test convergent by the Root Test; in principal (i.e., aside from computational considerations) the D'Alembert test contributes nothing new. In practice, however, it is frequently easy and effective.

For the proof, putting $\overline{\lim}\, a_{n+1}/a_n = L$, we have, $\varepsilon > 0$ implies there exists N such that

$$n \geq N - 1 \quad \text{gives} \quad \frac{a_{n+1}}{a_n} < L + \varepsilon.$$

But for $n \geq N$ we can estimate $\sqrt[n]{a_n}$ using ratios as follows:

$$a_n = \frac{a_n}{a_{n-1}} \cdot \frac{a_{n-1}}{a_{n-2}} \cdot \ldots \cdot \frac{a_{N-1}}{a_N} \cdot a_N < (L + \varepsilon)^{n-N} a_N,$$

$$\sqrt[n]{a_n} < (L + \varepsilon) \left[\frac{a_N}{(L + \varepsilon)^N}\right]^{1/n}.$$

Thus,

$$\overline{\lim}\, \sqrt[n]{a_n} \leq (L + \varepsilon) \lim_n \left[\frac{a_N}{(L + \varepsilon)^N}\right]^{1/n} = L + \varepsilon.$$

The claimed inequality follows because $\varepsilon > 0$ is arbitrary.

EXERCISE 9. Construct a convergent series $\sum_{k=1}^{\infty} a_k$ of positive terms for which $\overline{\lim}\sqrt[n]{a_n} < 1$ and $\overline{\lim}\, a_{n+1}/a_n = 1$.

A more subtle condition on the ratios a_{n+1}/a_n (which we may suppose all < 1) than that which gave D'Alembert's test would allow some subsequence to approach 1, but "slowly" in some sense, to achieve convergence:

"Test": "$(1 - a_{n+1}/a_n) \to 0$ *slowly" implies* $\sum_{k=1}^{\infty} a_k$ *converges.*

Turning to the Harmonic Series for a reference level for "slowly" leads to a conjecture:

$$\text{Convergence if} \quad \left(1 - \frac{a_{n+1}}{a_n}\right) > \left(1 - \frac{1/n}{1/(n-1)}\right) \qquad \text{for} \quad n \geq \text{some } N.$$

This simplifies to

$$\text{convergence if} \quad \underline{\lim}\, n\left(1 - \frac{a_{n+1}}{a_n}\right) > 1.$$

We shall, in fact, prove

Raabe's Test. *The series of (strictly) positive terms* $\sum_{k=1}^{\infty} a_k$ *converges if* $\underline{\lim}\, n(1 - a_{n+1}/a_n) > 1$. *It diverges if* $\overline{\lim}\, n(1 - a_{n+1}/a_n) < 1$.

Proof. If $\underline{\lim}\, n(1 - a_{n+1}/a_n) > 1$, there is an $l > 1$ and an index $N > 1$ such that

$$n \geq N \quad \text{gives} \quad n\left(1 - \frac{a_{n+1}}{a_n}\right) > l,$$

or

$$l a_n < n(a_n - a_{n+1}) \qquad \text{for} \quad n \geq N.$$

Summing over $n = N, \ldots, N + k$,

$$\begin{aligned}
l \sum_{n=N}^{N+K} a_n &< N(a_N - a_{N+1}) + (N+1)(a_{N+1} - a_{N+2}) + (N+2)(a_{N+2} - a_{N+3}) \\
&\quad + \cdots + (N+k)(a_{N+k} - a_{N+k+1}) \\
&= N\{(a_N - a_{N+1}) + (a_{N+1} - a_{N+2}) + \cdots + (a_{N+k} - a_{N+k+1})\} \\
&\quad + (a_{N+1} - a_{N+2}) + 2(a_{N+2} - a_{N+3}) + \cdots + k(a_{N+k} - a_{N+k+1}) \\
&= N(a_N - a_{N+k+1}) + (a_{N+1} - a_{N+2}) + 2(a_{N+2} - a_{N+3}) + \cdots \\
&\quad + k(a_{N+k} - a_{N+k+1}) \\
&= N(a_N - a_{N+k+1}) + a_{N+1} + a_{N+2} + \cdots + a_{N+k} - k a_{N+k+1} \\
&= (N-1)a_N + \sum_{n=N}^{N+k} a_n - (N+k)a_{N+k+1}.
\end{aligned}$$

Thus,

$$(l-1)\sum_{n=N}^{N+k} a_n < (N-1)a_N - (N+k)a_{N+k+1} < (N-1)a_N,$$

$$\sum_{n=N}^{N+k} a_n < \frac{N-1}{l-1}a_N, \qquad \text{all } k.$$

It follows that the series converges, the partial sums being bounded.

EXERCISE 10. Prove the assertion about divergence.

EXERCISE 11. Show that, for a and b fixed, the series

$$1 + \frac{1+a}{1+b} + \frac{1+a}{1+b}\cdot\frac{2+a}{2+b} + \frac{1+a}{1+b}\cdot\frac{2+a}{2+b}\cdot\frac{3+a}{3+b} + \cdots$$

converges when $b > a + 1$, and diverges when $b < a + 1$.

EXERCISE 12. Show that the series

$$1 + (\tfrac{1}{2})^p + (\tfrac{1}{2}\cdot\tfrac{3}{4})^p + (\tfrac{1}{2}\cdot\tfrac{3}{4}\cdot\tfrac{5}{6})^p + \cdots$$

converges for $p > 2$, and diverges for $p < 2$.

Further discussions of delicate convergence tests are found in Bromwich [6] and Lick [21].

§4. Computation with Series

The operations of addition and multiplication by a fixed factor, when applied term-by-term to the terms of two convergent series, yield the terms of a convergent series whose sum is the result of performing the same operations on the two sums. This long-winded observation is stated concisely as the

Proposition. *If* $s = \sum_{i=1}^{\infty} a_i$ *and* $t = \sum_{j=1}^{\infty} b_j$ *then, for any* $\alpha, \beta \in \mathbb{R}$,

$$\alpha s + \beta t = \sum_{i=1}^{\infty} (\alpha a_i + \beta b_i).$$

EXERCISE 1. Give the proof and discuss absolute convergence. In particular, if $\mathscr{C}$ denotes the vector space of all convergent series under "term-wise" addition and scalar multiplication, do the absolutely convergent series form a subspace?

It is not obvious how to combine two series term-by-term in order to create a series converging to the *product* of their sums. (For the quotient, it is even less obvious!) One approach is to consider $\sum_{i=1}^{\infty} a_i$ as the "value" at

$x=1$ of the "polynomial of infinite degree" $\sum_{i=1}^{\infty} a_i x^{i-1}$, and imitate the multiplication of polynomials; "putting $x = 1$" in the formula

$$(a_1 + a_2 x + a_3 x^2 + \cdots)(b_1 + b_2 x + b_3 x^2 + \cdots)$$
$$= a_1 b_1 + (a_1 b_2 + a_2 b_1)x + (a_1 b_3 + a_2 b_2 + a_3 b_1)x^2 + \cdots,$$

suggests the series formula

$$\left(\sum_{i=1}^{\infty} a_i\right)\left(\sum_{j=1}^{\infty} b_j\right) = \sum_{k=2}^{\infty} (a_1 b_{k-1} + a_2 b_{k-2} + \cdots + a_{k-2} b_2 + a_{k-1} b_1). \quad (*)$$

Notice that this is the sum of all the products $a_i b_j$, $i, j = 1, 2, 3, \ldots$, taken in a certain *order*. (The distributive law for finite sums makes it clear that the product $(\sum_{i=1}^{\infty} a_i)(\sum_{j=1}^{\infty} b_j)$ *should be* the sum of these terms in *some* order.)

The product defined by (*) is called the **Cauchy product** of the two series. It is the sum $\sum_{i,j=1}^{\infty} a_i b_j$ ordered by following the successive *diagonals* of the array of $a_i b_j$'s:

Index sum 2----$a_1 b_1$ $\quad a_1 b_2 \quad a_1 b_3 \quad a_1 b_4 \quad \cdots$
Index sum 3----$a_2 b_1$ $\quad a_2 b_2 \quad a_2 b_3 \quad \cdots$
Index sum 4----$a_3 b_1$ $\quad a_3 b_2 \quad \cdots$
Index sum 5----$a_4 b_1$ $\quad \vdots$
$\vdots \qquad \vdots$

If the Cauchy product converges absolutely then the series $\sum_{i,j=1}^{\infty} a_i b_j$ converges in all of its rearrangements and the product $(\sum_{i=1}^{\infty} a_i)(\sum_{j=1}^{\infty} b_j)$ is unambiguously defined. But the general partial sum of $\sum_{i,j=1}^{\infty} |a_i b_j|$ in the Cauchy order is readily estimated, in order to investigate this, as follows:

$$|a_1 b_1| + |a_1 b_2| + |a_2 b_1| + \cdots + |a_1 b_{n-1}| + |a_2 b_{n-2}| + \cdots + |a_{n-1} b_1|$$
$$= |a_1|(|b_1| + |b_2| + |b_{n-1}|) + |a_2|(|b_1| + \cdots + |b_{n-2}|) + \cdots + |a_{n-1}||b_1|$$
$$\le |a_1|\left(\sum_{j=1}^{n-1} |b_j|\right) + |a_2|\left(\sum_{j=1}^{n-1} |b_j|\right) + \cdots + |a_{n-1}|\left(\sum_{j=1}^{n-1} |b_j|\right)$$
$$= \left(\sum_{i=1}^{n-1} |a_i|\right)\left(\sum_{j=1}^{n-1} |b_j|\right).$$

There follows at once the first part of the

Proposition. *If* $\sum_{i=1}^{\infty} a_i$ *and* $\sum_{j=1}^{\infty} b_j$ *converge absolutely then* $\sum_{i,j=1}^{\infty} a_i b_j$ *converges absolutely. Moreover, the sum of this series is given by the formula*

$$\sum_{i,j=1}^{\infty} a_i b_j = \left(\sum_{i=1}^{\infty} a_i\right)\left(\sum_{j=1}^{\infty} b_j\right).$$

EXERCISE 2. Show that $\sum_{i,j=1}^{\infty} a_i b_j$ has the indicated sum by using the order $a_1 b_1 + a_1 b_2 + a_1 b_3 + \cdots + a_2 b_1 + a_2 b_2 + a_2 b_3 + \cdots + a_3 b_1 + \cdots$, i.e., taking the "rows" of the array

$$\begin{array}{cccc} a_1 b_1 & a_1 b_2 & a_1 b_3 & \cdots \\ a_2 b_1 & a_2 b_2 & a_2 b_3 & \cdots \\ a_3 b_1 & a_3 b_2 & a_3 b_3 & \cdots \\ \vdots & \vdots & \vdots & \end{array}$$

in order.

This result serves for most computations of products.

It is interesting that the convergence of the *Cauchy* product (possibly conditional) can be inferred under weaker hypotheses. To see this, we compute the partial sum as follows, using the notations

$$A_n = \sum_{i=1}^{n} a_i, \qquad B_n = \sum_{j=1}^{n} b_j, \qquad s = \sum_{i=1}^{\infty} a_i, \qquad t = \sum_{j=1}^{\infty} b_j:$$

$$\begin{aligned} & a_1 b_1 + a_1 b_2 + a_2 b_1 + a_1 b_3 + a_2 b_2 + a_3 b_1 + \cdots + a_1 b_n + a_2 b_{n-1} + \cdots \\ & \quad + a_{n-1} b_2 + a_n b_1 \\ & = a_1(b_1 + b_2 + \cdots + b_n) + a_2(b_1 + b_2 + \cdots + b_{n-1}) + \cdots + a_n b_1 \\ & = a_1 B_n + a_2 B_{n-1} + \cdots + a_n B_1 \\ & = a_1(B_n - t) + a_2(B_{n-1} - t) + \cdots + a_n(B_1 - t) + A_n t. \end{aligned}$$

Since $A_n t \to st$, the convergence of the Cauchy product to the sum st is seen to be equivalent to

$$a_1(B_n - t) + a_2(B_{n-1} - t) + \cdots + a_n(B_1 - t) \to 0.$$

Observe that for large n, both $|a_n|$ and $|B_n - t|$ are small ... at least one of the factors in each term of this sum is small; initially, the second factor and ultimately, the first. This suggests splitting the sum at some N, to estimate it:

$$\left| \sum_{k=1}^{n} a_k(B_{n+1-k} - t) \right| \le \sum_{k=1}^{N} |a_k| |B_{n+1-k} - t| + \sum_{k=N+1}^{n} |a_k| |B_{n+1-k} - t|.$$

In the second sum, the crude estimate

$$|B_{n+1-k} - t| \le M = \sup_n |B_n - t|$$

is appropriate and gives

$$\left| \sum_{k=1}^{n} a_k(B_{n+1-k} - t) \right| \le \sum_{k=1}^{N} |a_k| |B_{n+1-k} - t| + M \sum_{k=N+1}^{n} |a_k|,$$

in which the second part can be made small if $\sum_{i=1}^{\infty} a_i$ is assumed to converge *absolutely*.

Assuming this, and putting $\sum_{i=1}^{\infty} |a_i| = A$ (we suppose the nontrivial case $A \neq 0$), the Cauchy condition yields an N, being given $\varepsilon > 0$, so that

$$\sum_{k=N+1}^{n} |a_k| < \frac{\varepsilon}{2M} \qquad \text{for all} \quad n > N.$$

Then $N^* > N$ can be chosen to give

$$n \geq N^* \quad \text{implies} \quad |B_{n+1-k} - t| < \frac{\varepsilon}{2A} \qquad \text{for} \quad k = 1, 2, \ldots, N.$$

Thus, for $n \geq N^*$

$$\left| \sum_{k=1}^{n} a_k (B_{n+1-k} - t) \right| < \frac{\varepsilon}{2A} \sum_{k=1}^{N} |a_k| + M \cdot \frac{\varepsilon}{2M} < \varepsilon.$$

We have proved the

Theorem (Mertens). *The Cauchy product of two series converges to the product of their sums provided the convergence of at least one of the factor series is absolute.*

EXERCISE 3. Consider the Cauchy product of the series $\sum_{k=1}^{\infty} (-1)^{k+1}/\sqrt{k}$ with itself, to show that the conclusion can fail when neither factor series converges absolutely. (Hint: The nth term of the Cauchy product is

$$c_n = (-1)^n \sum_{k=1}^{n-1} \frac{1}{\sqrt{k(n-k)}}, \qquad n = 2, 3, 4, \ldots.$$

Find the minimum of $1/\sqrt{x(n-x)}$, $1 \leq x \leq n-1$, to estimate c_n from below.)

We shall not examine the question of the quotient of two series systematically. In simple cases, especially when one (or both) series is finite, a series for the quotient is readily found. For example, imitating the long division process suggests the formula

$$\frac{1}{1-x} = 1 + x + x^2 + \cdots,$$

which is then easily validated for $|x| < 1$.

Next, we turn our attention to the case in which the terms of the series are *functions* of a real variable x:

$$f(x) = \sum_{i=1}^{\infty} a_i(x) \quad \text{and} \quad g(x) = \sum_{j=1}^{\infty} b_j(x).$$

The convention, that a function given by a formula is to have as domain all x for which the formula makes sense, now demands *convergence* of the numerical series of values at that x.

The computations that are of interest include the pointwise arithmetic combinations, composition, and the computations of Calculus: integration, differentiation, and summation of series.

Any combination $\alpha f + \beta g$, $\alpha, \beta \in \mathbb{R}$, has the series representation

$$(\alpha f + \beta g)(x) = \sum_{i=1}^{\infty} [\alpha a_i(x) + \beta b_i(x)]$$

on its domain, $\operatorname{dom} f \cap \operatorname{dom} g$.

The product function has the series representation

$$(fg)(x) = \sum_{i,j=1}^{\infty} a_i(x) b_j(x) \qquad \text{(any ordering)}$$

at least on the subset of $\operatorname{dom} f \cap \operatorname{dom} g$ consisting of all x at which both given series converge absolutely. Alternatively, we have

$$(fg)(x) = \sum_{i,j=1}^{\infty} a_i(x) b_j(x) \qquad \text{(Cauchy ordering)}$$

at least on the set of all x at which at least one of the series converges absolutely, by Mertens' theorem.

Certain simple composite functions yield new series formulas, e.g.,

$$\frac{1}{1+x} = 1 - x + x^2 - x^3 + \cdots, \qquad |x| < 1$$

(putting $-x$ for x in the geometric series)

$$\frac{1}{1+x^2} = 1 - x^2 + x^4 - x^6 + \cdots, \qquad |x| < 1$$

(putting x^2 for x).

Such ad hoc results are adequate for most purposes. We accept them in lieu of a systematic study of composition.

Integration is the simplest calculus computation to treat. The term-by-term integration of a series amounts to an interchange of two limit operations, and we know that the uniformity of the convergence on the range of integration suffices to justify such a step. We start with the obvious.

Definition. The series $\sum_{i=1}^{\infty} a_i(x)$ **converges uniformly on a set** S in case the sequence of partial sums $\{\sum_{i=1}^{n} a_i(x)\}$ converges uniformly on S.

The theorem on p. 184 translates as the

Theorem. *If* $f(x) = \sum_{i=1}^{\infty} a_i(x)$, *with the convergence uniform on* $[a, b]$, *and if*

$$a_i \in R[a, b], \qquad i = 1, 2, 3, \ldots,$$

then $f \in R[a, b]$ and

$$\int_a^b f(t)\,dt = \sum_{i=1}^{\infty} \int_a^b a_i(t)\,dt.$$

Remark. The Lebesgue theory of integration yields this term-by-term integration formula under much weaker hypotheses than the present uniformity. Indeed, the need to improve this result was a major motivation for the search for a better theory of integration.

EXERCISE 4. Show that the series

$$\frac{1}{1+t} = 1 - t + t^2 - t^3 + \cdots$$

is absolutely convergent uniformly on $[0, x]$ if $0 \le x < 1$, by comparison with $1 + x + x^2 + \cdots$. Then justify the formula

$$\log(1 + x) = x - \frac{x^2}{2} + \frac{x^3}{3} - \frac{x^4}{4} + \cdots, \qquad 0 \le x < 1.$$

In the same spirit, a result about taking derivatives term-by-term in a series follows from the theorem on p. 188:

Theorem. *If*

$$a_i(x) \in C^1(a, b), \qquad i = 1, 2, 3, \ldots,$$

and if

$$f(x) = \sum_{i=1}^{\infty} a_i(x) \quad \text{and} \quad g(x) = \sum_{i=1}^{\infty} \frac{da_i}{dx}(x),$$

with both series convergent uniformly on every closed subinterval of (a, b), then f is differentiable and $f' = g$, i.e., term-by-term differentiation is valid:

$$\frac{df}{dx}(x) = \sum_{i=1}^{\infty} \frac{da_i}{dx}(x), \qquad a < x < b.$$

If these theorems are to be useful, a test for uniform convergence on a closed interval is wanted, to verify the hypotheses. The most obvious insight is that, if the pointwise convergence of $\sum_{i=1}^{\infty} a_i(x)$ is inferred from comparisons and the same comparison series works at every $x \in [a, b]$, then the convergence is uniform on $[a, b]$. This gives the

Weierstrass *M*-Test. *If $0 \le a_i(x) \le M_i$ for all $x \in [a, b]$, $i = 1, 2, 3, \ldots$, and if $\sum_{i=1}^{\infty} M_i$ converges then $\sum_{i=1}^{\infty} a_i(x)$ converges uniformly on $[a, b]$.*

EXERCISE 5. Supply a detailed proof. Hint: The absence of any reference to the sums suggests that the Cauchy criterion be employed.

Remarks. 1. The condition $0 \le a_i(t)$ is often the result of considering absolute values.

2. The differentiation-term-by-term theorems feature the "convergence, uniformly on every closed subinterval" of a series or sequence. This kind of convergence is of central importance in more advanced topics in Analysis in the form "convergence, uniform on compact subsets."

Exercise 6. Show that $\{U_n(x)\}$ converges (pointwise in (a, b)), uniformly on every closed subinterval, if and only if it converges, uniformly on every compact subset.

3. The summation-of-series computation, which we shall not investigate, i.e., $\sum_{i=1}^{\infty} a_i(x)$ where each $a_i(x)$ has a series expansion

$$a_i(x) = \sum_{j=1}^{\infty} a_{ij}(x), \qquad i = 1, 2, 3, \ldots,$$

leads to the double sum $\sum_{i,j=1}^{\infty} a_{ij}(x)$ in the order

$$\sum_{i=1}^{\infty} \sum_{j=1}^{\infty} a_{ij}(x),$$

in which the rows of the array

$$\begin{matrix} a_{11}(x) & a_{12}(x) & a_{13}(x) & \cdots \\ a_{21}(x) & a_{22}(x) & a_{23}(x) & \cdots \\ a_{31}(x) & a_{32}(x) & a_{33}(x) & \cdots \\ \vdots & \vdots & \vdots & \end{matrix}$$

are added, then their sums added in order; it is this "row ordering" which arises naturally.

Finally, we mention a result that allows uniformity to be inferred in a different way.

Theorem. *If a series of positive terms has sum f:*

$$f(x) = \sum_{i=1}^{\infty} a_i(x) \quad \text{with } a_i(x) \ge 0 \quad \text{for all } x \text{ and } i = 1, 2, 3, \ldots,$$

then the convergence is uniform on any set S with the properties:

(i) *f and every a_i is continuous on S; and*
(ii) *S is compact.*

Proof. The sequence of remainders

$$r_n(x) = f(x) - \sum_{i=1}^{n} a_i(x)$$

is decreasing and converges to zero. We show this convergence to be uniform on S.

Fix $\varepsilon > 0$. For each $x \in S$ choose an index $n(x)$ so that $r_{n(x)}(x) < \varepsilon/2$, using the convergence at x.

The continuity of $r_{n(x)}$ at x provides an open interval J_x such that

$$y \in S \cap J_x \quad \text{implies} \quad r_{n(x)}(y) < r_{n(x)}(x) + \frac{\varepsilon}{2} < \varepsilon.$$

The open covering $\{J_x: x \in S\}$ has a finite subcovering $\{J_{x_1}, J_{x_2}, \ldots, J_{x_m}\}$ of S, and so any index N exceeding all of $n(x_1), n(x_2), \ldots, n(x_m)$ gives

$$r_N(x) < \varepsilon \qquad \text{for every} \quad x \in S.$$

The desired conclusion,

$$n \geq N \quad \text{implies} \quad r_n(x) < \varepsilon \quad \text{for all} \quad x \in S,$$

holds because the sequence of remainders $\{r_n(x)\}$ is decreasing at all $x \in S$.

§5. Power Series

A series of the form

$$a_0 + a_1(x - x_0) + a_2(x - x_0)^2 + \cdots = \sum_{k=0}^{\infty} a_k(x - x_0)^k$$

is called a **power series in powers of** $(x - x_0)$ or **with center of expansion** x_0, where $\{a_k\}$ is a sequence of constants, **the coefficients**.

Of course, if only finitely many coefficients are nonzero, the power series is just a polynomial. Beyond these simple ones, one wonders what kind of functions are given by power series, as the sum. We shall see that they are functions whose domains are intervals with x_0 as center, understood to include two extreme "intervals": all of $\mathbb{R}$ and the singleton $\{x_0\}$. Moreover, these functions will be seen to be infinitely differentiable at interior points.

These results pose the question whether an arbitrary C^∞ function on an interval is the sum function of some power series (can be **represented** by a power series). One sees at once that the only possible such series, with center of expansion some x_0 in the interior of the interval, is the Taylor series of f at x_0, i.e., the coefficients *must* be

$$a_k = \frac{f^{(k)}(x_0)}{k!}, \qquad k = 0, 1, 2, \ldots \quad (f^{(0)} \text{ means } f).$$

But, *surprisingly*, a Taylor series may fail to have as its sum function the function whose Taylor series it is.

This means that we must distinguish, among all C^∞ functions, those which *are* the sum of their Taylor series; the so-called analytic functions.

The elementary functions are analytic. This follows by way of Taylor's Theorem.

Finally, these observations show that the sum functions of power series share an important property with polynomials: two of them can only be equal throughout an interval;

$$\begin{aligned} &a_0 + a_1(x - x_0) + a_2(x - x_0)^2 + \cdots \\ &\quad = b_0 + b_1(x - x_0) + b_2(x - x_0)^2 + \cdots, \qquad |x - x_0| < A, \end{aligned}$$

if they have the same coefficients;

$$a_0 = b_0, \qquad a_1 = b_1, \qquad a_2 = b_2, \ldots.$$

This, in turn, is the basis for a method for solving differential equations. We shall only illustrate the method by an example, referring to Simmons [25] for an exposition of the method and the important Special Functions of Mathematical Physics to which it leads: another extension of the class of familiar functions.

EXERCISE 1. Find the expression of

$$f(x) = 1 - x + 2x^3 - 5x^4$$

as a power series in powers of $(x - 1)$.

By thinking of $x - x_0$ as a new variable u, any series in powers of $(x - x_0)$ becomes a series in powers of u, so the general properties of power series need only be set forth in the simple case where 0 is the center of expansion. Here is the fundamental result.

Theorem. *If $\sum_{k=0}^{\infty} a_k x^k$ converges when $x = x_1$ then, for any r with $0 < r < |x_1|$, the series converges absolutely, uniformly on $[-r, r]$. If the series diverges at some x_1 then it diverges at any x with $|x| > |x_1|$.*

Proof. The case $x_1 = 0$ is trivial, so we may suppose $x_1 \neq 0$ and write $|x| = |x_1||x/x_1|$, in which $|x/x_1| < 1$ for $x \in [-r, r]$, and compute the Cauchy condition:

$$\sum_{k=n}^{n+m} |a_k x^k| = \sum_{k=n}^{n+m} |a_k x_1| \left|\frac{x}{x_1}\right|^k \leq \sum_{k=n}^{n+m} |a_k x_1^k|.$$

Given $\varepsilon > 0$, the latter sum is $< \varepsilon$ provided n exceeds some N, whatever $m \in \mathbb{N}$ and $x \in [-r, r]$ may be. The absolute convergence, uniformly on $[-r, r]$ follows.

For the divergence statement, notice that convergence at x would entail convergence at x_1 by the statement just proved.

Any power series is convergent, trivially, at the center of expansion. Some are convergent *only* there:

EXERCISE 2. Show that $\sum_{k=0}^{\infty} k!\, x^k$ diverges if $x \neq 0$.

At the other extreme, some power series converge for *all* $x \in \mathbb{R}$:

EXERCISE 3. Show that $\sum_{k=0}^{\infty} (1/k!)x^k$ converges (absolutely) for all $x \in \mathbb{R}$.

The familiar geometric series illustrates the in-between case:

$$\sum_{k=0}^{\infty} x^k = \frac{1}{1-x} \qquad \text{for} \quad |x| < 1.$$

EXERCISE 4. Modify this example to create a series that converges on (a, b) and diverges outside $[a, b]$, where $a < b$.

In view of the theorem, the largest interval on which a given power series is convergent (centered at the center of expansion, of course) has "radius" given as in the

Definition. The **radius of convergence** of $\sum_{k=0}^{\infty} a_k(x - x_0)^k$ is

$$\rho = \sup\left\{x_1 - x_0 \colon \sum_{k=0}^{\infty} a_k(x_1 - x_0)^k \text{ converges}\right\},$$

(under the convention that the case of convergence everywhere is expressed as $\rho = \infty$). The **interval of convergence** of the series is the open interval $\{x \colon |x - x_0| < \rho\}$.

Remark. This term is not standard. A series may diverge, converge conditionally or converge absolutely at either of the endpoints $x_0 - \rho$ and $x_0 + \rho$ in case $\rho < \infty$. Thus the set of *all* points of convergence is always an interval, possibly semi-open or closed. The term "interval of convergence" should perhaps refer to that interval, and that is the usual usage. However, the term "set of convergence" is available for that interval, while a term is needed for the *open* interval because *it* is of the most interest. The usage is simplified by taking the above definition.

EXERCISE 5. Give an example of a series with interval of convergence $(-1, 1)$ which:

(a) converges at -1 and diverges at 1; and
(b) converges absolutely at both -1 and 1.

The fact that the convergence is uniform on closed subintervals of the interval of convergence implies that the sum function inherits the continuity of the partial sum functions on any closed subinterval, hence on the interval of convergence.

In fact, the sum function is even continuous on the *set* of convergence, according to

Abel's Theorem. *If a power series converges at an endpoint of its interval of convergence, its sum there* must be *the value that gives the (one-sided) continuity there of the sum function.*

EXERCISE 6. Show that the general case can be reduced to the special case expressed by the statement

$$\sum_{k=0}^{\infty} a_k x^k = f(x) \quad \text{for } x \in [0, 1] \quad \text{implies} \quad f(1) = \lim_{\substack{x \to 1 \\ x < 1}} f(x).$$

Proof of the special case. For $n = 0, 1, 2, \ldots$, put

$$f_n(x) = \sum_{k=0}^{n} a_k x^k.$$

The coefficients, when expressed as differences:

$$a_k = f_k(1) - f_{k-1}(1), \quad k = 1, 2, 3, \ldots, \quad \text{and} \quad a_0 = f_0(1),$$

allow these partial sums to be rewritten ("summation-by-parts"):

$$\begin{aligned} f_n(x) &= \sum_{k=1}^{n} [f_k(1) - f_{k-1}(1)]x^k + f_0(1) \\ &= \sum_{k=1}^{n} f_k(1)x^k - x \sum_{k=1}^{n} f_{k-1}(1)x^{k-1} + f_0(1) \\ &= f_n(1)x^n + \sum_{k=0}^{n-1} (1 - x)f_k(1)x^k. \end{aligned}$$

In case $0 \le x < 1$, we may pass to the limit, to obtain

$$f(x) = \sum_{k=0}^{\infty} (1 - x)f_k(1)x^k, \qquad 0 \le x < 1.$$

The geometric series,

$$1 = \sum_{k=0}^{\infty} (1 - x)x^k, \qquad 0 \le x < 1,$$

gives, upon multiplication by $f(1)$, a compatible expression for $f(1)$,

$$f(1) = \sum_{k=0}^{\infty} (1 - x)f(1)x^k, \qquad 0 \le x < 1,$$

to use in estimating $|f(1) - f(x)|$.

Thus

$$|f(1) - f(x)| \le \sum_{k=0}^{\infty} (1 - x)|f(1) - f_k(1)|x^k, \qquad 0 \le x < 1,$$

and, for $\varepsilon > 0$ fixed, N may be chosen so that $|f(1) - f_k(1)| < \varepsilon/2$ when $k \ge N$ ($f(1)$ being, by definition, $\lim_k f_k(1)$) and so

$$\begin{aligned} |f(1) - f(x)| &\le \sum_{k=0}^{N-1} (1 - x)|f(1) - f_k(1)|x^k + \frac{\varepsilon}{2} \sum_{k=N}^{\infty} (1 - x)x^k \\ &\le \sum_{k=0}^{N-1} (1 - x)|f(1) - f_k(1)|x^k + \frac{\varepsilon}{2}, \qquad 0 \le x < 1. \end{aligned}$$

Since the sum here is continuous and vanishes at $x = 1$, there is a $\delta > 0$ such that it is $< \varepsilon/2$ when $1 - \delta < x < 1$. Consequently, for $\varepsilon > 0$ there is a δ such that

$$1 - \delta < x < 1 \quad \text{implies} \quad |f(1) - f(x)| < \varepsilon,$$

as was to be shown.

Exercise 7. Assuming the expressions

$$\log(1 + x) = x - \frac{x^2}{2} + \frac{x^3}{3} - \frac{x^4}{4} + \cdots, \qquad |x| < 1,$$

and

$$\tan^{-1} x = x - \frac{x^3}{3} + \frac{x^5}{5} - \frac{x^7}{7} + \cdots, \qquad |x| < 1,$$

find the sums of the two series:

$$1 - \tfrac{1}{2} + \tfrac{1}{3} - \tfrac{1}{4} + \cdots \quad \text{(alternating harmonic)}$$

and

$$1 - \tfrac{1}{3} + \tfrac{1}{5} - \tfrac{1}{7} + \cdots.$$

The Root Test yields a formula for the radius of convergence of a power series.

Exercise 8. Apply the test to $\sum_{k=0}^{\infty} a_k(x - x_0)^k$, to prove the next result.

Proposition. *The radius of convergence of a power series with coefficients* $\{a_k\}$ *is*

$$\rho = \frac{1}{\overline{\lim}\ \sqrt[n]{|a_n|}},$$

where the meaning, by convention, is

$$\rho = \begin{cases} 0 \\ \infty \end{cases} \quad \textit{in case} \quad \overline{\lim}\ \sqrt[n]{a_n} = \begin{cases} \infty \\ 0 \end{cases}.$$

Remark. For any sequence $\{s_n\}$, it can be shown that

$$\underline{\lim}\left|\frac{s_{n+1}}{s_n}\right| \le \underline{\lim}\sqrt[n]{|s_n|} \le \overline{\lim}\sqrt[n]{|s_n|} \le \overline{\lim}\left|\frac{s_{n+1}}{s_n}\right|$$

(see p. 235). Thus, if $\lim|a_{n+1}/a_n|$ exists, it can replace $\overline{\lim}\sqrt[n]{|a_n|}$ in the formula for ρ, giving an alternative computation for ρ which is often to be preferred.

EXERCISE 9. Find the radius of convergence:

(a) $\sum_{n=1}^{\infty} \frac{2^n}{n^2} x^n$; (b) $\sum_{n=0}^{\infty} \frac{(n!)^2}{(2n)!} x^n$; (c) $\sum_{n=1}^{\infty} \frac{1}{(2^n + 1)!} x^n$.

Remark. At the endpoints of the interval of convergence (if any) the root test and the ratio test fail (a corollary to the proposition!). At such points, other tests must be used. The alternating series test is frequently applicable at one of the ends.

EXERCISE 10. Find the *set* of convergence

(a) $\sum_{n=1}^{\infty} n^\alpha x^n$, α real; (b) $\sum_{n=0}^{\infty} \frac{3^n + 2^n}{4^n + 5} x^n$.

The operations of term-by-term differentiation and integration of a power series $\sum_{k=0}^{\infty} a_k x^k$ *yield power series,*

$$\sum_{k=0}^{\infty} k a_k x^{k-1} \quad \text{and} \quad \sum_{k=0}^{\infty} \frac{a_k}{k+1} x^{k+1},$$

respectively. The fact that $n^{1/n} \to 1$, and the formula for radius of convergence, show that the new series have the same radius of convergence as the original series.

EXERCISE 11. State a theorem to this effect. Give the proof for the differentiated series. Infer the result for the integrated series.

In view of the fact that the convergence is uniform on any closed subinterval of the interval of convergence, the hypotheses of our theorems on term-by-term calculations in the previous section are satisfied. The next two theorems result.

Theorem. *If* $f(x) = \sum_{k=0}^{\infty} a_k x^k$ *on* $(-R, R)$ *with* $0 < R \le \infty$, *then* f *is differentiable with*

$$f'(x) = \sum_{k=0}^{\infty} k a_k x^{k-1} \quad \text{on } (-R, R).$$

Corollary. *The sum function of a power series is infinitely differentiable on its interval of convergence.*

Theorem. *If $f(x) = \sum_{k=0}^{\infty} a_k x^k$ on $(-R, R)$ and $t \in (-R, R)$ then*

$$\int_0^t f(x)\,dx = \sum_{k=0}^{\infty} \frac{a_k}{k+1} t^{k+1}.$$

EXERCISE 12. Show that the sum function $y(x)$ of $\sum_{n=0}^{\infty} x^n/n!$ satisfies $y' = y$ and $y(0) = 1$, the initial-value problem which is satisfied by e^x. Prove that if the sum function of a power series

$$f(x) = \sum_{k=0}^{\infty} a_k x^k$$

with a positive radius of convergence satisfies this initial-value problem then the coefficients must be given by $a_n = 1/n!$, $n = 0, 1, 2, \ldots$. This proves that

$$e^x = \sum_{k=0}^{\infty} \frac{x^k}{k!} \qquad \text{for all } x,$$

provided the initial-value problem is known to have a *unique* solution.

The repetition of the following calculations: differentiate term-by-term then set $x = x_0$, yields the next result:

Theorem. *If a function f is the sum of a power series on a nontrivial interval*

$$f(x) = \sum_{k=0}^{\infty} a_k (x - x_0)^k, \qquad x_0 - R < x < x_0 + R, \quad \text{with } R > 0,$$

then the coefficients can only be the Taylor coefficients of f at x_0:

$$a_n = \frac{f^{(n)}(x_0)}{n!}, \qquad n = 0, 1, 2, \ldots.$$

EXERCISE 13. Give the details of the proof.

EXERCISE 14. Taylor's Theorem (p. 159) shows that e^x and the nth partial sum of its Taylor series about $x_0 = 0$, its **Maclaurin Series**, differ by

$$\frac{e^\xi}{n!} x^n$$

for some ξ between 0 and x. Prove that this "remainder" approaches 0 as $n \to \infty$ for any x, thus establishing the fact that e^x *is* the sum of its Maclaurin series without using the uniqueness result needed in Exercise 12.

Taylor's Theorem, with various forms of the remainder, permits the direct verification of power series representations of known functions, as is illustrated by this exercise.

Various tricks can be used to derive new expansions from known ones. For example, putting $-x$ for x in the geometric series gives

$$\frac{1}{1+x} = 1 - x + x^2 - x^3 + \cdots, \qquad |x| < 1.$$

Integrating from 0 to x gives an expansion for the logarithm function

$$\log(1+x) = x - \frac{x^2}{2} + \frac{x^3}{3} - \frac{x^4}{4} + \cdots, \qquad |x| < 1.$$

Putting x^2 for x in the expansion of $1/(1+x)$ gives

$$\frac{1}{1+x^2} = 1 - x^2 + x^4 - x^6 + \cdots, \qquad |x| < 1.$$

Then an integration yields the Maclaurin expansion of $\tan^{-1} x$:

$$\tan^{-1} x = x - \frac{x^3}{3} + \frac{x^5}{5} - \frac{x^7}{7} + \cdots, \qquad |x| < 1.$$

These observations validate the conclusions of Exercise 7.

The sum of the series $\sum_{k=0}^{\infty} (k+3)x^k$ is found by noticing that, calling it $f(x)$,

$$x^2 f(x) = \sum_{k=0}^{\infty} (k+3)x^{k+2} = \sum_{k=0}^{\infty} \frac{d}{dx} x^{k+3} = \frac{d}{dx}\left[x^3 \sum_{k=0}^{\infty} x^k\right]$$

$$= \frac{d}{dx}\frac{x^3}{1-x} = \frac{3x^2 - 2x^3}{(1-x)^2}, \qquad \text{hence} \quad f(x) = \frac{3-2x}{(1-x)^2}.$$

EXERCISE 15. Prove the validity of the formula

$$\sin x = x - \frac{x^3}{3!} + \frac{x^5}{5!} - \frac{x^7}{7!} + \cdots, \qquad \text{all } x,$$

and then establish the Maclaurin series representation of $\cos x$ in one step.

EXERCISE 16. Find a series for the number

$$\int_0^1 \frac{\sin x}{x}\, dx$$

and give an approximate value that is in error by no more than 1/600.

The function

$$f(x) = \begin{cases} e^{-1/x^2}, & x \neq 0, \\ 0, & x = 0, \end{cases}$$

is the standard example of a C^∞ function that is *not* represented by its Maclaurin series. This follows from the fact that all the coefficients of this series are zero,

$$f^{(n)}(0) = 0, \qquad n = 0, 1, 2, \ldots, \quad \text{(see next exercise)},$$

hence the series converges to the zero function, *not to f!*

EXERCISE 17. Observe that for $x \neq 0$ the derivatives of f all have the form: a sum of terms like

$$\frac{\text{constant}}{x^l} e^{-1/x^2}, \qquad l \in \mathbb{N}.$$

The derivatives at 0 are then seen to be sums of terms like

$$\text{constant} \cdot \lim_{h \to 0} \frac{e^{-1/h^2}}{h^{l+1}}, \qquad l \in \mathbb{N}.$$

In the form

$$\lim_{h \to 0} \frac{h^{-(l+1)}}{e^{1/h^2}}$$

this has the indeterminate form ∞/∞, and a version of L'Hôpital's Rule applies. Show that these limits, hence all $f^{(n)}(0)$, are zero as claimed.

Definition. A function f is **analytic at** x_0 if it is defined and C^∞ on some neighborhood of x_0 *and* its Taylor series about x_0 converges to $f(x)$ for all x in some neighborhood of x_0.

Remarks. 1. The above f is not analytic at the origin.

2. If f is analytic at x_0 then it is determined throughout the interval of convergence of its Taylor series about x_0, however large that interval may be, by the sequence of numbers $f^{(n)}(x_0)$, $n = 0, 1, 2, \ldots$. Being derivatives, these numbers are completely determined in turn by the values of f on any neighborhood of x_0, however small.

3. The natural setting for the study of analytic functions is that of complex-valued functions of a complex variable. There, a neighborhood of a point is two dimensional, a power series has a "disk" of convergence and the radius of convergence is seen to be the distance from the center of expansion to the nearest point where the function is "singular." Thus, for example, the mystery of why the expansion

$$\frac{1}{1 + x^2} = 1 - x^2 + x^4 - x^6 + \cdots$$

is valid only for $|x| < 1$ is clarified by the fact that $x = \pm i$ are "singular points" of $1/(1 + x^2)$.

To conclude this section, we illustrate the use of power series for the solution of differential equations, using the example

$$(1 + x)y' = py, \qquad p \neq 0.$$

If there is a solution function $y(x)$ that is analytic at $x_0 = 0$, then it has a Maclaurin expansion

$$y(x) = \sum_{k=0}^{\infty} a_k x^k, \qquad |x| < R,$$

for some $R > 0$.

The problem is to determine the coefficients from the assumption that y is a solution. But

$$y'(x) = \sum_{k=0}^{\infty} k a_k x^{k-1}, \qquad |x| < R,$$

hence

$$\begin{aligned}(1 + x)y' - py &= \sum_{k=0}^{\infty} k a_k x^{k-1} + \sum_{k=0}^{\infty} k a_k x^k - \sum_{k=0}^{\infty} p a_k x^k \\ &= \sum_{k=0}^{\infty} \{(k + 1)a_{k+1} + (k - p)a_k\} x^k, \qquad |x| < R.\end{aligned}$$

The differential equation requires this series to give the zero function on $|x| < R$, and this can only hold if all the coefficients are zero:

$$a_{k+1} = \frac{p - k}{k + 1} a_k, \qquad k = 0, 1, 2, \ldots.$$

This "recursion formula" for the a_k's is the manifestation of the differential equation in coefficient language.

Since $a_0 = f(0)$ can be assigned any value c, after which the remaining a_k's are determined, we are finding a family of solutions $\{f_c(x): c \in \mathbb{R}\}$; a "one-parameter family." Taking $c = 1$ gives

$$a_0 = 1, \qquad a_1 = p, \qquad a_2 = \frac{p - 1}{2} a_1 = \frac{p(p - 1)}{2},$$

$$a_3 = \frac{p - 2}{3} a_2 = \frac{p(p - 1)(p - 2)}{3 \cdot 2}, \ldots.$$

In general,

$$a_n = \frac{p(p - 1)(p - 2)\ldots(p - n + 1)}{n!}$$

and so

$$f_1(x) = 1 + px + \frac{p(p - 1)}{2!} x^2 + \cdots + \frac{p(p - 1)\ldots(p - n + 1)}{n!} x^n + \cdots$$

is the only function that can be analytic at 0 and for which $f_1(0) = 1$ and $(1 + x)f_1' = pf_1$ on a neighborhood of 0. Clearly, $f_c = cf_1$ for $c \in \mathbb{R}$.

The ratio test shows that the radius of convergence is 1, f_1 *is* analytic at 0, and term-by-term differentiation verifies that f *is* a solution—as it must be in view of the procedure.

Remark. This proves the existence (by showing it!) and uniqueness (among the class of functions analytic at 0) of the solution of the initial-value problem ($p \neq 0$)

$$(1 + x)y' = py, \qquad y(0) = 1.$$

One might have noticed in this simple example that $(1 + x)^p$ is a solution and that the above series is just the Maclaurin series of $(1 + x)^p$, which amounts to the general **Binomial Theorem**

$$(1 + x)^p = 1 + px + \frac{p(p-1)}{2!}x^2 + \cdots, \qquad |x| < 1, \quad p \in \mathbb{R}.$$

More complicated differential equations can give rise to power series solutions which have no simple "closed form" expression as elementary functions. Mathematical Physics generates a variety of such "special functions," e.g., Bessel, Legendre, etc.

EXERCISE 18. Solve $y'' + y = 0$, $y(0) = 0$, $y'(0) = 1$, by power series.

§6. Fourier Series

If a **trigonometric series**

$$\frac{a_0}{2} + \sum_{k=1}^{\infty} \{a_k \cos kt + b_k \sin kt\}$$

is convergent for all real t, the sum function f has the property (it is **periodic of period** 2π)

$$f(t + 2\pi) = f(t) \qquad \text{for all} \quad t \in \mathbb{R}.$$

A function having this property is completely determined by its values on *any* closed interval of length 2π; it is the unique **extension by periodicity** of its restriction to any such interval. Note that every such restriction must take the same value at both ends of its closed interval of definition. Thus the periodic functions and the functions on the closed interval that have equal values at the endpoints are in one-to-one correspondence via this extension/restriction. We may focus our attention on any particular interval of length 2π, to be specific. We choose $[-\pi, \pi]$.

Remarks. 1. The functions on *any* nontrivial interval $[a, b]$ satisfying $f(a) = f(b)$ are matched in the same way with the functions on $\mathbb{R}$ that are periodic of period $b - a$, defined by

$$f(t + (b - a)) = f(t) \qquad \text{for all} \quad t \in \mathbb{R}.$$

This more general case is easily reduced to the $[-\pi, \pi]$ case by changing variable.

EXERCISE 1. Verify that

$$\begin{Bmatrix} \sin \\ \cos \end{Bmatrix} \frac{2\pi(t - (a + b)/2)}{b - a}$$

is periodic of period $(b - a)$. Generalize the above idea of "trigonometric series."

2. The interval $(-\infty, \infty)$, and other open intervals, lead to further extensions of these ideas; so-called "Fourier Integrals"

$$\int_{-\infty}^{\infty} \{a(u) \cos ut + b(u) \sin ut\}\, du$$

replace the "Fourier series," which we now proceed to introduce.

Suppose that a trigonometric series is convergent on $[-\pi, \pi]$ in such a way that the sum function f is integrable on $[-\pi, \pi]$ and that term-by-term integration is valid in the equation

$$f(t) = \frac{a_0}{2} + \sum_{k=1}^{\infty} \{a_k \cos kt + b_k \sin kt\},$$

which results. Then, multiplying on both sides by any one of the functions 1, $\sin t$, $\cos t$, $\sin 2t$, $\cos 2t$, $\ldots$, say by $\begin{Bmatrix} \sin \\ \cos \end{Bmatrix} nt$ for some $n = 0, 1, 2, \ldots$, we observe that:

1. The function on the left is integrable;

$$\begin{Bmatrix} \sin \\ \cos \end{Bmatrix} nt \cdot f(t) \in R[-\pi, \pi]$$

   and
2. Term-by-term integration of the resulting equation gives equations for the coefficients $\{a_k\}$ and $\{b_k\}$: for $n = 0, 1, 2, \ldots$,

$$\int_{-\pi}^{\pi} f(t) \begin{Bmatrix} \cos \\ \sin \end{Bmatrix} nt\, dt = \frac{a_0}{2} \int_{-\pi}^{\pi} \begin{Bmatrix} \sin \\ \cos \end{Bmatrix} nt\, dt + \sum_{k=1}^{\infty} a_k \int_{-\pi}^{\pi} \cos kt \begin{Bmatrix} \sin \\ \cos \end{Bmatrix} nt\, dt$$
$$+ \sum_{k=1}^{\infty} b_k \int_{-\pi}^{\pi} \sin kt \begin{Bmatrix} \sin \\ \cos \end{Bmatrix} nt\, dt.$$

Given f, this is a "system of ∞-many equations in ∞-many unknowns a_0, $a_1, \ldots, b_1, b_2, \ldots$." The integrals on the right are the "coefficients" in the system. They are evaluated by use of trigonometric identities like

$$\sin kt \cos nt = \tfrac{1}{2}[\sin(k+n)t + \sin(k-n)t].$$

The results are called the **orthogonality relations** for the sequence of functions $\{1, \sin t, \cos t, \sin 2t, \cos 2t, \ldots\}$. Namely, for $n, k \in \mathbb{N}$

$$\int_{-\pi}^{\pi} 1\,dt = 2\pi, \qquad \int_{-\pi}^{\pi} \sin nt\,dt = 0, \qquad \int_{-\pi}^{\pi} \cos nt\,dt = 0,$$

$$\int_{-\pi}^{\pi} \sin kt \sin nt\,dt = \begin{cases} \pi & \text{if } k = n, \\ 0 & \text{if } k \neq n, \end{cases}$$

$$\int_{-\pi}^{\pi} \sin kt \cos nt\,dt = 0,$$

$$\int_{-\pi}^{\pi} \cos kt \cos nt\,dt = \begin{cases} \pi & \text{if } k = n, \\ 0 & \text{if } k \neq n, \end{cases}$$

EXERCISE 2. Verify the orthogonality relations. Restate them in words.

The "system of equations" is actually *solved as it stands*, in view of these computations;[3] it reduces to

$$\begin{cases} a_n = \dfrac{1}{\pi}\displaystyle\int_{-\pi}^{\pi} f(t) \cos nt\,dt, \\ b_n = \dfrac{1}{\pi}\displaystyle\int_{-\pi}^{\pi} f(t) \sin nt\,dt. \end{cases} \qquad n = 0, 1, 2, \ldots. \qquad (*)$$

(Here, we have put $b_0 = 0$ and used the artificially introduced $\frac{1}{2}$ in the a_0 term of the general trigonometric series, as devices to include the $n = 0$ case under the general formulas.)

Definition. For any integrable f on $[-\pi, \pi]$, the numbers $(*)$ are the **Fourier coefficients of** f. The **Fourier series** of an integrable f is that trigonometric series in which the coefficients are the Fourier coefficients of f.

The first task of the subject is to identify those integrable functions which are "represented" by their Fourier series in some useful sense. Only the integrability is required for the Fourier series to be defined. Whether that series converges and, if so, whether the original f is its sum, are the issues.

[3] This phenomenon occurs more generally for any "orthogonal" sequence $\{u_0, u_1, u_2, \ldots\}$ of functions. See Goffman [9] for this approach to our subject. The final generalization is "Hilbert Space Theory." Goffman and Pedrick [12], chapter 5, gives a brief introduction.

Remark. The possibility presents itself, to generalize the definitions of integral and integrability in a way that yields representation results for a more general class of functions f than might be found using the Riemann integral. The desirability of achieving a strong term-by-term integration theorem is also apparent. These were additional motivations for the investigations which culminated in the Lebesgue theory of the integral; it yields a more satisfactory Fourier theory.

EXERCISE 3. If a Riemann integrable f is represented by a trigonometric series and the convergence is uniform on $[-\pi, \pi]$, then the series *must be* the Fourier series of f.

Observe that the values of the Fourier coefficients are unchanged if the values of f are altered arbitrarily at the points of a subset of $[-\pi, \pi]$ that is negligible for integration, i.e., that has content zero. Thus, many distinct functions share the same Fourier series, and the distinction between them should be ignored in the study of representation by Fourier series. We accomplish this by accepting a convention: any function f may be freely replaced by any other function that differs from it only on a set of content zero. In particular, the appropriate notion of "pointwise representation" is that the Fourier series of f at t should converge to $f(t)$ *for all t except* those of some subset of $[-\pi, \pi]$ whose content is zero.

EXERCISE 4. Define f by

$$f(t) = \begin{cases} t, & -\pi \le t < \pi, \\ -\pi, & t = \pi. \end{cases}$$

Show that the Fourier series of f is

$$2 \sin t - \sin 2t + \tfrac{2}{3} \sin 3t - \tfrac{1}{2} \sin 4t + \cdots + \frac{(-1)^{n+1}2}{n} \sin nt + \cdots.$$

Observe that the series gives the value zero for $t = \pm\pi$. Nevertheless, the "pointwise representation" of f by its Fourier series remains a possibility.

The student should calculate the Fourier coefficients of some functions. The calculations are sometimes simplified by recalling that:

(1) an **odd function** g, $g(-u) = -g(u)$ for all u, has $\int_{-a}^{a} g(u)\, du = 0$ for any a; and
(2) an **even function** g, $g(-u) = g(u)$ for all u, gives $\int_{-a}^{a} g(u)\, du = 2\int_{0}^{a} g(u)\, du$ for any a.

In particular, the Fourier series of an $\begin{Bmatrix}\text{odd}\\ \text{even}\end{Bmatrix}$ function has only $\begin{Bmatrix}\text{sine}\\ \text{cosine}\end{Bmatrix}$ terms.

Exercise 5. Supply the details.

Exercise 6. The periodic extension of the function in Exercise 4 is a "sawtooth wave" (indeed, a "ripsaw").

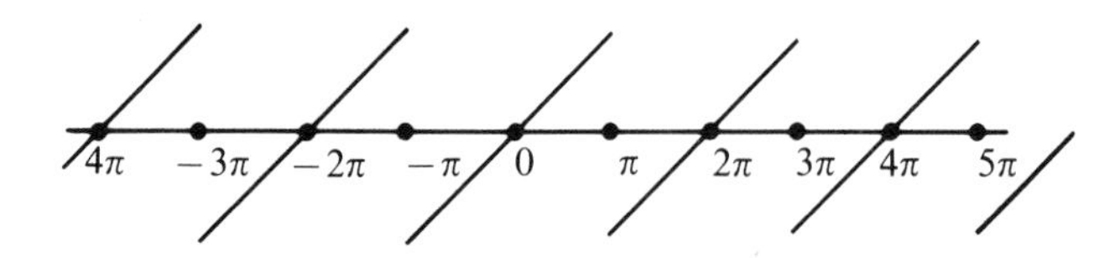

Another sawtooth wave ("cross-cut") results by extending by periodicity a function like

$$g(t) = \begin{cases} t + \pi, & -\pi \le t \le 0, \\ -t + \pi, & 0 \le t \le \pi. \end{cases}$$

(Recall the "plucked string," p. 219.) Find the Fourier series of g.

Exercise 7. Find the Fourier series of the "square wave" generated by extending

$$h(t) = \begin{cases} -\pi, & -\pi \le t \le 0, \\ \pi, & 0 < t < \pi, \\ -\pi, & t = \pi. \end{cases}$$

Exercise 8. Plot points on graph paper, to obtain fairly accurate superimposed graphs of the square wave and the first few partial sums of its Fourier series.

To enter upon the study of the convergence of a Fourier series, we need a manageable expression for its partial sums. We compute:

$$\begin{aligned} S_n(t) &= \frac{a_0}{2} + \sum_{k=1}^{n} \{a_k \cos kt + b_k \sin kt\} \\ &= \frac{1}{2\pi}\int_{-\pi}^{\pi} f(u)\,du + \frac{1}{\pi}\sum_{k=1}^{n}\left\{\int_{-\pi}^{\pi} f(u)\cos ku\,du\cdot\cos kt\right. \\ &\qquad\qquad \left. + \int_{-\pi}^{\pi} f(u)\sin ku\,du\cdot\sin kt\right\} \\ &= \frac{1}{\pi}\int_{-\pi}^{\pi} f(u)\left\{\frac{1}{2} + \sum_{k=1}^{n}[\cos ku\cos kt + \sin ku \sin kt]\right\}du \\ &= \frac{1}{\pi}\int_{-\pi}^{\pi} f(u)\left\{\frac{1}{2} + \sum_{k=1}^{n}\cos k(t-u)\right\}du. \end{aligned}$$

At this stage, one sees the desirability of a closed-form expression for

$$\tfrac{1}{2} + \cos\theta + \cos 2\theta + \cdots + \cos n\theta.$$

It happens that multiplication by $2\sin(\theta/2)$ *introduces telescoping*:

$$\begin{aligned}
&2\sin\frac{\theta}{2}\left[\frac{1}{2} + \cos\theta + \cos 2\theta + \cdots + \cos n\theta\right] \\
&\quad = \sin\frac{\theta}{2} + 2\sin\frac{\theta}{2}\cos\theta + 2\sin\frac{\theta}{2}\cos 2\theta + \cdots + 2\sin\frac{\theta}{2}\cos n\theta \\
&\quad = \sin\frac{\theta}{2} + \left[\sin\frac{3\theta}{2} - \sin\frac{\theta}{2}\right] + \left[\sin\frac{5\theta}{2} - \sin\frac{3\theta}{2}\right] + \cdots \\
&\qquad + \left[\sin\frac{2n+1}{2}\theta - \sin\frac{2n-1}{2}\theta\right] \\
&\quad = \sin\frac{2n+1}{2}\theta.
\end{aligned}$$

Thus, we may write, for $-\pi \le \theta \le \pi$ and $\theta \ne 0$,

$$\tfrac{1}{2} + \cos\theta + \cos 2\theta + \cdots + \cos n\theta = \frac{\sin(n+\frac{1}{2})\theta}{2\sin\frac{1}{2}\theta}$$

and obtain

$$S_n(t) = \int_{-\pi}^{\pi} f(u)\frac{\sin(n+\frac{1}{2})(t-u)}{2\pi\sin\frac{1}{2}(t-u)}\,du, \qquad n = 0, 1, 2, \ldots.$$

The partial sums are thus representable as "integral transforms" of f with fractions of sines as "kernel functions," called the Dirichlet kernels.

Exercise 9. Show that the Dirichlet kernels are periodic of period 2π, even though the numerator and denominator have period 4π. (Note: The formulas are meaningless when $t - u$ is an even multiple of 2π. Take the value 0 for all such points, to obtain periodicity.)

The study of the convergence of the series at t to a value S calls for an examination of $S_n(t) - S$, which indicates that a compatible expression for S should be found. But such an expression results from putting $f(u) =$ the constant function with value S in the above formula, for that function has all $S_n(t)$ also the constant function S. We obtain

$$S_n(t) - S = \int_{-\pi}^{\pi} [f(u) - S]\frac{\sin(u+\frac{1}{2})(t-u)}{2\pi\sin\frac{1}{2}(t-u)}\,du.$$

The substitution of $u + t$ for u gives

$$S_n(t) - S = \int_{t-\pi}^{t+\pi} [f(t+u) - S]\frac{\sin(n+\frac{1}{2})u}{2\pi\sin\frac{1}{2}u}\,du$$

and the range of integration can be changed back to $[-\pi, \pi]$ in view of the periodicity of the integrand (Exercise 9).

To isolate the dependence on n, we write the result as

$$S_n(t) - S = \int_{-\pi}^{\pi} \frac{[f(t+u) - S]}{2\pi \sin \frac{1}{2}u} \sin(n + \tfrac{1}{2})u \, du.$$

We obtain an intuition about this as follows. Put

$$F(u) = \frac{f(t+u) - S}{\sin \frac{1}{2}u}, \qquad u \neq 0,$$

and think of the expression as a "weighted average" of the values of F in $[-\pi, -\delta] \cup [\delta, \pi]$ for a small $\delta > 0$ (to avoid the possible singularity of F at $u = 0$, for now), where the "weighting factor" $\sin(n + \frac{1}{2})u$ embodies the dependence on n:

$$S_n(t) - S \cong \frac{1}{2\pi} \int_{\delta < |u| < \pi} F(u) \sin(n + \tfrac{1}{2})u \, du.$$

The dominant feature here is the highly oscillatory nature of the weighting factor $\sin(n + \frac{1}{2})u$, when n is large. This means that the "weights" at "nearby" values u and $u + \pi/(n + \frac{1}{2})$ cancel:

$$\sin(n + \tfrac{1}{2})u = -\sin(n + \tfrac{1}{2})\left(u + \frac{\pi}{n + \frac{1}{2}}\right).$$

Thus, if F does not change values abruptly, we can expect cancellations; $F(u)$ and $F(u + \pi/(n + \frac{1}{2}))$ with opposite weights, to yield weighted averages near zero when n is large.

This intuition suggests that the tendency of $S_n(t) - S$, hence whether the Fourier series of f at t converges with the sum S, is completely determined by the integral over $[-\delta, \delta]$, i.e., by the behavior of f *in a neighborhood of* t, however small. We say that the convergence (and sum) of a Fourier series of f is a *local property of* f, when this holds. We can prove the

Riemann Localization Theorem. *The convergence (and sum) of the Fourier series of the integrable function f is a local property of f.*

The proof consists of showing that

$$\lim_{n\to\infty} \int_{\delta < |u| < \pi} F(u) \sin(n + \tfrac{1}{2})u \, du = 0 \qquad \text{if} \quad 0 < \delta < \pi,$$

giving substance to the above intuition.

In fact, it is useful to express the intuition, and prove it, in the more general form of this basic lemma of the subject:

The Riemann–Lebesgue Lemma. *For any sequence of positive numbers $\{\lambda_n\}$*

approaching infinity,

$$g \text{ integrable on } [a, b] \quad \textit{implies} \quad \lim_{n\to\infty} \int_a^b g(u) \sin \lambda_n u \, du = 0.$$

(Riemann and Lebesgue gave proofs in the contexts of their respective meanings of integral and integrable.)

Proof (Riemann case). An integration-by-parts *reveals* the result whenever g *has* an integrable derivative, so that the computation is valid. Namely,

$$\int_a^b g(u) \sin \lambda_n u \, du = \frac{-1}{\lambda_n}\left\{g(u) \cos \lambda_n u]_a^b - \int_a^b g'(u) \cos \lambda_n u \, du\right\},$$

and the expression in braces is a bounded function of n, while $\lambda_n \to \infty$.

An integrable function being bounded, there is a constant M such that $g + M$ is nonnegative. The result is then easily seen to follow from the special case of a nonnegative integrable g, which is proved as follows.

There is a lower sum for g, say L_ε, $\varepsilon > 0$ having been fixed, such that

$$\int_a^b g - L_\varepsilon < \frac{\varepsilon}{2}.$$

L_ε has a canonical expression as $\int_a^b g_\varepsilon$, for a step function g_ε. Thus

$$\int_a^b g - L_\varepsilon = \int_a^b (g - g_\varepsilon) = \int_a^b |g - g_\varepsilon| < \frac{\varepsilon}{2}.$$

Since g_ε has derivative zero at all but finitely many points, the above proof applies to it:

$$\lim_n \int_a^b g_\varepsilon(u) \sin \lambda_n u \, du = 0.$$

Now

$$\begin{aligned}\left|\int_a^b g(u) \sin \lambda_n u \, du\right| &\le \left|\int_a^b (g(u) - g_\varepsilon(u)) \sin \lambda_n u \, du\right| + \left|\int_a^b g_\varepsilon(u) \sin \lambda_n u \, du\right| \\ &\le \int_a^b |g(u) - g_\varepsilon(u)| \, du + \left|\int_a^b g_\varepsilon(u) \sin \lambda_n u \, du\right| \\ &< \frac{\varepsilon}{2} + \left|\int_a^b g_\varepsilon(u) \sin \lambda_n u \, du\right|.\end{aligned}$$

Choosing an index N that makes the second term $< \varepsilon/2$ when $n \ge N$ gives the result.

EXERCISE 10. Show that the two sequences $\{a_n\}$ and $\{b_n\}$ of Fourier coefficients of an integrable f approach zero.

The Riemann Localization Theorem is a consequence of applying the lemma to the integrable function $F(u)$ on $[-\pi, -\delta]$ and $[\delta, \pi]$, respectively. Thus, the question whether the Fourier series of f at t has a sum S is reduced to that of whether

$$\lim_{n\to\infty} \int_{|u|<\delta} F(u)\sin(n+\tfrac{1}{2})u\,du = 0, \qquad F(u) = \frac{f(t+u)-S}{\sin\frac{1}{2}u}, \qquad u \neq 0.$$

Once more, the lemma is relevant, and yields the result under any assumptions that make F integrable on $[-\delta, \delta]$. This, in turn, is determined by the behavior of F in the neighborhood of $u = 0$, where a singularity of F *could* exist and prevent the integrability.

The existence of $\lim_{u\to 0} F(u)$, for example, insures the integrability, so we seek conditions on f that insure this property of F, to obtain a theorem.

Write

$$F(u) = \frac{f(t+u)-S}{u}\cdot\frac{u}{\sin\frac{1}{2}u}, \qquad u \neq 0,$$

and invoke the fact that $\lim_{u\to 0} u/\sin\frac{1}{2}u = 2$, to transfer the question to whether $\lim_{u\to 0}[f(t+u) - S]/u$ exists, which suggests a difference quotient for f at t.

We infer that if f is differentiable on $[a, b]$ then the Fourier series of f at t converges to $f(t)$ for all $t \in [a, b]$, simply by taking $S = f(t)$.

Taking account of our convention, we can say that f is represented pointwise by its Fourier series if f is differentiable except at most on a set of content zero.

This representation theorem covers the cross-cut saw functions and provides a satisfactory conclusion to our discussion of the plucked string. However, other applications call for a somewhat more general representation result, one in which f is allowed some jump discontinuities (ripsaw, square wave). An adequate generalization is close at hand.

Denote the one-sided limits and derivatives of f at t, assuming they exist, as follows:

$$f_+(t) = \lim_{\substack{u\to 0\\ u>0}} f(t+u), \qquad f_-(t) = \lim_{\substack{u\to 0\\ u<0}} f(t+u),$$

$$f'_+(t) = \lim_{\substack{u\to 0\\ u>0}} \frac{f(t+u) - f_+(t)}{u}, \qquad f'_-(t) = \lim_{\substack{u\to 0\\ u<0}} \frac{f(t+u) - f_-(t)}{u}.$$

Suppose all these limits exist at t. We shall prove that the Fourier series of f at t converges to the "middle of the jump"

$$\frac{f_+(t) + f_-(t)}{2}$$

(hence to $f(t)$ at points of continuity). We need only examine a suitable form of the expression

$$S_n(t) - \tfrac{1}{2}[f_+(t) + f_-(t)]$$

$$= \frac{1}{2\pi}\int_{-\pi}^{\pi} \frac{f(t+u) - \frac{1}{2}[f_+(t) + f_-(t)]}{\sin\frac{1}{2}u} \sin(n+\tfrac{1}{2})u\,du$$

$$= \frac{1}{4\pi}\int_{-\pi}^{\pi} \left\{\frac{f(t+u) - f_+(t)}{u}\cdot\frac{u}{\sin\frac{1}{2}u} + \frac{f(t+u) - f_-(t)}{u}\cdot\frac{u}{\sin\frac{1}{2}u}\right\}$$

$$\times \sin(n+\tfrac{1}{2})u\,du.$$

The only possibility for the expression in braces to fail to be integrable would arise at $u = 0$. But our assumptions insure the existence of the limit as $u \to 0$, hence the integrability, of this expression, so that the lemma applies and yields the result.

These observations allow us to claim the

Representation Theorem. *The integrable function f on $[-\pi, \pi]$ is represented pointwise by its Fourier series provided it has one-sided limits and one-sided derivatives at all points except at most those of some set of content zero.*

Remarks. 1. Dirichlet met the needs of the early applications by achieving such a result in the case of finite exceptional sets.

2. The negligible sets for Lebesgue integration are the sets of *measure* zero. In particular, countable sets are negligible. The existence of the one-sided limits except on a countable set is known for any monotonic, or, more generally, bounded variation, function f. One speculates that a function of bounded variation is represented pointwise almost everywhere by its Fourier series. The issue is the existence, almost everywhere, of the one-sided derivatives. The striking fact that a monotonic (or bounded variation) function is differentiable almost everywhere is proved as the first theorem in Riesz–Nagy [22]. The proof is accessible to the reader of this book.

The representation theorem for functions of bounded variation is also proved in Titchmarsh [26], but drawing upon some knowledge of the Lebesgue integral.

3. Our approach led to the use of differentiability hypotheses in a natural way, but one is left to wonder if such a strong hypothesis is required; does pointwise representation almost everywhere follow just from supposing f continuous almost everywhere?

This question remained open for more than a century. It was settled in 1966 by L. Carleson, in the affirmative.

An interesting earlier result (1904) due to Fejér achieves a weaker kind of representation when f is continuous:

Definition. A series $\sum_{k=0}^{\infty} u_k$ is **summable** (in the C-1 or Cesaró sense) if the successive averages of its partial sums

$$\sigma_n = \frac{1}{n}\{S_0 + S_1 + \cdots + S_{n-1}\}, \qquad S_m = \sum_{k=0}^{m} u_k, \qquad m, n \in \mathbb{N},$$

form a convergent sequence. The **Cesaró sum** is then $\lim \sigma_n$.

Fejér's Theorem. *If f is continuous on $[-\pi, \pi]$ then the Fourier series of f at t is summable to $f(t)$ for all t. Indeed,*

$$\lim \sigma_n(t) = f(t) \qquad \textit{uniformly on } [-\pi, \pi].$$

Remarks. 1. Every convergent series is summable with the same sum (p. 89, Exercise 27). There are divergent series that are summable, e.g., $1 - 1 + 1 - 1 + 1 - 1 + \cdots$.

2. Since the continuous f can be approximated uniformly by the elementary function $\sigma_n(t)$ and $\sigma_n(t)$ can be approximated uniformly by a polynomial (a partial sum of its Taylor series, say), we obtain a new proof of the Weierstrass Approximation Theorem.

We conclude with a short series of exercises which constitute a proof of Fejér's Theorem (see also Goffman [9] or Widder [27]).

Exercise 11. Using the formulas for the partial sums $S_k(t)$ of the Fourier series of f, obtain

$$\sigma_n(t) = \frac{1}{n\pi} \int_{-\pi}^{\pi} \frac{f(t+u)}{\sin \frac{1}{2}u} \sum_{k=0}^{n-1} \sin(k + \tfrac{1}{2})u \, du, \qquad n \in \mathbb{N}.$$

Exercise 12. Find a closed-form expression for

$$\sin \tfrac{1}{2}\theta + \sin \frac{3\theta}{2} + \cdots + \sin(n - \tfrac{1}{2})\theta$$

to prove that

$$\sigma_n(t) = \frac{1}{2n\pi} \int_{-\pi}^{\pi} f(t+u) \left[\frac{\sin(n/2)u}{\sin \frac{1}{2}u}\right]^2 du, \qquad n \in \mathbb{N}.$$

Exercise 13. By considering constant functions in $[-\pi, \pi]$ obtain the integral representation

$$\sigma_n(t) - f(t) = \frac{1}{2n\pi} \int_{-\pi}^{\pi} [f(t+u) - f(t)] \left[\frac{\sin(n/2)u}{\sin \frac{1}{2}u}\right]^2 du.$$

Exercise 14. Show that

$$\frac{1}{2n\pi} \int_{-\pi}^{\pi} \left[\frac{\sin(n/2)u}{\sin \frac{1}{2}u}\right]^2 du = 1, \qquad n \in \mathbb{N}.$$

Exercise 15. For any $0 < \delta < \pi$,

$$|\sigma_n(t) - f(t)| \leq \frac{1}{2n\pi} \int_{|u|<\delta} |f(t+u) - f(t)| \left[\frac{\sin(n/2)u}{\sin(u/2)}\right]^2 du.$$

$$+ \frac{1}{2n\pi} \int_{\delta<|u|<\pi} |f(t+u) - f(t)| \left[\frac{\sin(n/2)u}{\sin \frac{1}{2}u}\right]^2 du.$$

Exercise 16. Complete the proof of Fejer's Theorem, using the uniform continuity of f and Exercise 14 to estimate the first term, and the boundedness of the second integral to estimate the second.

Bibliography

1. Agnew, Ralph Palmer. *Calculus.* New York, NY: McGraw-Hill, 1962.
2. Artin, Emil. *The Gamma Function.* New York, NY: Holt, Rinehart, and Winston, 1964.
3. Apostol, Tom M. *Mathematical Analysis,* Second Edition. Reading, MA: Addison-Wesley, 1974.
4. Boas, Ralph P., Jr. *A Primer of Real Functions,* Third Edition. Washington, DC: Mathematical Association of America, 1972, 1981.
5. Boas, Ralph P., Jr. *Invititation to Complex Analysis.* New York, NY: Birkhauser, 1987.
6. Bromwich, Thomas J. L'Anson. *An Introduction to the Theory of Infinite Series.* New York, NY: Macmillan, 1965.
7. Clark, Colin W. *Elementary Mathematical Analysis,* Second Edition. Pacific Grove, CA: Brooks/Cole, 1982.
8. Gelbaum, B. and Olmsted, J. *Theorems and Counter-examples in Mathematics.* San Francisco, CA: Holden-Day, 1964; New York, NY: Springer-Verlag, 1990.
9. Goffman, Casper. *Introduction to Real Analysis.* New York, NY: Harper and Row, 1966.
10. Goffman, Casper. *Calculus of Several Variables.* New York, NY: Harper and Row, 1965.
11. Goffman, Casper. *Real Functions,* Revised Edition. Boston, MA: Prindle, Weber, and Schmidt, 1967.
12. Goffman, Casper and Pedrick, George. *A First Course in Functional Analysis,* Second Edition. Englewood Cliffs, NJ: Prentice-Hall, 1965; New York, NY: Chelsea, 1983.
13. González-Velasco, Enrique A. Connections in Mathematical Analysis: The Case of Fourier Series. *American Mathematical Monthly,* vol. 99, no. 5, May 1992, p. 427.
14. Hille, Einar. *Analysis,* Vols. I and II. New York, NY: Blaisdell Publishing Company, 1966.
15. Hardy, G. H. *Divergent Series.* New York, NY: Oxford University Press, 1949.
16. Halmos, Paul. *Measure Theory.* New York, NY: Springer-Verlag, 1974.

17. Kaplan, Wilfred. *Operational Methods for Linear Systems.* Reading, MA: Addison-Wesley, 1962.
18. Kline, Morris. *Mathematical Thought From Ancient to Modern Times.* New York, NY: Oxford University Press, 1972.
19. Landau, E. *The Foundations of Analysis*, Third Edition. New York, NY: Chelsea, 1951, 1966.
20. Lebesgue, Henri. *Measure and the Integral.* San Francisco, CA: Holden-Day, 1966.
21. Lick, Dale R. *The Advanced Calculus of One Variable.* New York, NY: Appleton-Century-Crofts, 1971.
22. Riesz, Frigyes and Nagy, Bela Sz. *Functional Analysis.* New York, NY: Frederick Unger, 1955; Mineola, NY: Dover, 1990.
23. Ross, Kenneth A. *Elementary Analysis: The Theory of Calculus.* New York, NY: Springer-Verlag, 1980.
24. Rudin, Walter. *Principles of Mathematical Analysis*, Third Edition. New York, NY: McGraw-Hill, 1976.
25. Simmons, George F. and Robertson, John S. *Differential Equations with Applications and Historical Notes*, Second Edition. New York, NY: McGraw-Hill, 1972, 1991.
26. Titchmarsh, Edward C. *Theory of Functions*, Second Edition. New York, NY: Oxford University Press, 1939.
27. Widder, David V. *Advanced Calculus*, Second Edition. Englewood Cliffs, NJ: Prentice-Hall, 1947, 1961; Mineola, NY: Dover, 1992.

Index

(Defining occurrences of a term are cited in boldface.)

(continued)

Lidl/Pilz: Applied Abstract Algebra.
Macki-Strauss: Introduction to Optimal Control Theory.
Malitz: Introduction to Mathematical Logic.
Marsden/Weinstein: Calculus I, II, III. Second edition.
Martin: The Foundations of Geometry and the Non-Euclidean Plane.
Martin: Transformation Geometry: An Introduction to Symmetry.
Millman/Parker: Geometry: A Metric approach with Models. Second edition.
Moschovakis: Notes on Set Theory.
Owen: A First Course in the Mathematical Foundations of Thermodynamics.
Palka: An Introduction to Complex Function Theory.
Pedrick: A First Course in Analysis.
Peressini/Sullivan/Uhl: The Mathematics of Nonlinear Programming.
Priestley: Calculus: An Historical Approach.
Protter/Morrey: A First Course in Real Analysis. Second edition.
Protter/Morrey: Intermediate Calculus. Second edition.
Ross: Elementary Analysis: The Theory of Calculus.
Samuel: Projective Geometry.
Readings in Mathematics.
Scharlau/Opolka: From Fermat to Minkowski.
Sigler: Algebra.
Silverman/Tate: Rational Points on Elliptic Curves.
Simmonds: A Brief on Tensor Analysis. Second edition.
Singer/Thorpe: Lecture Notes on Elementary Topology and Geometry.
Smith: Linear Algebra. Second edition.
Smith: Primer of Modern Analysis. Second edition.
Stanton/White: Constructive Combinatorics.
Stillwell: Mathematics and Its History.
Strayer: Linear Programming and Its Applications.
Thorpe: Elementary Topics in Differential Geometry.
Troutman: Variational Calculus with Elementary Convexity.
Valenza: Linear Algebra: An Introduction to Abstract Mathematics.